Actes du XIVème Congrès UISPP, Université de Liège, Belgique, 2-8 septembre 2001

Acts of the XIVth UISPP Congress, University of Liège, Belgium, 2-8 September 2001

UNIVERSITÉ de Liège

SECTION 11
ÂGE DU BRONZE EN EUROPE ET EN MÉDITERRANÉE
BRONZE AGE IN EUROPE AND THE MEDITERRANEAN

Colloque / Symposium 11.2

Le problème de l'étain à l'origine de la métallurgie

The Problem of Early Tin

Édité par / Edited by

Alessandra Giumlia-Mair
Fulvia Lo Schiavo

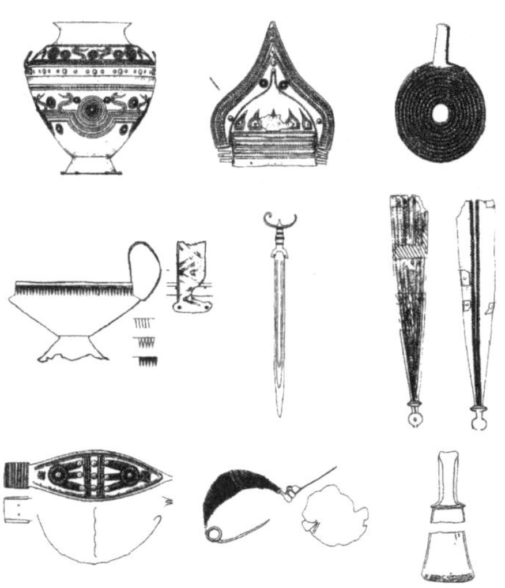

BAR International Series 1199
2003

Published in 2016 by
BAR Publishing, Oxford

BAR International Series 1199

Acts of the XIVth UISPP Congress, University of Liège, Belgium, 2-8 September 2001
Colloque / Symposium 11.2
Le problème de l'étain à l'origine de la métallurgie / The Problem of Early Tin

ISBN 978 1 84171 564 3

Avec la collaboration du Ministère de la Région Wallonne
Direction générale de l'Aménagement du territoire, du Logement et du Patrimoine
Subvention n° 02/16341
Mise en page / Editing: Rebecca MILLER

Marcel OTTE, Secrétaire général du XIVème Congrès de l'U.I.S.P.P.
Université de Liège, Service de Préhistoire
7, place du XX août, bât. A1
4000 Liège Belgique

Tél. 0032/4/366.53.41 Fax 0032/4/366.55.51 Email: prehist@ulg.ac.be
Web: http://www.ulg.ac.be/prehist

Typesetting and layout : Darko Jerko

BAR Publishing is the trading name of British Archaeological Reports (Oxford) Ltd.
British Archaeological Reports was first incorporated in 1974 to publish the BAR
Series, International and British. In 1992 Hadrian Books Ltd became part of the BAR
group. This volume was originally published by Archaeopress in conjunction with
British Archaeological Reports (Oxford) Ltd / Hadrian Books Ltd, the Series principal
publisher, in 2003. This present volume is published by BAR Publishing, 2016.

Printed in England

BAR
PUBLISHING

BAR titles are available from:

BAR Publishing
122 Banbury Rd, Oxford, OX2 7BP, UK
EMAIL info@barpublishing.com
PHONE +44 (0)1865 310431
FAX +44 (0)1865 316916
www.barpublishing.com

TABLE DES MATIÈRES / TABLE OF CONTENTS

1ère Section – Géologie et sédimentologie
Ist Section – Geology and sedimentology

1 - Tin in the Mediterranean Area: History and Geology 3
R. Valera & P. Valera (Italy)

2 - Natural Tin-Bronze Alloy in Iberian Peninsula Metallurgy: Potentiality and Reality 15
S. Rovira & I. Montero (Spain)

3 - Bronze Age Tin Mines in Central Asia 23
J. Cierny & G. Weisberger (Germany)

IIème Section – Métallurgie et métallographie - Archéométallurgie en Europe
IInd Section – Metallurgy and metallography - Archaeometallurgy in Europe

4 - Casting and metallography of Tin Bronzes in Clay Mould 35
Q. Wang & B. Ottaway (GB)

5 - The "Etruscan Tin": A Preliminary Contribution from Researches at Monte Valerio and Baratti – Populonia (Southern Tuscany, Italy) 55
M. Benvenuti, L. Charantini, L. Norfini, A. Casini, S. Guideri, G. Tanelli (Italy)

6 - Analyse élémentaire des dépôts Lorrains : Essai de caractérisation d'une production métallique de la fin de l'Âge du Bronze 67
C. Veber, B. Mille, D. Bourgarit (France)

7 - L'étain dans les examens de quatre ensembles villanoviens. Premiers résultats. 77
A. Lehöerff (France)

8 - The Use of Tin and Lead in Bronze Age Metallurgy 87
M. Primas (Switzerland)

9 - Iron Age Bronze and Tin in the Oriental Alps 93
A. Giumlia-Mair (Italy)

IIIème Section – Archéométallurgie dans les grandes iles de la Méditerranée Chypre, Sardaigne, Sicile
IIIrd Section – Archaeometallurgy on major Mediterranean Islands: Cyprus, Sardinia, Sicily

10 - The Trade of Tin and the Island of Copper 109
V. Kassianidou (Cyprus)

11 - The problem of early tin from the point of view of Nuragic Sardinia. 121
F. Lo Schiavo (Italy)

12 - The discovery of tin near the "megaron temple" of S'Arcu 'e is Forros (Villagrande Strisaili, Nuoro) 133
M. A. Fadda (Italy)

13 - La métallurgie du bronze en Sicile 139
R. M. Albanese Procelli (Italy)

IVème Section – Problèmes de métrologie et le commerce de métaux
IVth Section – Metrology and metal trade problems

14 - Investigating weight systems in Nuragic Sardinia 149
M. Ruìz-Galvez Priego (Spain)

****

Articles qui n'a pas été présenté au Colloquium et qui est hors du programme original / Papers not presented at the Colloquium and out of the original program:

15 - Potential Tin Sources in the Aegean 159
Nikos Skarpelis (Greece)

16 - The Problem of Tin in the Ancient World 165
J.E. Dayton (Spain)

1ère Section – Géologie et sédimentologie

Ist Section - Geology and sedimentology

TIN IN THE MEDITERRANEAN AREA: HISTORY AND GEOLOGY

Roberto G. VALERA & Paolo G. VALERA

Résumé : Récentes trouvailles d'inattendus fragments d'étain dans le sanctuaire nuragique de S'Arcu 'e is Forros, Sardaigne, ont soulevé des nouvelles vives discussions sur la provenance de cet métal et spécialement sur l'origine de son minéral père : la cassitérite. L'ancien débat sur le problème de l'étain dans la région méditerranéenne peut aussi recevoir un nouveau aiguillon, qui devrait aussi être discuté à la lueur de l'actuelle connaissance sur la présence d'étain en Sardaigne et en Toscane. Pour donner une contribution concrète à la discussion, nous proposons une approche à partir d'une soignée consideration des plus anciennes informations disponibles dans la littérature, qui a fourni très fréquemment des données et des descriptions au sujet de « plumbum album » et «calculi nigri ». Une comparaison entre cettes descriptions et la connaissance d'aujourd'hui de la présence et de la géologie de l'étain dans la région méditerranéenne est fructueux, et peut encore nous amener à de découvertes très interessantes. Entre beaucoup d'auteurs qui directement ou indirectement ont parlé d'étain, en référant aussi à ses anciennes sources, nous pouvons rappeler des noms bien connus : Homère, Strabon, Plin l'Ancien. Cet approche a été suivi par plusieurs entre les auteurs les plus recents, mais notre proposition suggère de donner un poids plus important à un traitment pluridisciplinaire des données : historique, archéologique et géologique.

Abstract: Recent findings of unexpected tin scraps in the Nuragic sanctuary of S'Arcu 'e is Forros, Sardinia, have promoted new lively discussions about the provenance of this metal and especially about the origin of its parent ore: cassiterite. The ancient debate on the problem of tin in the Mediterranean area can thus receive new input, which should also be discussed in the light of the present knowledge about tin occurrences in Sardinia and Tuscany. In order to give a positive contribution to such a discussion, we propose an approach starting from a careful consideration of the earliest information available in literature, which has very often given data and descriptions about "plumbum album" and "calculi nigri". A comparison between these descriptions and the present knowledge of tin occurrences and geology in the Mediterranean area is fruitful, and may still lead to some very interesting discoveries. Among the many authors who directly or indirectly spoke about tin, referring also to their older sources, we may remember well known names such as Homer, Strabo, and Pliny the Elder. Actually, this approach has been followed by several recent authors, but our proposal suggests to give greater weight to a multidisciplinary processing of the data, i.e. a historical, archaeological, and geological approach.

INTRODUCTION

The *tin mystery*, as it was defined by Bob Maddin in Matsue (1998) is still in discussion, and many authors had contributed to the debate long before that date. The terms of the question are rather clear, and the geological side seems to be sufficiently ascertained. Actually, the distribution of tin-bearing mineralisation in the Mediterranean area is well outlined, and new entries are not be expected, because of the today's known geological constraints.

A new event brings additional substance to the discussion: the recent find of tin scraps in a Nuragic sanctuary in Sardinia (S'Arcu 'e is Forros: Valera et al., 2002; Fadda, this vol.; Lo Schiavo, this vol.). Particular interest is provided by the arising suggestion of tin supply from the island itself, since tin bearing mineralisation occurrences are mentioned in the literature (Penhallurick, 1986), and a foundry activity is supposed to have developed in this sanctuary area. Therefore the question is open, and we are going to discuss the data and their possible contribution to the "tin mystery" debate in the light of the present, and especially the historical, knowledge.

THE SARDINIAN TIN ORE DEPOSITS

It is convenient to clarify the reality of the tin occurrences in Sardinia. Most of them are known for their strict mineralogical, rather than economic, value. In the past century a moderate interest was only devoted to two deposits, both in the SW of the island: Perdu Cara (Monte Linas) and Canali Serci (Villacidro). They belong to a rich metallogenic province, well known for its long lasting production of various commodities, mainly lead, zinc, silver, fluorite, and barite, but also for the presence of molybdenum, cobalt, and copper.

Perdu Cara (or Sedda Planuceddu, Monte Linas: Zuffardi, 1958; Biste, 1979) is a NE-SW trending vein mineralisation, 10-100 cm thick, 1500 m in length, consisting of quartz-cassiterite in the northern section, changing to quartz-arsenopyrite in the south. Cassiterite was irregularly dispersed in quartz, and also in the wall rock for few decimetres. Some small trenches and pits of indecipherable age were visible before a recent exploitation activity, which totally exhausted the deposit. Two outstanding questions arise: was BA technology adequate to the Perdu Cara primary deposit exploitation? Was a cassiterite placer concentration possible in the local morphology? The answer to the first question is negative, because of the difficult composition of the vein (cassiterite specks scattered in massive quartz). The second question is open: the formation of small placers was certainly possible in a downstream creek, and if so it was totally exploited, because no traces are visible now.

Canali Serci (Piepoli & Collari, 1936; Biste, 1979) is a quartz-siderite lode with sphalerite, galena, pyrite, marcasite, pyrrhotite, cassiterite, and rare stannite and tetrahedrite. Minor cassiterite is finely intergrown with major zinc and lead sulfides, and it is only visible under microscope. Again the same questions arise: was a primitive tin metallurgy viable for such a primary ore association? Was a cassiterite placer concentration possible in the local morphology? Both answers are negative, because only a modern flotation process can separate and concentrate each ore mineral from such a primary mixture, and only a weak geochemical increase may be expected in the stream sediments derived from the erosion of the Canali Serci lode.

In conclusion, we may say that the only hypothetical possibility, though very weak, of a Sardinian tin beneficiation was offered by an eventual small placer from the Perdu Cara mineralisation.

THE NEW SARDINIAN TIN SCRAPS

The new tin fragments, weighing some tens of grams each, were found in association with scraps of Cu-Pb alloys, suggesting the existence of a local metallurgical activity, referred to the LBA (Fadda, this vol.). Some fragments were analised to detect the Pb content and lead isotope ratios. The results are given in Table 1.

The analyses of lead isotope ratios in one of the tin scraps were kindly performed by Prof. E. Pernicka, and the results clearly show the presence of a foreign lead. In fact, if we compare the lead isotope ratios of the tin scrap (fig. 3) with those of the lead coming from the two families of Sardinian

Table 1 – S'Arcu 'e is Forros: Pb content and Pb isotope ratios in tin and copper-lead alloy scraps.

Sample	Pb %	208/206Pb	207/206Pb	206/204Pb
*Tin scrap**	0.3	2.05735	0.847325	18.5195
*Cu-Pb alloys***				
AVR 123	2.5	2.12746	0.86845	17.9639
AVR 124	0.17	2.12174	0.86734	17.9491
AVR 125	5.4	2.11629	0.87482	17.7645
AVR 126	3.5	2.10119	0.87534	17.6467

[*Analytical data: E. Pernicka, 2000, pers. comm.] [**Analytical data by the authors]

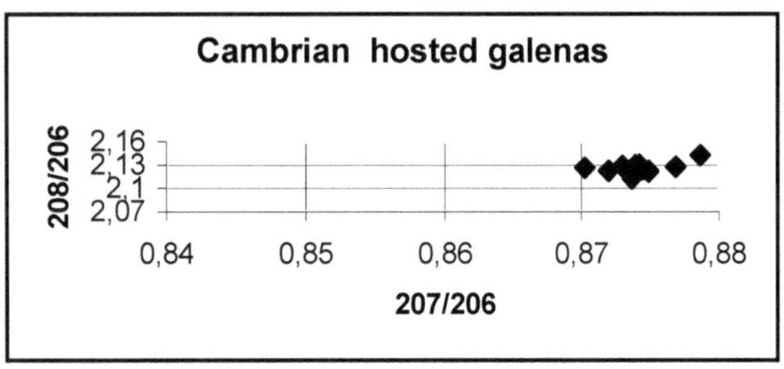

Figure 1. The lead isotope ratios of Sardinia Cambrian hosted galenas (modified from Valera et al., 2002).

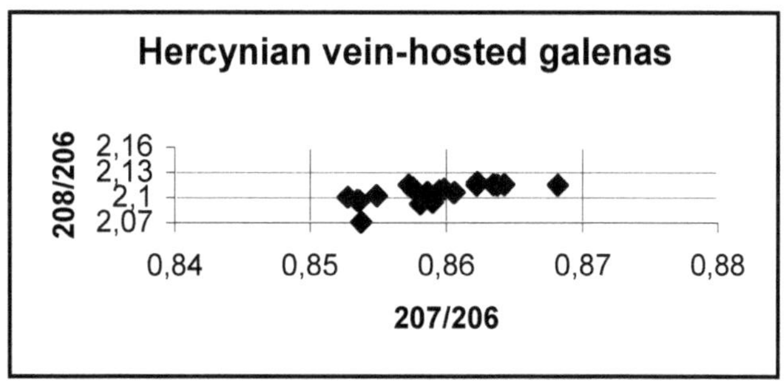

Figure 2. The lead isotope ratios of Sardinia Hercynian vein-type hosted galenas (modified from Valera et al., 2002).

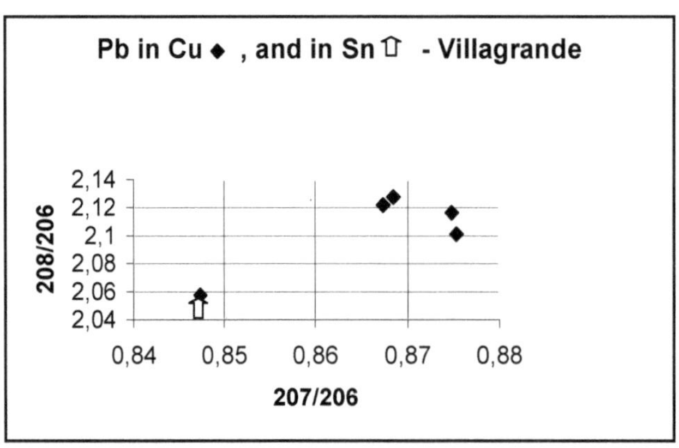

Figure 3. The lead isotope ratios of the metallic scraps of Villagrande.

Table 2: Trace element grades in some tin scraps of S'Arcu 'e is Forros.

PPM	AVR-113	AVR-114	AVR-115	AVR-117	AVR-118
1 - Li	-	-	-	-	-
2 - Sc	1,2	1	-	-	-
3 - V	23,6	27,1	28,4	31	20,9
4 - Cr	3,4	5,2	5,4	4	3,5
5 - Co	42,7	58,9	29,6	127	12,5
6 - Ni	50,6	51,1	43,3	53,3	43,6
7 - Cu	51,3	118	55	38,3	10,9
8 - Zn	10,6	19,5	9,6	11,8	10
9 - Ga	0,1	0,2	0,2	0,3	0,2
10 - Y	-	-	-	-	-
11 - Zr	-	-	-	-	-
12 - Nb	2	1,6	1,6	2,1	1,8
13 - Cd	9	50	33	48	42
14 - In	7	3	19	11	12
15 - Mo	3,3	4,5	2,2	3,6	1,9
16 - Sb	63,7	43,1	44	42,2	48,4
17 - Cs	4,8	4,9	5,2	5,2	6,2
18 - La	0,3	0,5	0,3	0,4	0,5
19 - Ce	2,3	2,6	2,5	2,6	3,1
20 - Pr	-	-	-	-	0,2
21 - Nd	-	-	-	-	-
22 - Sm	0,7	0,8	0,9	0,9	0,9
23 - Eu	0,5	0,7	0,8	0,8	0,9
24 - Gd	0,3	0,3	0,2	0,3	0,3
25 - Tb	-	-	-	-	-
26 - Dy	-	-	-	-	-
27 - Ho	-	-	-	-	-
28 - Er	-	-	-	-	-
29 - Tm	-	-	-	-	-
30 - Yb	-	-	-	-	-
31 - Lu	-	-	-	-	-
32 - Hf	-	-	-	-	-
33 - Ta	3,8	3,1	3	3,1	3
34 - W	6,8	4	3,4	3,8	4,6
35 - Tl	0,5	0,4	0,4	0,3	0,3
36 - Pb	3100	2000	1300	500	8700
37 - Bi	10,4	1,4	14,6	1	4
38 - Th	-	-	-	-	-
39 - U	-	-	-	-	-

galenas (fig. 1 and fig. 2; identified as Cambrian hosted and Hercynian vein-hosted: Valera et al., 2002), it is absolutely clear that the lead isotope ratios from the tin scrap fall far apart from the fields of the two families, whereas the lead isotope ratios of the Cu-Pb alloys are well consistent with the field of the Cambrian hosted galenas, once more according to the trend of the Nuragic leads.

We also analised[1] the trace elements in different tin scraps, and the results show a rather homogeneous composition, with the exception of the Pb behaviour (fig. 4a-e: the element numbers correspond to the list in table 2).

The fluctuation of the lead values can be explained by a variable addition of, or pollution by, the same metal during the metallurgic process. As cassiterite samples were not found in the sanctuary, and because of the foreign lead isotope ratios of the analysed tin scrap, we assume that its metallurgy was developed in its mother land. The actual challenge is the identification of such a mother land, a major concern for all scholars facing similar problems (Maddin, 1998). In our case, a simple geological approach suggested the following procedure:

- check of the trace element patterns obtained by the analyses of different tin scraps;
- choice of a couple of elements typical and common to the trace element patterns. The selected couple must behave in a similar way in the different scraps, so that the two elements can be defined as markers. Since the Cd-Sb couple in the Villagrande tin scraps is in good agreement with such requirements, we decided to use it as the marker couple for our case;
- comparison between the trace element patterns of the Villagrande tin scraps and the trace element patterns of cassiterites from different metallogenic provinces. Only

[1] All measurements were made by quadropole ICP-MS on sample solutions, obtained by ultrapure nitric acid digestion. The uncertainties (RSD%) for individual measurements are <0,5%, and accuracy <1%. The values are based on the NBS "Common Lead Isotopic Standard" (NIST SRM-981).

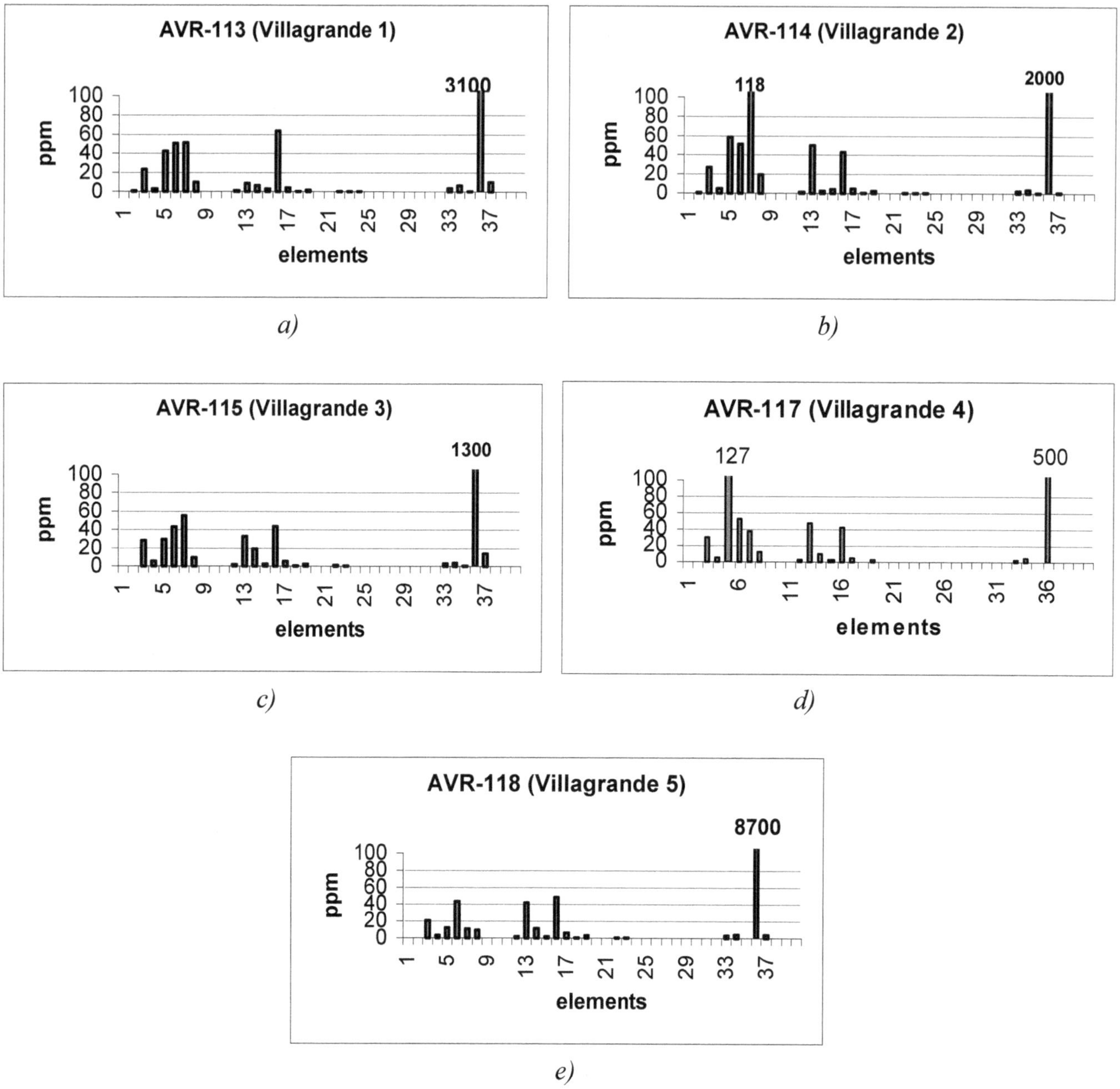

Figure 4 (a-e) – Trace element patterns in five tin scraps from Villagrande (element numbers in Table 2 and 3).

for a preliminary outline, we analysed some available samples of cassiterites from Sardinia, Tuscany, Spain, Erzgebirge, Nigeria, and China. Nigeria and China are of course only for external reference, whereas the other districts could not be discarded *a priori* as options for the origin of the Villagrande tin. The comparison between metal tin and tin oxide is reasonable, if we can exclude the risk of pollution by, or intentional addition of the selected markers during the metallurgic refining process. Actually the refining process should be characterised by a more or less appreciable loss of most of the trace elements contained in the starting cassiterites, rather than a concentration;

– the last step of our procedure is the possible recognition of a metallogenic province, whose cassiterites show a similar outline as per the selected markers.

The results of this preliminary approach are rather interesting. Each analised cassiterite showed a fairly distinct "personality" with peculiar trace element patterns. However only one sufficiently reliable conclusion can be drawn: i.e., the lack of any link between the Villagrande tin and the Sardinian cassiterites. In fact the Cd-Sb couple has very low values in the Sardinian (Perdu Cara) cassiterites, and the same result is shown by the other cassiterites (Erzgebirge, Nigeria, Spain, China), except for the Monte Valerio sample (Fig. 5a-i, tab. 3).

In conclusion, the above data bring additional support to the exclusion of Sardinia as mother land of the Villagrande tin. The few samples of other districts we analysed seem to point rather to Tuscany, but the problem is completely open. We need a larger population of samples from the considered tin deposits, but also have to extend the sampling procedure to as many deposits as possible, which might be representative, belonging to the important tin metallogenic provinces of Erzgebirge, Hiberian Peninsula, Brittany, and Cornwall.

Table 3 – Trace element grades in various cassiterites (in ppm). Cas-1: China; Cas-2:Erzgebirge; Cas-3,4,5,6 : Sardinia (Perdu Cara); Cas-7: Monte Valerio; Cas-12: Nigeria ; Cas-14b : Spain.

	Cas-1	Cas-2	Cas-3	Cas-4	Cas-5	Cas-6	Cas-7	Cas-12	Cas-14b
1 - Li	-	99,8	2,8	49	5,7	5,7	11	7,3	27,1
2 - Sc	25,7	864	-	-	-	-	39	-	-
3 - V	52,6	-	-	-	-	-	46	37,2	10,4
4 - Cr	13,3	27,2	8,5	-	2,5	-	173	6,3	6,2
5 - Co	5,3	5,3	2	4,2	2,3	6,5	255	5,2	-
6 - Ni	26,6	43,9	16,7	21,4	11,5	12,2	1999	12,8	6,5
7 - Cu	58,6	204	314	105	124	169	350	123	31,2
8 - Zn	48,9	148	79	110	108	78,2	544	23,4	-
9 - Ga	nd	48,6	6,1	10,4	3,8	1,3	19,5	5,9	2,7
10 - Y	0,6	-	-	-	-	-	17	3,1	-
11 - Zr	19,5	111	1,5	2,2	1	1,6	22,4	23,7	-
12 - Nb	26,1	648	8,5	12,5	8,9	9,4	13,5	170	14,1
13 - Cd	2	-	2	2	19	11	43	1,4	0,77
14 - In	19	20	25	26	16	22	27	-	-
15 - Mo	11,8	16,9	9,5	9,3	9	8,7	11,1	-	-
16 - Sb	-	2,3	1	3,7	2,7	8,5	228	-	-
17 - Cs	2,8	5	-	1,5	1	-	3	-	3,3
18 - La	1,2	1,1	0,5	1,6	-	0,5	8,2	16	-
19 - Ce	1,3	1,8	0	3,4	-	0,2	10,2	45	0,64
20 - Pr	0,7	0,5	0,3	0,7	0,3	0,4	2,5	2	-
21 - Nd	-	-	-	-	-	-	7,1	5,7	-
22 - Sm	0,3	-	-	0,3	-	-	2,4	0,8	-
23 - Eu	1	0,8	0,8	0,9	0,7	0,7	1,5	-	-
24 - Gd	1,2	1	0,8	1,3	0,8	0,9	3,8	1,2	-
25 - Tb	0,24	-	-	0,1	-	-	0,6	-	-
26 - Dy	-	-	-	-	-	-	3,3	0,59	-
27 - Ho	-	-	-	-	-	-	0,7	-	-
28 - Er	0,4	0,3	-	0,2	-	-	2,5	0,38	-
29 - Tm	-	-	-	-	-	-	-	-	-
30 - Yb	0,2	0,4	-	-	-	-	2,1	0,47	-
31 - Lu	0,6	0,4	0,4	0,4	0,3	0,3	0,6	-	-
32 - Hf	1,3	35,2	-	-	-	-	-	4,1	-
33 - Ta	15,8	325	10,9	11,4	10,4	10,9	11,5	73,9	-
34 - W	25,5	570	103	181	107	145	448	42,8	-
35 - Tl	1,1	0,8	0,6	0,8	0,5	0,6	1	-	-
36 - Pb	205	558	223	988	158	1080	2072	-	67
37 - Bi	3,7	124	1	0,8	0,8	0,7	87,2	-	-
38 - Th	-	18,9	-	0,4	-	-	2,2	-	-
39 - U	1,9	88,6	1,2	1,3	0,6	0,5	2,3	7,8	1,6

EARLY TIN IN SARDINIA

Bronze Age metallurgy in Sardinia is well known, and its high level of activity is upheld by a great deal of products found in several sites and hoards. The first bronze artifacts appear in the so called "Bonnannaro Culture" (XVIII-XVII century B.C.: BA Phase I, Lo Schiavo 1997). Imported models are dominant at the beginning, rapidly followed by the birth and expansion of a local metallurgic industry. Copper, lead, and minor silver were already known in the previous Neolithic Ozieri culture, which is considered the first Sardinian metal-using culture, spanning from the end of the 4th millennium B.C. to the beginning of the 3rd millennium B.C. (Lo Schiavo, loc cit.). However, before the discovery of the s'Arcu 'e is Forros' tin, only three sites had been mentioned for the presence of tin by several authors (full references and detailed analysis in Lo Schiavo, this volume): the sites are Abini (Teti) and Forraxi Nioi (Nuragus), where tin ore (cassiterite) was recognized, and La Maddalena (Silanus-Lei) where the only certain metal tin described in Sardinia to date (Vivanet, 1890) was found. Actually Tylecote et al. (1984) do not accept the classification of the Forraxi

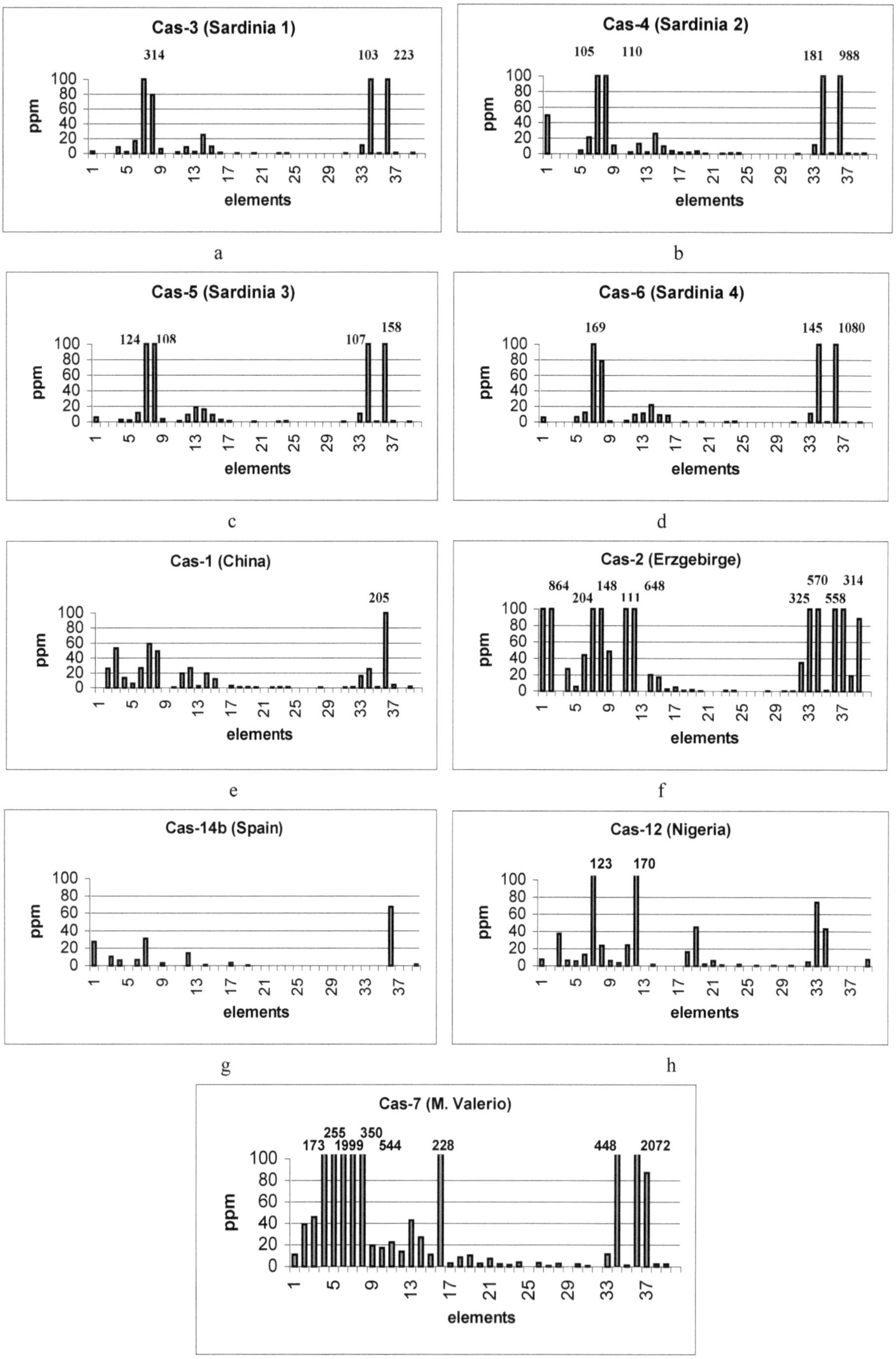

Figure 5. Trace element patterns of cassiterites from different areas. a-d: Sardinia (Perdu Cara); e: China (unknown deposit); f: Erzgebirge (Cinovec); g: Spain (Mina Casualidad, Zamora); h: Nigeria (unknown placer); i: Tuscany (Monte Valerio) (element numbers in Table 2 and 3).

Nioi materials analysed by Cambi (1959) as cassiterite, and state that they are 90% "corroded tin", probable remnants of metallurgic activity. In any case the question is open, because it is not completely sure which specimens were studied by Cambi, and because they are not easily available for a check-up. Therefore the only certain and available metal tin nowadays, is the one we are discussing, described by Fadda (this vol.), and already announced by Valera et al. (2002). The material is stored at the Soprintendenza Archeologica per le province di Sassari e Nuoro (Nuoro Territorial Service).

FACTS, PROBLEMS, AND OBSERVATIONS

At this point we can list a number of factual data dealing with the ancient tin occurrences and metallurgy in Sardinia. These are emphasized by the S'Arcu 'e is Forros field and laboratory studies, and by a careful consideration of the data already mentioned in the literature. We shall here list the factual data that appear particularly important and meaningful to our approach:

1. S'Arcu 'e is Forros sanctuary:
   - chronology: LBA to EIA (Fadda 1989a, 1989b, 1991)
   - the Pb isotope ratios of the tin scrap analysed by Prof. Pernicka are foreign to Sardinia
   - the Pb isotope ratios of the Cu-Pb alloy scraps are consistent with the Sardinian Cambrian hosted leads
   - the trace element analyses of the tin scraps are not consistent with those of known Sardinian cassiterites (Perdu Cara)
   - among the large number of collected scraps no cassiterite was observed
   - the local metalworkers clearly employed only metal tin for bronze alloying.
2. Ancient metallurgy in Sardinia:
   - the only metal tin found in Sardinia to date (before the discovery of S'Arcu 'e is Forros) is that of Silanus-Lei (La Maddalena)
   - cassiterite was certainly used in Abini, and probably in Forraxi Nioi
   - whether it be ore or metal, the tin employed in Sardinian metallurgy is to be assigned to a supply by import
   - copper and bronze objects analysed by Begemann et al. (2001) show a variable composition in Sn, Pb, As, Sb, Ag, and Fe content. Such variability can be frequently observed in other published analytical data. Especially lead is well represented (Tylecote et al., 1984: copper-lead ingot of table 6.7)
   - an Ag-bronze, tin-free figurine, exhibited in the Cagliari Archaeological Museum as the *"Bronzetto with long hair-plaits"*, has been analysed and described by Atzeni et al. (1992)
   - silver, or silver-bearing artifacts are very rare in Nuragic age sites
   - the trend of silver grade in the lead scraps of Nuragic age is consistent with that of most of the Sardinian lead ores (cf. the high silver grades of Nuragic lead scraps in Lo Schiavo & Valera, 1998). Should any desilveration by cupellation have been performed, it would have to be done only on selected rich ores, though obtaining an imperfect recovery.

These facts are to be carefully considered, in discussing issues that can be summarized as follows:

- From a more general point of view, was the tin actually a true "base metal" for the BA cultures?
- What about tin ore deposits in the Mediterranean area?
- Where did the general metal supply come from? In the S'Arcu 'e is Forros sanctuary only the lead of the copper scraps is of certain local origin, and according to isotope ratios it belongs to the Cambrian-hosted deposits of SW Sardinia (Valera et al., 2002)
- In particular, what about the origin of the tin ore and that of the metal tin found in the Nuragic sites?
- What about the origin of the silver used for the *"Bronzetto with long hair-plaits"*?
- Was the silver-bronze an experiment, or a first answer to foreign suggestions? Might it be a possible reference to a *"candidum Corinthium aes"* (Giumlia-Mair, 2000)?
- Why does metal tin occur in S'Arcu 'e is Forros and La Maddalena, and why tin ore in Abini and Forraxi Nioi? Did this depend on different sources, and/or different traders, and/or different metallurgic styles or advancements, considering that the origin of tin was foreign to Sardinia?
- In the S'Arcu 'e is Forros scraps, lead was intentionally added to copper, and was of Sardinian origin, as said above. Was it a matter of lively metallurgic experimentation? Was there a rule for different uses? Was it a "salting" trick to increase the metal value? When the lead content is particularly high, might it be referred to the general technological evolution that occurred all over Europe in the EIA, which was marked by hoards of as-cast metal artifacts with high lead contents (Huth, 1998)?

The above selection of problems outlines some of the major features of the BA metal industry, where tin origin and metallurgy seem to be the outstanding open question: the *tin mistery*. But: why a mistery? A great deal of studies and analyses that were carried out in the past decades all point towards one main conclusion: ancient metalworkers were highly skilled and capable of obtaining exceptional results. Their attitude to metals was not only matter of advanced technology, but it was also clearly based on a kind of philosophic approach. Each ore, each metal had a "meaning", not only a "value", and we should recover the same attitude if we wish to understand their mentality and behaviour, so as to solve what we have called *problems*, or *the mistery*. In short, only a modern scientist can refer to TIN as a *base metal* (Penhallurick, 1986): according to a BA point of view, which was obviously substantially different from ours, the position of tin was high up in the metal hierarchy, i.e. it was considered a *precious metal*, characterized by a special behaviour (how else could we interpret the *tin disease*? cf below), that had to be used only in particular conditions, because it gives a particular meaning to the artifacts. Its indispensable presence in the alloys for weapons gave it additional preciousness: i.e. the power, or rather, the status symbol of power. Only through the recovery of this primeval approach, turning the *tin mistery* into *tin philosophy*, may we try and give a contribution to many of the open questions. No doubt the ancient authors, whose words are the proper key to such an interesting world, are of great help.

DISCUSSION: TIN MISTERY OR TIN PHILOSOPHY?

It is well known that a long, deep, and lively debate developed in the past decades among scientists about tin, which is still considered an important, rare, and valuable metal. We shall bring some contribution to the discussion on the case of tin from a number of particular angles, i.e. the geological and the historical, while leaving the archaeological conclusions open.

A serious discussion on the tin case must be based on its occurrence and behaviour in the earth's crust. This means, tin in nature. As regards the average tin crustal abundance, conflicting data are reported in the specialized scientific literature: from 2 to 40 p.p.m. (parts per million) according to different authors, the former being recently the most credited figure. But let us compare tin with some other metals. Generally accepted figures are: iron 50.000, copper 70, lead 16, arsenic 5, silver 0,1, and gold 0,005 (Routhier, 1963). Whatever the correct value, tin crustal abundance is low, and is mirrored by the scarce occurrence of tin ore deposits. Actually an important concept must be taken into consideration in order to assign an effective weight to such figures: the relative abundance of a given element is not the only controlling factor of its efficiency in giving rise to mineral concentrations of economic value, i.e. to ore deposits. An essential role is also played by its behaviour in the geological processes, and in general in the processes which develop in, and on, the earth's crust: we refer to the physical-chemical leading constraints concerning solubilities, complexing, sensibility to various reactions, etc., not least, its preferred primary genetic environment, igneous (felsic or mafic), sedimentary, and metamorphic. Such a picture will help us understand why lead, whose abundance falls in the same range as tin, is much more frequent as an occurrence of ore mineral. On the other hand, the value of lead compared to tin also depends on its economic evaluation, which is in turn derived from the aforementioned meanings that were given to the metals in antiquity. Unlike lead, tin had a generally restricted use, devoted to artifacts of particular importance or sense. Therefore, in the case of metals, the meaning of *economy* in antiquity does not only refer to evaluation simply from a material point of view: it also has an undeniable *philosophic* component. All along human history, different types of *geologic bodies* (e.g. obsidian) assumed the "honour" of *ore deposits* due to their economic value as considered at a particular stage when their intrinsic value was assessed also according to a more philosophic approach.

From a genetic point of view valuable tin ore deposits are mostly associated to a group of granites that had long been called *tin granites* in the Old World (Lehmann, 1990). Tin granites occur only in some particular lithospheric sections, and characterize few metallogenic districts. The Mediterranean area is very poor, almost lacking tin granites. Nevertheless, many cassiterite occurrences are mentioned, especially in the Balkan region (Ilavsky, 1979), unfortunately of no practical interest, even in a BA framework. Actually, the nearest tin districts are those known since antiquity: NW Iberian peninsula, Brittany, Erzgebirge, and Cornwall.

Tin minerals are few, the most important and abundant being the tin oxide *cassiterite* (tin content 78%). Cassiterite is a very stable mineral, with high specific gravity (7.0). It is mostly affected by the mechanics of the weathering processes, so it can give rise to secondary deposits or "placers" which are of course always associated with primary deposits related to tin granites. Only the tin sulfide *stannite* (tin content 27%) can undergo chemical weathering processes, but its rare occurrence does not have a great influence on the geochemistry of tin.

Lead has a crustal abundance comparable to that of tin, but its occurrence in the earth crust, and its behaviour in the geochemical processes is absolutely lively and various. Its atomic radius, ionic charge, ionization potential etc. allow Pb^{2+} to replace diadochically some important elements such as Ca^{2+} and K^{+} (Rankama & Sahama, 1950), and this occurs in igneous rocks. But Pb^{2+} may also proceed to the replacement of Ca^{2+} in minerals formed at lower temperatures. Therefore lead participates in many natural processes in the exogenous environment, and can be concentrated in so many different conditions, that we may frequently find it either in hydrothermal processes at high to low temperatures, or in sedimentary processes, etc. Many lead minerals are known in nature, the most important being the lead sulfide *galena* (lead 86%), but sulphosalts and oxides are not rare. Actually galena is one of the most frequent, and economically important minerals occurring in base metal ore deposits, both of the hydrothermal and the sedimentary type, and lead ore deposits are widely distributed all over the world. Therefore lead has never been considered a precious metal, even if through its association with silver in galena it became a very important ore since the earlyest times of metallurgy, when the cupellation process was set up. Following lead, silver is also common, though in much lower quantities. Due to its brightness, aesthetics, and physical properties, silver has been considered a precious metal since the very beginning of time. Many silver-bearing ore deposits are known in the Mediterranean area, whereas tin is practically absent. So when tin first appeared in the Near East EBA cultures, it found an already well developed metallurgic tradition, especially based on arsenical copper production, and on cupellation to obtain silver. It is worth quoting that the crustal abundance of arsenic (cf above) is also comparable to that of tin: but it had long been an essential component of the metallurgic know-how.

All existing finds and documents that may be examined point to the exploitation of the *experimental method* by the metalworkers, which made it possible to produce different types of alloys according to the different needs, and to process pure metals in the best possible way. The appearance of tin was a revolutionary shake in the metalworking world, because it gave a new, exciting possibility to create precious artifacts and weapons that looked very different and had unexpected physical properties. At that time the classification of known metals was very simple: gold and silver were precious, scarce but common (in the sense that they occurred in many countries); arsenic was useful but scarce although common; copper and lead were the base metals, frequent and common. Tin suddenly became essential, but it was rare, scarce and unknown in the then "developed countries". According to the Ebla tablets (Pettinato, 1979) it is possible to estimate the metal ratio to 1 shekel of silver in Ebla, 2500 B.C.: gold is 4.75, tin 0.66, copper 0.02. Copper, though important, is well apart from tin, which is closer to silver.

Tin was so highly esteemed that from the beginning and for a long time, it was even purchased in remote countries, together with other valuable goods like lapislazuli, amber, and nephrite. The metallogenic province of Fergana – Tien Shan (Afghanistan; Penhallurik, 1986) is one of the most accepted sources of tin, and is perfectly suitable from a geologic point of view. However, exploration soon started in countries that were then known for their mineral wealth. Unfortunately the Mediterranean area is particularly poor in tin granites, so the first positive results arrived after quite a long time, when products of the western Hiberian peninsula, Brittany, Cornwall, and later of Erzgebirge became available. Perhaps a contribution was also given by Tuscany, and recent indications (Gillis, 1991) assign a possibilitiy to Turkey (Bolkardag) and Macedonia (Cer). But it cannot be denied that if something fruitful such as exploitable tin ore had been found in the Mediterranean area, its memory would have been recorded and left traces today, as happened in all other known mining districts of antiquity: Ergani Maden, Cyprus, Lavrion, Sardinia etc.

In any case, tin was considered a sort of "royal metal": not only precious, but *precious and indispensable.* It was essential in the preparation of the bronze weapons - a status symbol of power as stated above. It is always mentioned in ancient literature as a highly valuable metal. A very famous passage in Homer's Iliad describes the work of Hephaestus to make Achille's shield. There tin, together with gold, is quoted as a true precious metal ("the oxen were made of gold and tin": XVIII, 574)[2]. Another reference is the arrival of the Diomede's chariot "inlaid with gold and tin" (XXIII, 504)[3].

A very authoritative voice is that of Herodotus, who describes some features of the northern European regions, when speaking vaguely about islands named Cassiterides "where tin should come from" (III: 115, 1). However, a few lines later Herodotus says that "tin and amber arrive from the extreme borders of the world", and "it seems certain that the greatest quantities of gold occur in northern Europe" (III: 115, 2; 116,1)[4], concluding that "actually, the extreme regions, surrounding the other countries, own the things we consider the finest and rarest" (III: 116,3)[5]. Therefore, according to Herodotus tin is to be classified among "the finest and rarest things".

Furthermore, describing the mineral wealth of the Iberian peninsula, Strabo quotes Posidonius saying that "among the Artabrois, the inhabitants of the north-westernmost Lusitania, the land flourishes with silver, tin, and white gold (actually mixed with silver)" (III, 2.9)[6].

We may also interpret as a reference to tin (and gold) the words of Job in the Scriptures (XXVIII, 11): "They dam up the trickling streams and bring to light the hidden treasures". However, the most conclusive definition occurs in the well known passage by Pliny the Elder (XXXIV, 47): "Sequitur natura plumbi, cuius duo genera, nigrum atque candidum. pretiosissimum in hoc candidum, Graecis appellatum cassiterum fabuloseque narratum...". Tin is *apertis verbis* declared "most precious".

This brief anthology, which could be much expanded, should bring us to a basic change of approach, to a sort of new philosophy, we would say, of the processing of the tin concept. Keeping in mind a BA point of view, tin is one of the THREE precious metals, and certainly not the least. Many authors dealt with tin, and important and well documented studies have been published in the past decades. Among them Gillis (1991) starts revealing a new feeling, when she says that "tin must have been an extremely valuable commodity", and also suggests that the tin-covered vessels in the Minoan-Mycenean world "were at least the equivalent of silver in value and status".

So tin was rare, scarce, and precious. But it was also weak, its weakness being given by the so called "tin disease", a structural instability that controls crystallisation in the pure metal state (Maddin, 1998). It is the transformation β tin – α tin, which starts developing when the metal undergoes decreasing temperatures below 13 °C, and is accompanied by an irreversible architectural transformation into powder, then easily dispersed. The effects of this process must soon

---

2 'Εν δ' ἀγέλην ποίησε βοῶν ὀρθοκραιράων·
αἱ δὲ βόες χρυσοῖο τετεύχατο κασσιτέρου τε,
μυκηθμῷ δ' ἀπὸ κόπρου ἐπεσσεύοντο νομόνδε
πὰρ ποταμὸν κελάδοντα, παρὰ ῥοδανὸν δονακῆα.

3 αἰεὶ δ' ἡνίοχον κονίης ῥαθάμιγγες ἔβαλλον,
ἅρματα δὲ χρυσῷ πεπυκασμένα κασσιτέρῳ τε
ἵπποις ὠκυπόδεσσιν ἐπέτρεχον·

4 Ἐξ ἐσχάτης δ' ὦν ὅ τε κασσίτερος ἡμῖν φοιτᾷ καὶ τὸ ἤλεκτρον. Πρὸς δὲ ἄρκτου τῆς Εὐρώπης πολλῷ τι πλεῖστος χρυσὸς φαίνεται ἐών.

5 Αἱ δὴ ὦν ἐσχατιαὶ οἴκασι περικλήουσαι τὴν ἄλλην χώρην καὶ ἐντὸς ἀπέργουσαι τὰ κάλλιστα δοκέοντα ἡμῖν εἶναι καὶ σπανιώτατα ἔχειν αὐτά .

6 'Εν δὲ τοῖς 'Αρτάβροις, οἳ τῆς Λυσιτανίας ὕστατοι πρὸς ἄρκτον καὶ δύσιν εἰσίν, ἐξανθεῖν φασι τὴν γῆν ἀργυρίῳ, καττιτέρῳ, χρυσίῳ λευκῷ (ἀργυρομιγὲς γάρ ἐστι)·

have been discovered by early metalworkers at the beginning of the utilisation of tin, because of the dramatic constraints of this transformation. It is likely that the appearance of the first tin ingots and their trade coincided with the solution of the tin disease problem, which was simply the addition of even low quantities of other metals to the pure metal tin. Though rough, the aforementioned experimental method depended on the availability of various minerals, mainly Bi-, or Sb-, or Cu-, or Pb-sulphides or sulphosalts. Actually (cf figg. 4 a-e, table 2) the Villagrande tin scraps contain variable but substantial lead, and a little bit higher antimony than the average values quoted by Maddin (1998). Such Sb and Pb values could explain how the tin scraps have been preserved in a stable state, though for a long time in a climate with frequent winter temperatures below 0 °C.

Giving due weight to the practice of the *experimental method* by the BA metalworkers, we can now perceive how and when the tin trade shifted from cassiterite to metal ingots. This very likely happened as soon as metal tin was stabilised, and this was clearly achieved with the development and progressive improvement of smelting, which very often involved different ore minerals, and from slight to substantial changes in processing conditions.

The same interpretation should be given to many important events of the history of metals, e.g. the invention of bronze. Actually such an invention can be explained in two ways. One should assume the existence of metallogenic districts where originally both Cu and Sn ores contemporaneously occur. In this case the primitive metalworkers followed a natural suggestion by mixing coexisting minerals. Of course at the beginning they did not know that their mixture derived from two components, nor the relative content of each component. The other is simply provided by the experimental method, which led to the discovery of the mixture of well known copper ores with the strange metal hidden in the black sand and small pebbles which Pliny later called *calculi nigri* (XXXIV, 47). The experimental approach is clearly at the base of all the metallurgic activity in antiquity, and it is proved by the composition of metal objects occurring in extremely different sites. Of course, it had to be based on the availability of different types of metals and minerals. In the same way we may also explain the production of arsenical copper, and that of arsenical-antimonial copper, whose discovery and diffusion clearly started, and moved from fahlore-bearing districts, being soon followed by the improvement of the alloying method through the beneficiation of other As- and Sb-bearing ores. Further examples are given by copper-lead mixtures and by the addition of lead to tin. However, one of the best examples of the experimental method seems to be given by the above mentioned *bronzetto* of the Cagliari Museum, a bronze alloy with silver (6.5%) instead of tin.

The availability of different types of metals and minerals was guaranteed by the establishment of trade routes that reached all the known mineral districts, capable of providing all the required commodities to countries short of mineral wealth. On the other hand the flow of goods may also have acted as a suggestion for the beneficiation of local mineral resources previously unknown to the local people because of a lack of metallurgic experience. We must look inside this framework to get an idea of the origin of silver in the production of the Ag-bronze figurine, keeping in mind that silver was a rather uncommon metal in Nuragic culture, though well represented in Sardinian galenas.

CONCLUSIONS

The above discussion leads to some interesting conclusions, though it leaves many open questions. First of all we support the clear evidence of the outstanding position of tin in the BA metal hierarchy. Tin was maintained as a true precious metal for a long time, and was essential to the production of powerful weapons. It was so important, that we cannot expect to find a source of tin in the Mediterranean area or its immediate surroundings, forgotten by history for about two millennia and unknown even to Strabo and Pliny. Its supply from the Far-East was suddenly interrupted when Mari was destroyed, but it was soon replaced by the new western sources that became available and were beneficiated: NW Iberia, Brittany, and later Cornwall and Erzgebirge.

The metal trade routes were of strategic importance, and were certainly accompanied by an exchange of know-how. The skill and experience of the metalworkers were continuously improved through experiment aimed at obtaining the most useful and the newest products. F. Braudel (1998) says that itinerant metalworkers and smiths are well known characters in the Near East history, and their traveling since the beginning of the 3rd millennium is absolutely certain.: at about 2500 B.C. in the big towns of the Near East, metalworking is owned by guilds of foreigners who jealously preserve their secrets, and do not mix with the citizen communities. The results of this lively metallurgy are given by the metal mixtures and alloys found all along the history of the BA and later. In particular, the Nuragic "*Bronzetto with long hair-plaits*" could have been prepared as an experiment to improve the aesthetics approaching the *candidum Corinthium aes* (Giumlia-Mair, 2000), and the lead-copper mixtures of Villagrande could have been in line with the general technological evolution that occurred all over Europe in the EIA, and was marked by hoards of as-cast metal artifacts having high lead contents (Huth, 1998).

The evolution of the knowledge and use of tin was clearly marked by problems relating to the instability of its two phases. The refinement of the know-how to correct it was an important step, and was achieved through the deliberate (or unintentional?) addition of "impurities" like Pb, Sb, and Bi. Stabilising the precious phase was a brilliant result, and it was rapidly transformed into the improvement of the tin trade that changed from cassiterite to tin ingots.

The combined presence of both tin ore and metal tin in Sardinian bronze metallurgy may be explained in two ways: the simpler being the technological evolution from a smelting process based on the copper + tin ore couple ("cementation": Cambi, 1959; Herdits et al., 1995) to an evolved smelting process based on copper + metal tin, when tin ingots were made available and disease-free directly at the source of the ore deposit. Actually in the preliminary phase of the development of metallurgy, we may presume that the production of bronze was obtained everywhere by cementation. Of course, we may consider that some metalworkers preferred to transform the cassiterite into metal tin themselves, but it is difficult to imagine a possible prevalence of demand on supply in primitive trades.

Therefore the geological data suggest that the occurrence of metal tin scraps at Villagrande should be interpreted as the document of a second and more evolved trading phase, whose chronology is not yet possible to determine with a reliable precision, within the long time span of the sanctuary (from the Recent Bronze Age to the beginning of the EIA). Still disputable is the origin of the metal.

Acknowledgements

Discussions and interaction with Dr. F. Lo Schiavo is gratefully aknowledged. Prof. E. Pernicka is thanked for the analysis and comments on the Villagrande tin scrap. We also thank Sandro Rivoldini for the analytical procedures related to the ICP-MS methods on our difficult material. The research was carried out with grants from the Italian National Council of Research, through the Centro Studi Geominerari e Mineralurgici, University of Cagliari.

Author's address

Roberto G. VALERA

Paolo G. VALERA

Dipartimento di Geoingegneria e Tecnologie Ambientali (DIGITA)

Piazza D'Armi

09123 Cagliari, ITALY

Email: valera@unica.it

Bibliography

As regards the ancient authors we have relied upon the following editions:

HERODOTUS: Le storie (Aristide Colonna & Fiorenza Bevilacqua eds.), Classici UTET, Unione Tipografico-editrice Torinese, 2 voll., 1996

HOMER: Omero, Iliade, BUR Classici Greci e Latini , 2 voll., 2nd ed. 2000. © 2000 RCS Libri S.p.A., Milano

PLINY THE ELDER: C. Plini Secundi Naturalis Historiae (L. Ian – C. Mayhoff eds.), 6 voll., Stuttgart 1967

SCRIPTURES: Holy Bible. New Living Translation, © 1996 by *Tyndale Charitable Trust*

STRABO: Strabone, Geografia (Iberia e Gallia), BUR Classici Greci e Latini vol. III-IV, 2nd ed. 2000.

© 2000 RCS Libri S.p.A., Milano

***

ATZENI, C., MASSIDDA, L., SANNA, U., VIRDIS, P.F., 1992, Some metallurgical remarks on Sardinian bronzetti. In: Sardinia in the Mediterranean: A Footprint in the Sea (T.K. Andrews & R.H. Tykot eds), Sheffield, p. 347-354

BEGEMANN, F., SCHMITT-STRECKER, S., PERNICKA, E. & LO SCHIAVO, F., 2001, Chemical Composition and Lead Isotopy of Copper and Bronze from Nuragic Sardinia. European Journal of Archaeology Vol. 4(1) p. 43-85

BISTE, M., 1979, Die Anwendung geochemischer Indikatoren auf die Zinn-Höffigkeit herzinischer Granite in Süd-Sardinien. Berliner Geowiss. Abhandl. Rehie A/Band 18, Verlag von Dietrich Reimer in Berlin, 111 pp.

BRAUDEL, F., 1998, Les Mémoires de la Méditerranée (Roselyne de Ayala and Paul Braudel eds.). Éditions de Fallois, 427 pp.

CAMBI, L., 1959, Problemi della metallurgia etrusca. Studi Etruschi, XXVII, p. 415-432

FADDA, M.A., 1989a, Villagrande (Nuoro) – Tempio a "Megaron" di S'Arcu 'e is Forros. Nuovo Bollettino Archeologico Sardo, 2 (1986), p. 278-281

FADDA, M.A., 1989b, Villagrande Strisàili (Nuoro) – "*megaron*" temple at S'Arcu 'e is Forros, Appendix in Lo Schiavo, F. & Vagnetti, L., 1989, p. 233-237

FADDA, M.A., 1991, Villagrande Strisàili (Nuoro) Località S'Arcu 'e is Forros. Tempio a megaron. In: Bollettino d'Archeologia, 10, p. 108-110

FADDA, M.A., 2001, A Nuragic sanctuary of S'Arcu 'e is Forros, Villagrande Strisaili district, Nuoro Province (this volume)

GILLIS, C., 1991, Tin in the Aegean Bronze Age. Hydra, 8, p. 1-30

GIUMLIA-MAIR, A., 2000, Roman metallurgy: workshops, alloys, techniques and open questions. In: Ancient Metallurgy between Oriental Alps and Pannonian Plain (A. Giumlia-Mair ed.), Workshop, Trieste, p. 107-120

HERDITS, H., KEEN, J., & STEINBERGER, M., 1995, Wie kommt das Zinn in die Bronze? Ein Beitrag zur experimentellen Archaeologie. Archæologie, 6/1, p. 78-85

HUTH, C., 2000, Quality and quantity in late Bronze Age and Early Iron Age exchange systems. In: Ancient Metallurgy between Oriental Alps and Pannonian Plain (A. Giumlia-Mair ed.), Workshop, Trieste, p. 25-39

ILAVSKY, J., 1979, Metallogenèse de l'Europe Alpine Centrale et du Sud-Est. Institut Géologique Dionyz Stúr, Bratislava, 414 pp.

LEHMANN, B., 1990, Metallogeny of Tin. Lecture Notes on Earth Sciences (S. Bhattacharji, G.M. Friedman, H.J. Neugebauer, A. Seilacher eds.), 32, Springer-Verlag, 211 pp.

LO SCHIAVO, F., 1997, Voce: *Sardegna*, Enciclopedia dell'Arte Antica, II Supplemento, vol. V, p. 141-157

LO SCHIAVO, F., 2001, The Problem of Early Tin from the point of view of Nuragic Sardinia (this volume)

LO SCHIAVO, F. & VAGNETTI, L., 1989, Late Bronze Age long distance trade: the Role of the Cypriots. In: Early Society in Cyprus, Edinburgh, p. 217-243

LO SCHIAVO, F. & VALERA, R. G., 1998, Archaeomining and Archaeometallurgy: an essential multidisciplinary Approach. Silver in the Prehistory of Sardinia, Italy. Proc. Of the Fourth International Conference on the Beginning of the Use on Metals and Alloys (BUMA-IV). Kunibiki Messe, Matsue, Shimane, Japan, p. 185-190

MADDIN, R., 1998, Early Metallurgy: The Tin Mistery-1998. Proc. BUMA IV, Matsue, p. 1-14

PENHALLURICK, R.D., 1986, Tin in Antiquity. The Institute of Metals, London, 271 pp.

PETTINATO, G., 1979, Ebla. Un impero inciso nell'argilla. Arnoldo Mondadori Editore, 223 pp.

PIEPOLI, P. & COLLARI, N., 1936, Sul giacimento stannifero di Canali Serci (Villacidro: Iglesias). Periodico di Mineralogia, VII, 2, p. 147-172

RANKAMA, K. & SAHAMA, TH.G., 1950, Geochemistry. The University of Chicago Press, 912 pp.

ROUTHIER, P., 1963, Les Gisements Métallifères. Geologie et principes de recherche. Masson et C.ie Éd., Paris VIe, 2 voll., 1282 pp.

TYLECOTE, R.F., BALMUTH, M.S. & MASSOLI NOVELLI, R., 1984, Copper and bronze metallurgy in Sardinia. Studies in Sardinian Archeology, Ann Arbor, The University of Michigan Press, p. 114-162

VALERA, R.G., VALERA, P.G. & LO SCHIAVO, F, 2002, Lead in Nuragic Sardinia: Ores, Isotopy, and Archaeology. Martin Bartelheim, Ernst Pernicka & Rüdiger Krause (eds.), Die Anfänge der Metallurgie in der Alten Welt / The Beginnings of Metallurgy in the Old World. Archäometrie - Freiberger Forschungen zur Altertumswissenschaft, 1, 359-377

VIVANET, F., 1890, Lei. Avanzi di una fonderia dell'età del bronzo scoperti nel territorio del Comune. Notizie degli scavi 1890, p. 334-336

VODRET, F., 1959, Sui bronzi preistorici dell'epoca nuragica. Rendiconti del Seminario della Facoltà di Scienze dell'Università di Cagliari, XXIX, 3-4, p. 248-272

ZUFFARDI, P., 1958, Su una nuova segnalazione di cassiterite in Sardegna e sulla presenza di tracce di stagno in alcuni adunamenti idrotermali sardi. Resoconti Ass. Min. Sarda, LXII, 4, p. 27-43

NATURAL TIN-BRONZE ALLOY IN IBERIAN PENINSULA METALLURGY: POTENTIALITY AND REALITY

Salvador ROVIRA & Ignacio MONTERO

Abstract: The Iberian Peninsula is a rich area in tin resources, but they are not homogeneously distributed. Tin resources are concentrated in the western and north-western regions. It is not unusual to find copper-tin ores in those areas where tin resources are scarcer. So, in the eastern Pyrenees, the central ranges (Sierra de Guadarrama and Mounts of Toledo), Sierra Morena or in the Murcia region mixed copper-tin ores have been found. Some of these ores were used by prehistoric metallurgists as has been determined by the analyses of metallurgical debris from Bronze Age sites (Bauma del Serrat del Pont, Villaviciosa de Odón, Gravera Puente Viejo, Llanete de los Moros). A general discussion about natural tin alloy in the early metallurgy of the Iberian Peninsula based on the analytical study of some archaeological samples using XRF, SEM and metallography is carried out.

1. INTRODUCTION

An alloy is an advanced technological feature. The use of alloys in prehistoric metallurgy means a superior stage of knowledge and a better control over the final product. The possibility of alloying two different metals involves the knowledge of both as separate elements, and the resultant properties of the new metal are advantageous from an efficiency point of view. This is especially important in the case of bronze due to the fact that tin was not used as metal out of the alloy. So, tin alloy could not take place by recycling two different objects made of different metals, like gold and silver alloys, both alloys used in the Bronze Age too.

Modern metallurgy marks the limit of 1% to define the presence of a second element in the metal as an intentional alloy. Metal refining and the strict control on the production give effect to this limit, but it is not always suitable for the early metallurgy. Some discrepancy between scholars on this subject has produced a long debate not only in the case of the tin but also in relation to arsenic.

The so-called "arsenical bronze" was considered for a long time as a true alloy. The arsenic content over 1% in many copper objects from different cultures in the Near East and Europe was explained by the intentional addition of arsenic ores in the smelting process, although no ore sample were found in archaeological context. Nowadays, it is widely accepted (Craddock, 1995: 287-288) that arsenical copper is the result of reducing polymetallic Cu-As ores (arsenates). So, we are speaking about a natural alloy that can reach up to 8% As. A wide range of natural copper alloys could be obtained depending on the type of copper ore reduced (Cu-Sb, Cu-Pb or Cu-Ag), even ternary and quaternary alloys, such as Cu-As-Ni or Cu-As-Sb-Ag, are known in the European prehistoric metallurgy (Arnal *et al.*, 1979, Ambert, 1994).

Copper ores and partially reduced ores are common finds in the archaeological record at Chalcolithic and Bronze Age sites in the Iberian Peninsula. Most of them have a high arsenic content (Rovira *et al.*, 1997). The best example is Almizaraque (Almeria) where all the metallurgical process could be reproduced (Delibes *et al.*, 1991). The arsenic percentage in the metal objects was related directly to the copper ores found in the site.

The abundance of Cu-As ores through all the Iberian Peninsula explains the composition of metal objects from different regions, although the distribution of the mean arsenic content is irregular. Positive correlation is detected between the mean values of arsenic in objects and in ores from the same area: when the arsenic value in metals is low the copper resources studied show low values too (Ruiz-Taboada & Montero, 2000).

Some disagreement exists about the minimum tin percentage for an alloy to be considered bronze. Usually the 1% lower limit is used, but when the first alloys must to be described some researchers prefer to increase it up to 2%. This is one of the arguments in the polemic of Muhly versus Yener and Vandiver for accepting or not the Kestel mine as tin supplier in the third millennium BC, in the Anatolian Bronze Age. Muhly (1993: 240) considers that any alloy containing under 2% Sn can not be properly called bronze. Pare (2000: 2), in his recent paper about the full use of bronze in the European prehistory, suggests to raise the limit up to 4% Sn to define the intentional copper alloy. Anyway, all these limits demand a detailed study about the purpose for producing alloys with percentages of the alloyed element that do not change so much the characteristics and properties of the base metal. In other words, how to explain low percentages of tin (< 2%) in early metallurgy?

Poor bronzes could be the result of two very different circumstances[1], as Tylecote (1991: 213) pointed out:

[1] Pare (2000: 2) accept the intentional addition of very small amounts of tin (< 1%) based on Ebla texts. However, a low amount of tin or arsenic does not improve the copper enough and this recipe would not be probably applicable to the European metallurgy.

Table 1. Experimental smelting (% in weight) after Sarabia (1992).

	Cu	Sn	Pb	Zn	Ni	Fe	Rest
Base metal	84,82	9,50	2,06	3,00	0,35	0,12	0,14
First melting	86,54	8,78	1,55	1,40	1,05	0,07	1,21
After several melting	94,62	3,09	0,00	1,09	0,94	0,25	0,01

1.- A natural or unintentional alloy due to the use of polymetallic copper ores containing tin;

2.- A recycled bronze melted without adding tin. The tin percentage decreases quickly after some heat treatments (table 1).

Yener and Vandiver (1993: 256) reply to Muhly that the objects that "unintentionally" contain less than 2% Sn comes from ores rich in tin, giving a "natural tin-bronze". This would be evidence for the existence of some kind of tin in the Tarsus Mountains, against all the scholars that have been rejected such hypothesis. On the contrary, if small tin percentages come from bronze recycling it means that true bronze alloy was known in the area at that time.

The natural tin alloy has been poorly researched because nobody has paid attention to Cu-Sn ores, thinking they are very odd in Nature, apart from the fact that none of the well dated prehistoric copper mines contained them. However, descriptions of some important mining areas such as Cornwall (England) (Tylecote, 1991: 218), Erzgebirge (Germany) (Marechal, 1985: 35) and the Massif Central (France) (Leblanc, 1997: 23) mentioned this type of ore.

We find enough evidence of the use of Cu-Sn ores in the Iberian Peninsula archaeological record, although we do not think this natural alloy would have had a real influence on the development of tin-bronze metallurgy. The first tin alloys[2] usually have high tin percentage, so there is little doubt about their intentional production. Otherwise, we found in a later chronology some sites where poor bronzes (less than 5% Sn) are majority, like Monte Aguilar and other sites at the Bardenas Reales (Navarra), or some objects from the Southeast, Toledo and Huesca provinces (Rovira *et al.*, 1997).

To value the impact of tin alloy in the Iberian Bronze Age metallurgy we need to know if prehistoric technology could obtain natural bronzes, how and where it took place. Otherwise the explanation about the use of poor bronzes will lie on recycling and/or on the tin raw material supply. This paper deals firstly with the Cu-Sn ores distribution in the Iberian geography. Mineralogical data provided by geologist and the analysis of ore samples picked up from copper mines by the authors in the last years draw a general view about the frequency and accessibility of this type of ore. In a second step we handle the archaeological record: the analysis of ore samples and some reducing crucibles found in Bronze Age settlements. There, the comparison of these items with the rest of the materials from the same site gives information about the fact of natural bronze formation. Finally we present a detailed analytical study of one sample to describe the Bronze Age smelting conditions of this Cu-Sn ores.

2. COPPER-TIN ORES

There is a lot of copper mineralogical species; some of them are also tin compounds. Table 2 summarises all the species described by geologists with their chemical formula. Among all, only stannite is abundant enough, being the most cited. In fact, the Iberian Peninsula ore databases we have consulted contain only locations of stannite and in one case of cernýte (Barquilla, Salamanca). The distribution of stannite matches up in a general sense that of cassiterite, although some of the sites are too small to be worked nowadays. Stannite is located at Barquilla (Salamanca), Lumbrales (Salamanca), Otero de Herreros (Segovia), Bustarviejo (Madrid), Fuente Obejuna (Córdoba), Valle de la Serena (Badajoz), Calañas (Huelva), Valverde del Camino (Huelva), Sierra de Cartagena (Murcia) and Viella (Lerida) (Fig. 1: letters).

This geographic distribution is completed with the analysis of ore samples from some copper mines in central Spain made by the Archaeometallurgical Project of the Iberian Peninsula (fig. 1: numbers 2-10): Toledo province (Montero *et al.*, 1990), Madrid province (Rovira y Montero, 1994) and Avila (Fernández Manzano *et al.*, 1997). All of them are located in granitic formations.

Table 3 shows the mean and the standard deviation of some sampled mines. San Roque (Avila), Orgaz and San Pablo (Toledo) mines also contain zinc. Bismuth is present in some mines from Madrid and Toledo, and arsenic also appears in those containing bismuth, except San Luis mine (Toledo). The most complex ores come from Cerro de la Mina (Toledo): the samples contain tin, bismuth, lead and arsenic. Camuñas-3 is completely different, having high means of antimony and silver. This copper mine was probably worked in the Iberian period (V-III centuries BC.) (Montero, 2001).

More information comes from bibliography (fig. 1: numbers 1 and 11-14). Merideth (1998: 140-142) describes Cu-Sn ores from La Golpeja (Salamanca). These minerals are natural admixtures containing malachite and cassiterite. Some mixed copper-tin ores has been identified in the Pyrenees, in the north of Spain. The mines of La Ferrera and Manera (Girona), close to Bauma del Serrat del Pont, are located in granitic areas. X-Ray Diffraction analysis identify, next to copper oxides and arsenates, hidroromarchite ($Sn_3O_2(OH)_2$) from La Ferrera and cassiterite (SnO_2) and "bronze" (Cu_6Sn_5) from Manera (Alcalde *et al.*, 1998: 95). Copper ores from

2 At this moment the earliest bronzes are two awls from Guidoiro Areoso (Pontevedra) and a needle from the Early Bronze Age level at the Bauma del Serrat del Pont (Girona) (Alcalde *et al*, 1998: 96).

Table 2. Cu-Sn Mineralogical species

Cernýite	Cu2(Cd,Zn,Fe)SnS4	**'Bronze'**	Cu6Sn5
Cabriite	Pd2SnCu	**Chatkalite**	Cu6FeSn2S8
Colusite	Cu26V2(As,Sn,Sb)6S32	**Ferrokesterite**	Cu2(Fe,Zn)SnS4
Hemusite	Cu6SnMoS8	**IMA96.052**	Cu2HgSnS4
Kesterite	Cu2(Zn,Fe)SnS4	**Kiddcreekite**	Cu6SnWS8
Kuramite	Cu3SnS4	**Levyclaudite**	Pb8Sn7Cu3(Bi,Sb)3S28
Mawsonite	Cu6Fe2SnS8	**Mohite**	Cu2SnS3
Mushistonite	(Cu,Zn,Fe)Sn(OH)6	**Nekrasovite**	Cu26V2(Sn,As,Sb)6S32
'Oulankaite'	(Pd,Pt)5(Cu,Fe)4SnTe2S2	**Oulongaite**	(Pd,Cu,Fe)9SnTe2S2
Petrukite	(Cu,Fe,Zn,Ag)3(Sn,In)S4	**Rhodostannite**	Cu2FeSn3S8
Sakuraiite	(Cu,Zn,Fe)3(In,Sn)S4	**Sorosite**	Cu(Sn,Sb)
Stannite	Cu2FeSnS4	**Stannoidite**	Cu8(Fe,Zn)3Sn2S12
Stannopalladinite	(Pd,Cu)3Sn2	**Stibiocolusite**	Cu13V4(Sb,Sn,As)3S16
Taimyrite	(Pd,Cu,Pt)3Sn	**Tatyanaite**	(Pt,Pd,Cu)9Cu3Sn4
Unnamed(CuStannide)	Cu3Sn	**Velikite**	(Cu,Hg)11Sn4S16
Vinciennite	Cu10Fe4Sn(As,Sb)S16	**Volfsonite**	$Cu^{1+}_{10}Cu^{2+}Fe^{2+}Fe^{3+}_{2}Sn^{4+}_{3}S_{16}$

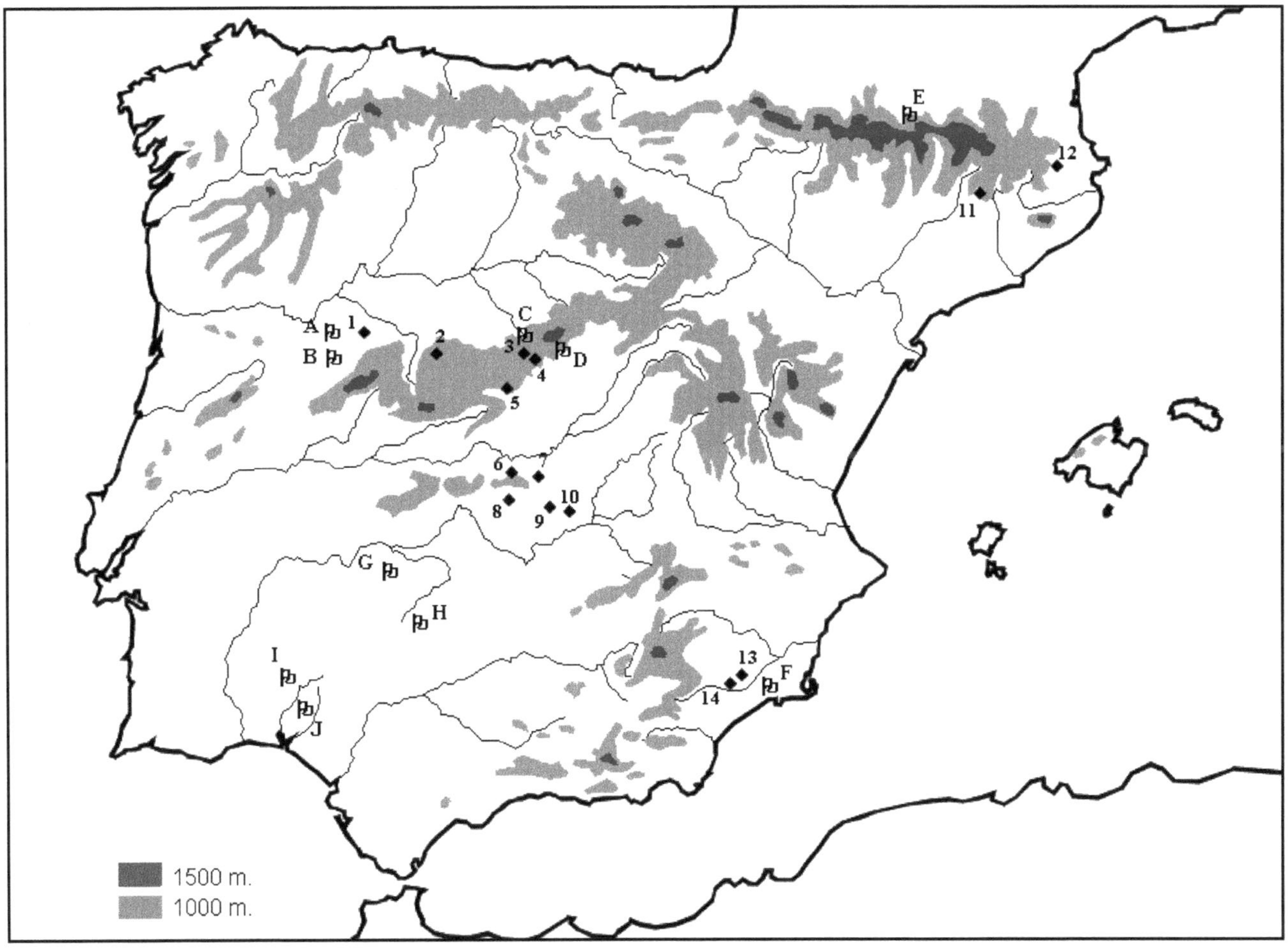

Figure 1. Cu-Sn ores (data from geological reports): A. Lumbrales (Salamanca), B. Barquilla (Salamanca), C. Otero de Herreros (Segovia), D. Bustarviejo (Madrid), E. Vilella (Lerida), F. Sierra de Cartagena (Murcia), G. Valle de la Serena (Badajoz), H. Fuente Obejuna (Córdoba), I. Calañas (Huelva), J. Valverde del Camino (Huelva); (data from field survey): 1. Golpejas (Salamanca), 2. Aldeagordillo and La Colilla (Avila), 3. Garganta de los Montes (Madrid), 4. Lozoyuela (Madrid), 5. Galapagar (Madrid), 6. San Martín de Montalban (Toledo), 7. Orgaz (Toledo), 8. San Pablo de los Montes (Toledo), 9. Madridejos (Toledo), 10. Camuñas (Toledo), 11. Riner (Lerida), 12. La Ferrera and Manera (Gerona), 13. Sierra de Tercia (Murcia), 14. Cerro de la Mina de Siscar (Murcia).

the Bronze Age mine of Riner (Lérida) were analysed at the beginning of XX century. Two published samples contain 12,8 % Sn and 4,7 % Sn respectively. The mine is a small horizontal gallery. The objects recovered were three sandstone moulds, pottery vessels, flint arrowheads and some human bones, probably the owner of all of them died under a fallen stone (Serra Vilaro, 1920).

Finally, there is some data from the Southeast region. Cu-Sn ores were first described by Rickard (1932), who mentioned

Table 3. XRF Analysis of Cu-Sn ores. Mean values (% in weight)

Mina	Prov.	Fe	Ni	Cu	Zn	As	Ag	Sn	Sb	Pb	Bi
La Colilla	AV	8,29	nd	82,7	nd	nd	0,22	**8,41**	0,04	nd	nd
STD		2,41		6,9			0,19	**6,45**	0,4		
Aldeagordillo	AV	17,7	0,18	74,5	nd	0,20	0,50	**4,99**	0,02	0,02	nd
STD		20,1	0,30	19,0		0,41	0,32	**2,77**	0,03	0,06	
San Roque	AV	5,50	0,47	87,16	**3,1**	nd	0,12	**2,47**	0,02	0,49	nd
STD		3,49	0,66	5,36	**1,5**		0,06	**0,70**	0,01	0,65	
Lozoyuela	M	2,52	0,27	80,60		**1,60**	0,07	**3,26**	0,07	**2,42**	**12,47**
STD		2,06	0,17	14,81		**1,08**	0,05	**3,54**	0,06	**2,31**	**11,18**
Orgaz	TO	13,13	1,29	74,81	**6,77**		0,038	**3,97**	0,017	**0,67**	
STD		5,34	0,82	8,25	**4,26**		0,012	**2,28**	0,008	**0,28**	
San Luis	TO	2,74	0,67	81,06			0,157	**5,14**	0,016	**1,69**	**8,40**
STD		1,10	0,51	9,56			0,129	**3,67**	0,013	**1,11**	**5,32**
San Pablo	TO	6,80	0,73	78,01	**7,35**	0,84	0,023	**5,86**	0,014	0,50	**1,23**
STD		6,09	0,36	8,52	**4,18**	0,62	0,033	**4,73**	0,010	0,14	**2,24**
Camuñas3	TO	3,24		86,94			**2,41**	**1,42**	**5,42**		
STD		3,04		3,36			**1,82**	**1,04**	**3,63**		

that this type of ores could give natural bronzes up to 2% Sn. Low percentages of tin appears in samples from Cerro de la Mina de Siscar and Sierra de la Tercia, both in Murcia province, the latter also giving high arsenic (Ayala, 1991: 383).

3. Cu-Sn ORES FROM ARCHAEOLOGICAL RECORD

It is not surprising that evidences of the use of similar ores have been found in Bronze Age sites close to the areas where Cu-Sn has been described before. Figure 2 shows geographical distribution of archaeological sites with this kind of Cu-Sn ores. In most cases there is no evidence of natural tin alloy development: artefacts or other metallurgical remains from these sites were usually made of copper or arsenical copper, even with very low tin impurities. A good example is Aldeagordillo (Avila). This site is a small settlement dated between 4115 ± 20 and 3685 ± 25 BP. Some metal objects, copper drops, a fragment of crucible and fragments of copper ores were recovered. The ores have the same composition that the near mine with the same name described before, altough the mean value is slightly lower (2,63 % Sn versus 4,99 % Sn). However metallic prills from the site (table 4), waste material of the melting process, do not show tin. They are pure coppers without arsenic. That is not strange since ores from the site and from the mines have very low content of As, Sb and Pb.

Only in the Southeast of the Iberian Peninsula very low tin bronze (< 2%) appears since Chalcolithic in sites such as Cueva Sagrada (Lorca, Murcia), a burial cave dated in the middle of III millenium cal BC. Bronze Age objects from Murcia province also show tin percentages between 0,5-1,5 (Montero, 1994: 200-202). However, at Bauma del Serrat del Pont (Girona), the earliest bronzes (6-8% Sn) dated in the first half of the 3th millenium cal BC could be the result of natural alloy due to the circumstance that Cu-Sn ores exist in the neighbouring area. Some tuyeres were found in the metallurgical workshop, being an exceptional find in the Iberian Peninsula archaeological record (Alcalde *et al.*, 1998). The tuyeres allow higher temperature in the smelting process, a key factor to get a natural alloy, as we will describe further.

4. A CASE STUDY

There are few doubts about the fact that early production of tin bronze in Prehistory was based on co-smelting of copper and tin minerals (malachite, azurite and cassiterite). The experiments made by Gowland (1912) did demonstrate that the method was viable. More recently, the experimental work by Rostoker and Dvorak (1991: 16-17) gave details on thermodynamic of co-smelting reactions.

More close and interesting to the aims of this paper are the experiments carried out by C. Merideth using small fragments of ores from the mine at Golpejas (Salamanca). These minerals are natural admixtures containing malachite and cassiterite. Sample no. 74 has a composition of 6.3% tin and 90.3% copper, and the bronze prill obtained after smelting maintains the same composition (!). The same happened to the sample no. 82, with a composition of 35.4-36.1% tin and

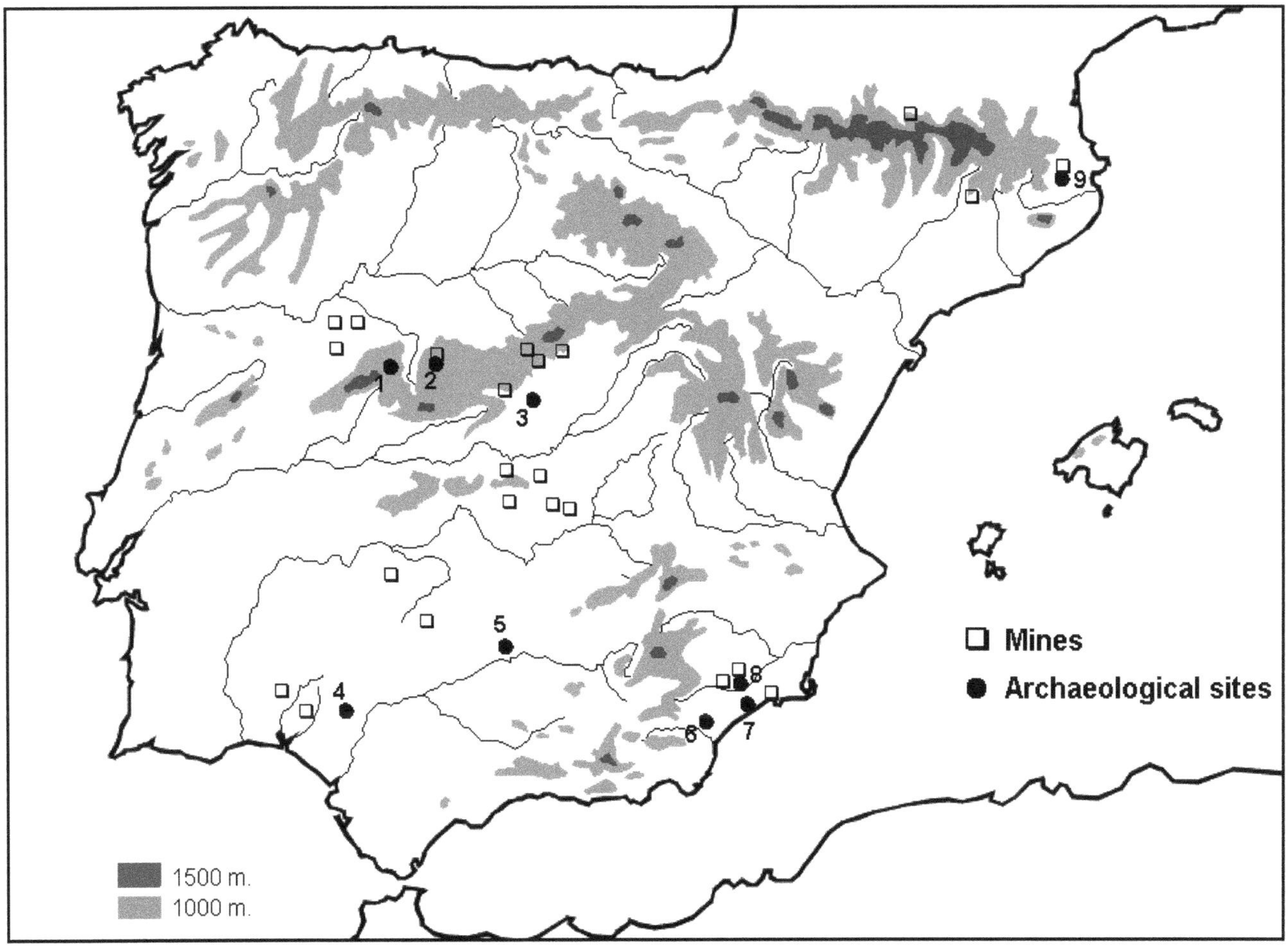

Figure 2. Bronze Age sites with evidences of Cu-Sn ores use: 1. La Corvera (Salamanca), 2. Aldeagordillo (Avila), 3. Villaviciosa de Odón, 4. Algarrobillo (Sevilla), 5. Llanete de los Moros (Córdoba), 6. Fuente Alamo (Almería), 7. Ifre (Murcia) 8. Rincón de Almendricos and Cueva Sagrada (Murcia), 9. Bauma del Serrat del Pont (Gerona).

Table 4. XRF Analysis from Aldeagordillo (Avila) (% in weigth; nd= non detected; tr=traces)).

Record	Object	Fe	Ni	Cu	Zn	As	Ag	Sn	Sb	Pb	Bi
PA6342	Prill	tr	tr	99,90	Nd	nd	0,043	nd	nd	nd	--
PA5297	Prill	tr	nd	99,93	Nd	nd	tr	nd	0,072	nd	nd
PA5293	Prill	0,08	0,08	99,82	Nd	tr	0,01	nd	0,015	nd	nd

66.8-67.1% copper. The resulting bronze is an alloy with 35% tin and 66% copper (Merideth 1998: 140-142).

However, these last experiments have been done under controlled conditions in the laboratory, very far from the working conditions of the open-mouthed reducing crucibles normally used by prehistoric metallurgists of the Iberian Peninsula. This difference has a dramatic importance, as we will discuss immediately after, for understanding the existing restrictions to get natural bronzes in early metallurgical times.

During the archaeological works in the Site 4 at Villaviciosa de Odon (Madrid) one fragment of a reducing crucible was found. This site is dated of Late Chalcolithic-Early Bronze Age by the archaeological assemblage (Rovira and Montero, 1994). The first results by XRF (Table 5) show that natural alloy was not developed, and a metal drop (PA3047A) from the smelting crucible (PA3407) only contains tin impurities.

The bowl shapped fragment exhibits the characteristic slagged surface in the inner side. Analysis of one sample of the slaggy material has given the results in tables 6 and 7.

We can infer from the analytical data that complex copper ore also containing appreciable amounts of arsenic and tin was smelted. This is a kind of mineral that we have found in some mines of the Central Range Mountains, not too far from the settlement.

A small metal prill partially surrounded by slag has been formed and remains adhered to the crucible wall (fig 3A). Its elemental composition (VO-1/1 in Tab. 6) indicates it is an arsenical copper containing 0.35% tin. However, the matrix in the slag show tin oxide amounts in the order of 1.3% and 2.2% that had not been reduced nor passed to the metal alloy. Why?

Phase structure of the slag is indicative of a primitive and low efficient reducing process. By the one hand, both the

Table 5. XRF Analysis from Site nº 4 (Villaviciosa de Odón). (% in weight).

NUM_ANALIS	TIPO	FE	NI	CU	ZN	AS	AG	SN	SB	PB
PA3046	Palmela arrowhead	0.04	nd	99.5	nd	0.20	0.017	0,004	0.154	nd
PA3047A	Metal drop	0.96	nd	96.0	nd	2.87	tr	0,17	nd	nd
PA3047	Smelting crucible	17.8	nd	65.3	nd	11.9	nd	5,08	nd	nd
PA3048	Smelting crucible	49.6	nd	24.4	nd	22.5	0.107	0,36	0.066	2.50

Table 6. Slag in smelting crucible (SEM microanalysis, weight %)

ANALYSI	PHASE	MgO	Al_2O_3	SiO_2	K_2O	CaO	FeO	CuO	As_2O_3	SnO_2
VO-1/2	Glassy matrix	0	5.11	16.9	1.3	20.7	17.3	11.3	10.6	2.2
VO-1/3	Pyroxene laths	3.6	4.4	32.7	0.1	25.5	24.3	1.5	2.7	3.6
VO-1/4	Delafossite needle	0	0	24.2	0	7.5	25.3	37.9	4.7	0
VO-1/5	Glassy matrix	3	4.6	37.2	1.8	18.7	8.7	12.8	10.6	1.3
VO-1/6	Ceramic (bulk)	1.1	16.1	64.1	4.8	5.3	7.2	0	0	0

Table 7. Metal prill in slag (SEM microanalysis, weight %)

ANALYSIS	PHASE	Cu	Sn	As	Fe
VO-1/1	Metal prill	95.58	0.35	4.07	0

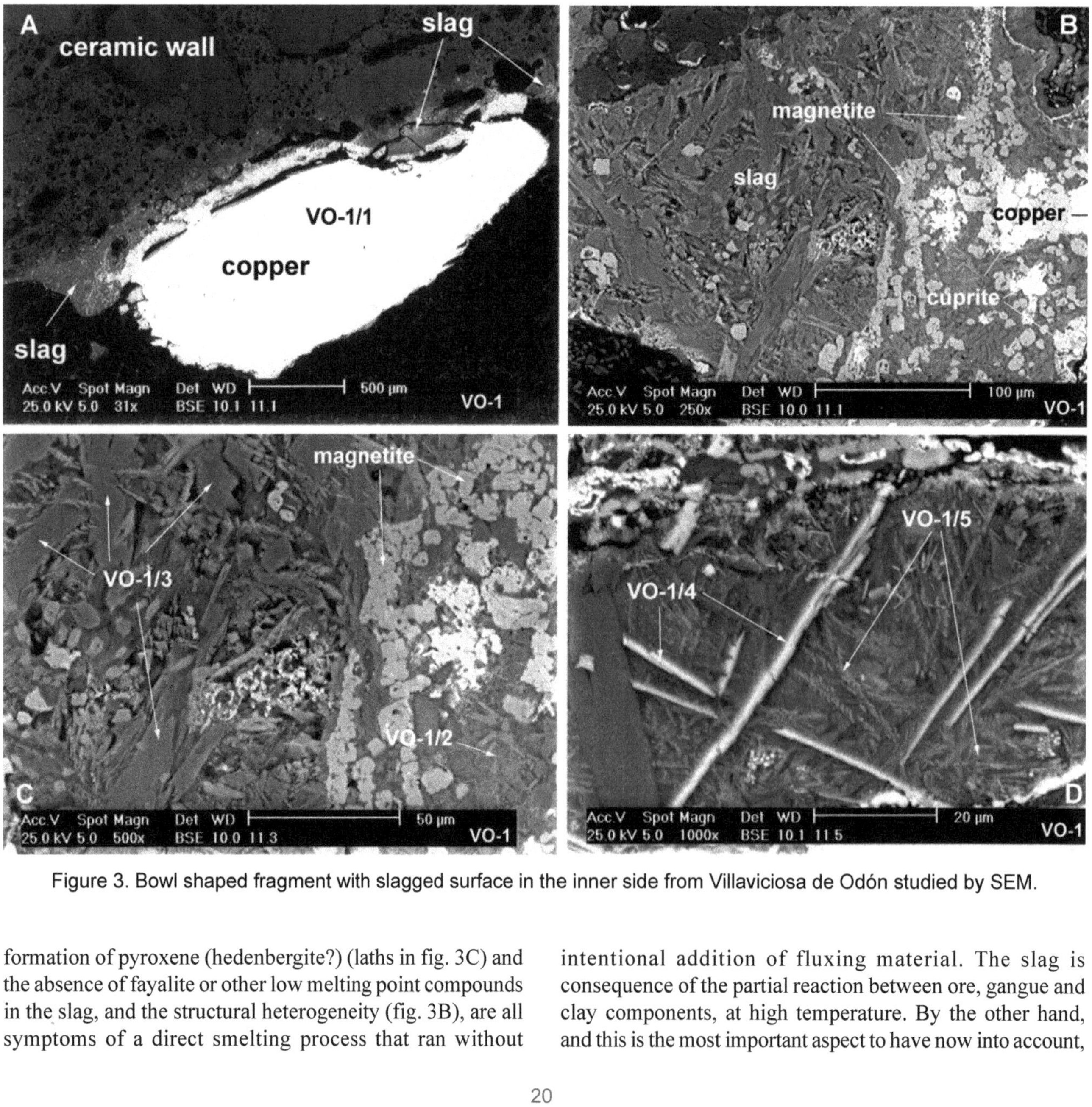

Figure 3. Bowl shaped fragment with slagged surface in the inner side from Villaviciosa de Odón studied by SEM.

formation of pyroxene (hedenbergite?) (laths in fig. 3C) and the absence of fayalite or other low melting point compounds in the slag, and the structural heterogeneity (fig. 3B), are all symptoms of a direct smelting process that ran without intentional addition of fluxing material. The slag is consequence of the partial reaction between ore, gangue and clay components, at high temperature. By the other hand, and this is the most important aspect to have now into account, the presence of abundant crystals of secondary magnetite (fig. 3C) and delafossite needles (fig. 3D) is due to the oxidizing atmosphere into the smelting bowl. Magnetite or, more generally speaking, Fe^{+3} oxides are formed by oxidation of Fe^{+2} oxides usually contained in the gangue or in the ceramic. Delafossite ($Cu^{+1}Fe^{+3}O_2$) is also one product formed and stable in oxidizing conditions.

It thus becomes obvious that the copper smelting conditions in the crucible from Villaviciosa de Odon have been rather oxidizing. In such conditions, efficiency of cassiterite reduction to metallic tin is very low because the reaction requires an atmosphere with a partial oxygen pressure much lower than the existing when delafossite and magnetite is formed or, in other words, a more restricted reducing atmosphere. Then, the greater part of tin remained in the slag as tin oxide.

Rostoker and Dvorak (1991: 8, Tab. 2) suggested that co-smelting would not necessarily imply an intermediate stage as metals of the alloyed materials. So, tenorite (CuO) can react with cassiterite and carbon to produce bronze and carbon monoxide. However, according to the archaeometallurgical finds, tenorite is very scarce (if any) in early copper slags. Only cuprite (Cu_2O) is usually present (Rovira 2000: 484). As Hauptmann *et al.* (1996: 6) state, tenorite decomposes to cuprite at 1026° C and this can be the reason why tenorite is absent in the final products. That being so, tenorite may be formed during the smelting process but alloying reaction could not take place because of the oxidizing conditions in the crucible.

The case of arsenic is quite different. The Spanish copper ores are frequently complex in composition: associations of malachite, azurite, olivenite, xenevixite, clinoclase and conicalcite have been reported by X-Ray diffraction analysis of archaeological samples. This leads to the production of natural arsenical copper at the established working conditions of reducing crucibles (Montero 1994, Rovira 2000).

5. CONCLUSIONS

The purpose of this paper was to investigate the natural tin-bronze alloy in the prehistoric metallurgy of the Iberian Peninsula. After this research several points are clear:

1. Mixed Cu-Sn ores are more frequent than expected.
2. Evidence of their use appears in the archaeological record.
3. These archaeological sites are usually close to the mining areas where Cu-Sn mineralogical species take place. This supports the hypothesis of local or regional metallurgical production without long distance raw material or final product exchange.
4. Natural tin alloy are generally not obtained. Sometimes bronze with less than 2% Sn could emerge, as in the Southeast examples. The exception could be the bronzes of Bauma del Serrat del Pont (Gerona).
5. Natural bronze obtention depends not only of copper-tin ore availability but of appropriate working conditions during the smelting process. Then, technological level plays here an important role. Given the fact that, by what we know, the reducing crucible method was used in the Iberian Peninsula until pre-Roman times, the only way to get a better reducing ambience inside the crucible is by covering it with abundant charcoal and controlling the oxygen input of the tuyeres, as Gowland did in his experiment. This strategy does not seem to have been normally practised in Prehistory until the end of Early Bronze Age or the beginning of Middle Bronze, the time when low tin bronzes start to be recorded (Rovira *et al.*, 1998: 157, fig. 10).
6. It is difficult, in that advanced period, to distinguish natural bronzes from the ones produced by recycling metal or by co-smelting. More investigation must be done in this field.

Author's address

Salvador ROVIRA
Museo Arquecologico Nacional
C) Serano, 13 28001 - Madrid (Spain)

Ignacio MONTERO
Dpk Prehistoria (CSIC)
C) Serano, 13 28001 - Madrid (Spain)

Bibliography

ALCALDE, G., MOLIST, M., MONTERO, I., PLANAGUMÁ, Ll., SAÑA, M. & TOLEDO, A. 1998, Producciones metalúrgicas en el nordeste de la Península Ibérica durante el III milenio cal. AC: el taller de la Bauma del Serrat del Pont (Tortellà, Girona). *Trabajos de Prehistoria* 55 (1), p. 81-100.

AMBERT, P. 1994: Les mines préhistoriques de Cabrières (Hérault): Bilan de quinze ans de recherches, en Archéologie en Languedoc. *Revue de la Fédération Archéologique de l'Hérault* 18, p. 89-100.

ARNAL, J., BOCQUET, A., ROBERT, A. & VERRAES, G. 1979, La naissance de la metallurgie dans le sud-est de la France. In *The origins of metallurgy in Atlantic Europe, Proceedings of the Fifth Atlantic Colloquium*, edited by M. Ryan, p. 35-63.

AYALA, M.M., 1991, El poblamiento argárico en Lorca. Estado de la cuestión. Lorca: Excmo. Ayuntamiento de Lorca.

CRADDOCK, P.T., 1995, *Early metal mining and production.* Edinburgh: Edinburgh University Press.

DELIBES, G., FERNÁNDEZ-MIRANDA, M., FERNÁNDEZ-POSSE, M. D., MARTÍN, C., MONTERO, I. & ROVIRA, S. 1991, Almizaraque (Almería, Spain): Archaeometallurgy during the chalcolithic in the southeast of the Iberian Peninsula. In *Découverte du métal*, edited by Ch. Eluere & J. P. Mohen. Paris: Picard, p. 303-315.

FERNÁNDEZ MANZANO, J., HERRÁN, J.I., OREJAS, A., HERNANSANZ, M. & PARADINAS, S. 1997, Minería y poblamiento Calcolítico en Ávila de los Caballeros. In *II Congreso de Arqueología Peninsular (Zamora, 24-27 septiembre 1996)*. Vol. II Neolítico, Calcolítico y Bronce, edited by R. de Balbín y P. Bueno. Zamora: Fundación Rei Afonso Enriques, p. 527-541.

GOWLAND, W., 1912, Copper and its alloys in early times. *Journal of the Institute of Metals* 4, p. 4-41.

HAUPTMANN, A., BACHMANN, H. G. & MADDIN, R., 1996, Chalcolithic copper smelting: new evidence from excavations at Feinan, Jordan. Archaeometry 1994, Ankara. *The Proceedings of the 29th International Symposium on Archaeometry*. Ankara, 9-14 May 1994. Ankara: Tübitak, p. 3-10.

LEBLANC, M., 1997, Gites et gisements de cuivre de la France meridionale: typologie et caracteristiques geochimiques. *Archéologie en Languedoc* 21, p. 21-26.

MARECHAL, J.R. 1985, Methods of ore roasting and the furnaces used. In *Furnaces and smelting technology in antiquity*, edited by P.T. Craddock & M.J. Hughes. London: Bristish Museum Occasional Paper 48, p. 29-41.

MERIDETH, C., 1998, *An Archaeological Survey for Ancient Tin Mines and Smelting Sites in Spain and Portugal.* BAR International Series 714. Oxford: BAR Publishing.

MONTERO, I., 1994, *El Origen de la Metalurgia en el Sureste Peninsular*. Almería: Instituto de Estudios Almerienses.

MONTERO, I., RODRÍGUEZ, S. & ROJAS, J. M., 1990, *Arqueometalurgia de la provincia de Toledo: Minería y recursos minerales de cobre*. Toledo: Diputación Provincial de Toledo.

MUHLY, J.D. 1993, Early Bronze Age tin and the Taurus. *American Journal of Archaeology* 97: 239-253.

PARE, Ch., 2000, Bronze andthe Bronze Age. In *Metals make the World go round*, edited by C.F.E. Pare. Oxford: Oxbow books, p. 1-38.

RICKARD, T. A. (1932): *Man and Metals*. London.

ROSTOKER, W. & DVORAK, J. R., 1991, Some experiments with co-smelting to copper alloys. *Archeomaterials* 5, p.: 5-20.

ROVIRA, S., 2000, Análisis de "escorias" calcolíticas de fundición de cobre utilizando la microscopía electrónica de barrido. *III Congreso Nacional de Arqueometría.* Sevilla: Universidad de Sevilla y Fundación El Monte, p. 477-486.

ROVIRA, S. & MONTERO, I., 1994, Metalurgia Campaniforme y de la Edad del Bronce en la Comunidad de Madrid. In *El Horizonte Campaniforme en la Región de Madrid en el centenario de Ciempozuelos*, edited by Mª C. Blasco . Patrimonio Arqueológico del Bajo Manzanares 2, p. 137-164

ROVIRA, S., MONTERO, I. & CONSUEGRA, S. 1997, *Las primeras etapas metalúrgicas en la Península Ibérica. I Análisis de materiales*. Madrid: Instituto Universitario Ortega y Gasset.

ROVIRA, S., MONTERO, I. & GÓMEZ RAMOS, P., 1998, The beginning of the use of metals in Spain. *Proceedings of the Fourth International Conference on the Beginning of the Use of Metals and Alloys* (BUMA IV). BUMA IV Organizing Committee & The Japan Institut of Metals. Sendai, p. 153-158.

RUIZ-TABOADA, A. & MONTERO, I., 2000, The pattern of use of stone and copper in Central Spain during the Bronze Age. *European Journal of Archaeology* 3(3), p. 350-369.

SARABIA, F.J., 1992, Arqueología experimental. La fundición de bronce en la Prehistoria Reciente. *Revista de Arqueología* 130, p. 12-22.

SERRA VILARO, J.1920, Mina i fundició d'aram del primer període de l'edat del Bronze de Riner. *Butlletí del Centre Excursionista de Catalunya*, 301. Any XXX, p. 33-39.

TYLECOTE, R.F., 1991, Early copper base alloys; natural aor man-made?. In *Découverte du métal*, edited by Ch. Eluere & J. P. Mohen. Paris: Picard, p. 213-221.

YENER, K.A. & VANDIVER, P.B. 1993, Reply to J.D. Muhly, "Early Bronze Age tin and the Taurus". *American Journal of Archaeology* 97: 255-262.

BRONZE AGE TIN MINES IN CENTRAL ASIA

Jan CIERNY & Gerd WEISGERBER

Abstract: After a general introduction to the problem of the provenience of Near Eastern tin in the Bronze Age 2nd millennium B.C., tin mines and mining techniques are described for Uzbekistan and Tajikistan. The enormous amount of tin produced by the Andronovo people could also have been for export.

Résumé : Commençant avec une introduction générale la provenance du métal étain suivit une description courte des mines et techniques de l'âge du bronze de l'Usbekistan et de Tadjikistan. Le grand surplus de la production d'étain par les gens de la culture d'Andronovo peut être était aussi pour l'exportation.

The use of metals by men reaches back until the 9th millennium B.C. By that time one started in Anatolia to form native copper bits by hammering them to beads and needles. But 3000 years later it became necessary to smelt copper ores to metal because natural copper finally was not available in sufficient quantities any more. Apart from copper also lead and antimony were smelted. Gold was washed out of the river sands and melted or cut or hammered together to form jewellery or idols, and finally precious objects were cast. One could separate silver from silver-bearing liquid lead very early. Until the production and use of the iron in the 2nd millennium B.C. all important metals were available in Anatolia before 3000 B.C. and are named in these few sentences above - except for one, tin. So the aim couldn't be to discover new metals but to optimise the characteristics of the materials available. The copper was just too soft and in comparison with gold it could not be cast particularly well and easily because air and gas bubbles often limited the quality of the product. Depending on the origin of the ores different characteristics and properties of the metals were remarked. One probably started to experiment with artificial additions to the smelt after the experience of different results with different proveniences of copper (with unknown natural additions), and had an epochal success. Both hardness and casting behaviour of copper could be improved decisively by alloying it with tin. This copper tin alloy — optimal at 90% copper and 10% tin — has been called «bronze» for the last 500 years only. But since 2.500 B.C. this bronze became most popular and metal finds like equipment, weapons and jewellery in the ancient Near East were produced of the new metal particularly in Mesopotamia[1]. Also monumental statues of rulers, gods or goddesses consist of this alloy. As this metal mixture appears only in the 2nd millennium in Central Europe[2] the 3rd millennium in the Near East already is described as Early Bronze Age. The metal compositions here in question are known since the first analyses over 130 years ago. Since the introduction of the spectral analyses into the twenties of the 20th century the question about the origin of the two metals copper and tin arose more and more because there was and there is no copper and no tin available in Mesopotamia. Copper could descend from numerous sources at that time, e.g. Anatolia, the Caucasus but also the ore bodies of Iran and Arabia were suppliers for copper. Within the last 20 years, today's sultanate of Oman analytically[3] and archaeologically[4] could be identified as the copper country "Magan" as it is known from Sumerian cuneiform tablets[5]. But for the four other regions mentioned analytical researches just have started, that's why they cannot yet be confirmed scientifically in the finds of Mesopotamia. However, nevertheless they are considered as suppliers because copper became reduced in these regions on a large scale from ores at that time too.

Metallurgy of tin requires the knowledge ore reduction of tin ores because native tin hardly occurs. The process can be carried out easily in small furnaces as ethnographic examples show[6]. Cassiterite (SnO_2) is considered as main ore for reduction processes. Cassiterite still was and is today washed out mainly from sedimental deposits of former river beds in placer mines. Such deposits are available mainly in East Asia; Malaysia and Thailand still today belong to the greatest tin producers of the world. In the ancient Near East namely in Anatolia[7], Syria, Lebanon, Palestine, Arabia and Iran tin placers are not available, there is no proof that the few in Egypt[8] were used in antiquity, the same for Afghanistan[9]. So, where the wider Mesopotamia got its tin from?[10] Despite many cuneiform information about commerce in tin during the first quarter of the second mill. BC[11] and despite finds of

1 Pernicka 1998.

2 For the European tin compare Roden 1985.

3 Hauptmann 1985; 1987; Prange 2001.

4 Weisgerber 1981; 1983;1991.

5 Heimpel 1987;Weisgerber 1983.

6 Bartelemy 1938.

7 Belli 1991;Muhly et al. 1991; for the hard discussion about the tin mines of Kestel in the Taurus mountains see Yener 2000.

8 Sabet *et al.* 1976.

9 Cleuziou,Berthoud 1982; Stech,Pigott 1986.

10 Dossin 1970; Muhly 1973 and 1976; Crawford 1974; Franklin *et al.* 1978.

11 Dercksen 1996.

many tin ingots in Mediterranean ship wrecks of the second half of that millennium[12], the sources of the tin remain quite unknown and mysterious.

As geology can give no positive answer, also the evaluation of the cuneiform texts delivers only minor hints to the provenience. There are many texts from early 2nd millennium trade stations referring to tin but they refer only to transport and exchange of tin already arrived in the wider Mesopotamian area[13]. Ships from a country named Meluhha (most probably India[14]) bring small quantities of tin to Sumer in the 3rd millennium BC. But India only has a few tin and the production is not proved for the time in question[15].

The 120 years old discussion to the origin of the tin[16] received a dramatic new urgency when, starting from the sixties, more and more ship wrecks were discovered in the Mediterranean and their freights sometimes included tin bars or their remains[17]. The sensational wreck find of Uluburun off to the southwest Turkish coast from the time of about 1300 B.C. carried more than 300 copper bars (10 t) and approx. 40 tin bars (1 t).[18] Of course for the origin of the Mesopotamian tin in the 3rd millennium these discoveries only can serve as background information from a younger period. But also for the 2nd millennium it is still completely unclear where the shipped tin originally came from. But the ships cargoes give a clear idea that the strange metal in question was available in large quantities and that somewhere there was a surplus to be traded over far distances. Geological surveys had discovered some tin occurrences in Afghanistan in the seventies and the archaeology started to accept that this tin had possibly been taken to the west from there via Iran. However, wars and civil wars in this country prevented every further research.

Curiously enough a communication of the Russian archaeologist Litvinskij from 1950 was nearly overlooked by the scientific community where he had reported apparently prehistoric tin mines in Central Asia[19]. The idea that Central Asian tin could have been available to the cultures of the ancient Near East during the 3rd millennium isn't so strange as it seems at first glance. Bronze Age finds abroad from Central Asia show a net of far relations, most probably based on local raw materials. Since a long time the lapis lazuli occurrence of Badakhschan[20] in north-eastern Afghanistan near the border to Tajikistan has been understood as the origin of this semiprecious stone so common in most valuable finds in Anatolia and Egypt, Mesopotamia and Iran. This long distance trade seems to have even started in 4th millennium B.C.[21] The routes of these far distance connections since earliest times between Central Asia and very far civilizations, from Troy to Abu Dhabi and Pakistan, were always the same, they were and are determined by the geomorphology of the landscapes between Karakum in the West und Taklamakan in the East. Since 1912 they are described as "silk road".

Recent research of the German Archaeological Institute Berlin, the University of Freiberg, and the Deutsches Bergbau-Museum (German Mining Museum) have revealed two major tin ore producing centres in Uzbekistan and Tajikistan, both in the surroundings of the Zeravshan valley. For the first time Bronze Age tin mines, large and in great number, were found and studied. Our own researches of the last years have shown that the Central Asian tin ore bodies had undergone a considerable surplus exploitation of these hard rock deposits, and it easily could have been that besides lapis lazuli tin also was an important export factor of the region.[22]

The aim of the project sponsored by the Volkswagen Foundation was to check the potential of Uzbekistan and Tajikistan with regard to the early exploitation of the tin occurrences. Besides the geological conditions both the miner's techniques of exploitation of the ore bodies and the archaeological remains of settlements in the mining areas should be examined. Under the leadership of the Eurasia Institute of the German Archaeological Institute (Prof. Dr. H. Parzinger), the chair of Archaeometallurgy of the Technical University Freiberg (Prof. Dr. E. Pernicka) and the Institute of Mining Archaeology of the German Mining Museum Bochum (Prof. Dr. G. Weisgerber) together with the Archaeological Institute of Samarkand of the Academy of Science of the Republic of Uzbekistan (Tashkent) and the Institute for History, Archaeology, and Ethnology of the Academy of Science of the Republic of Tajikistan (Duschanbe) one started to work on these problems.

In Uzbekistan the tin bearing hills are situated between Samarkand and Bokhara near the village of Karnab in a hilly steppe covered with wormwood (Fig. 1). The area with the old mines extends in Karnab on a sea-level between approx. 450 -500 m over an area of approx. 1500 m of length and 500 m of breadth. Today's phenotype of the ancient mining area is essentially formed by the prospection works of the Soviet geologists. The surface of the area is churned up very strongly and cut by hundreds of meters of prospection trenches perpendicular to the W-E ore veins. The ore bodies were described by Dr. Joachim Lutz, a geologist from Freiberg. The veins almost stand vertically in the granite and have 0.40 to 1.20 m of thickness. They reach to an unknown depth and could be easily discovered at the surface into the granite rock with which they are connected genetically due to their texture or unweathered bright colour. Cassiterite (SnO_2) occurs here primarily in quartz veins. The veins stand very steeply till vertically (Fig. 2). The thickness of the veins

12 Bass 1967; 1986; 1989; Güçur 1995; Kilian 1993; Maddin,Stech,Muhly 1977; Maddin 1989.

13 Dercksen 1996.

14 Weisgerber 1991.

15 Chakrabarti 1979.

16 Muhly 1973 and 1976;Franklin *et al.* 1978.

17 For tin remains in the Gelidonia wreck see Bass 1967.

18 Bass 1986 & 1989; Gülçur 1995.

19 Litvinskij 1950; Penhullurick 1986.

20 Kuhlke1976.

21 Herrmann 1968; Pinnock 1988. Tosi 1974.

22 Alimov *et al.* 1998 und 1999; Boroffka *et al.* 2000 and 2002; Ruzanov, Cierny, Weisgerber 1999.

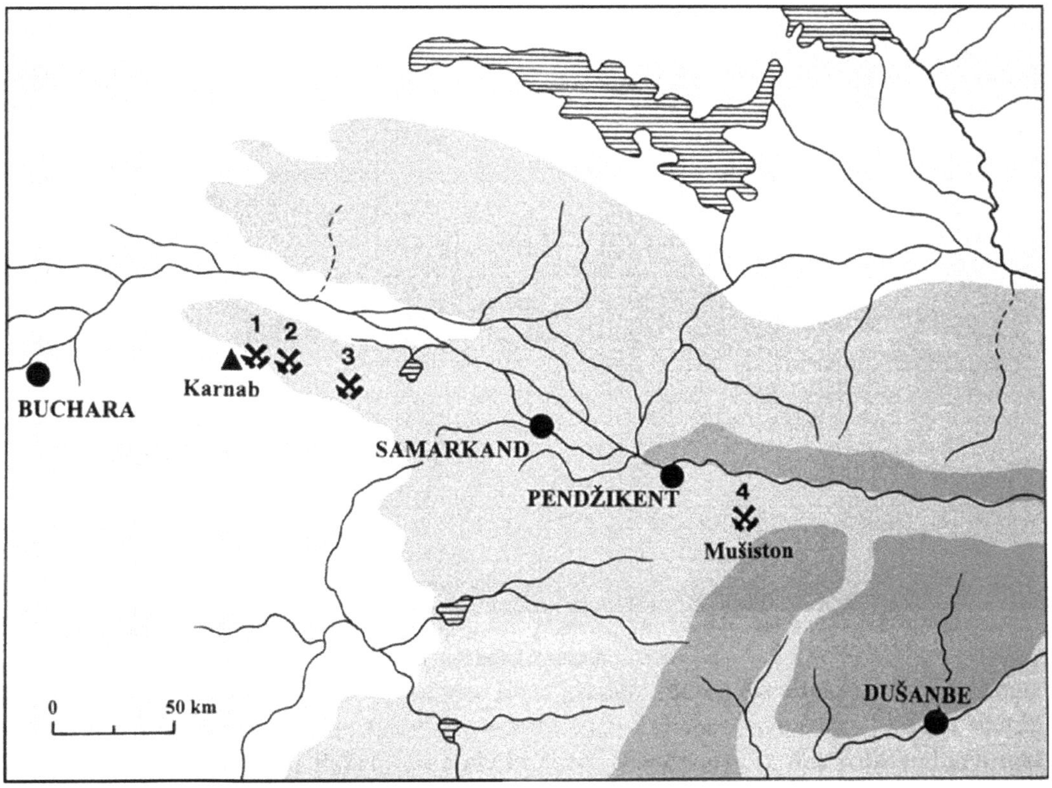

Fig.1: Geographical position of the tin mining areas in Central Asia (Uzbekistan and Tajikistan).

Fig. 2: Karnab, mine 6. Mining work left a completely exploited vertical vein.

Fig. 3: Karnab, mine 6. The dark vein which contains the tin ore can be seen in the remaining horizontal security pillar to the left.

seldom exceeds a meter, the real cassiterite mineralisation ranges between 5 to 10 centimetres (Fig. 3) . During the genesis of the ore body the granite got crumblier in the contact zones of the veins. This approached the miner's work a little, but nevertheless work was unbearable hard. As not one of the mines had remained open over the passed millennia, in Karnab several mines had to be excavated during the project by Bochum mining archaeologist Jan Cierny M.A. The mines

Fig. 4: Karnab, mine 5. At some places the uneven bottom of the vein was reached, at other places ground water stopped further penetrating.

Fig. 5: Karnab, mine 5. By exploiting the ore vein clearly the type of a trench mine was created.

can be described as trench mines, which followed the veins into the depth (Fig. 5). This goes by the width of the vein, therefore the mines are often extremely narrow. Some mining cavities are so narrow that no adult man could have been able to work effectively in prehistoric times (Fig. 6), one has to assume, that children or women were also engaged in the exploitation work as can be demonstrated for other prehistoric mining activities[23]. Nowadays Central Europeans only can move with great effort in these mines, there is no way for them to work there with heavy tools. Because all mines were totally filled and covered with sediments (loess) the prospecting trenches of the Soviet geologists are of a great help to discover and sometimes to enter the old mines. They had to be vacated bucket by bucket what took a long time. It was particularly interesting and helpful that for safety reasons the ancient miners sometimes did not take all of the mineralised vein out of the mine but left rock bridges, which still today contain the original ore (Fig. 3). Values of 4% tin were measured. Most mines couldn't be excavated completely since the ground water lined up after 9 respectively 17 m of depth. In one mine the bottom was reached after only 7 m of depth, it is irregular and uneven (Fig. 4). Climbing helps like kick holes or steps could be seen only in some places. Only where the hard vein material also contained ore and this could be won the vein was attacked by hammering and fire setting (Fig. 6). One heated rock and vein so much by fire that it got cracked by the arising tensions of the different minerals in the rock. Than the cracked material was hammered away

Fig. 6: Karnab, mine 6. Deep in the mine the curved foot wall clearly indicates traces of former firesetting.

[23] Extremely narrow Late Bronze Age galleries in Timna/Israel, see Conrad & Rothenberg 1980, Fig. 155, Beil. 14, gallery in direction Pkt. –328,90.

Fig. 7: Karnab, mine 3. Hammer stones and their fragments became either piled up or layed down according their find level.

Fig. 8: Changali, surface. Thousands of hammer stones once were fastened to a handle by a string fixed in the groove so that the bed rock could be smashed with a rather strong tool.

mainly with heavy hammers made of local limestone (Fig. 7. 8). They could not be made of local pebbles as it is usually done in prehistoric mining since there aren't any pebbles in the steppe.

One mine almost completely excavated is 30 m long, it formerly may have been considerably deeper, but today ground water was reached in 17 m of depth. At an average breadth of 60 cm a cavity of 270 m^3 was created by the miners what corresponds to about 730 t of rock broken out. If one accepts an ore thickness of only 5 cm, then at least 60 t of tin ore can be calculated for. Calculated at 2% tin alone this single mine delivered about one ton of tin. This corresponds by chance exactly to the tin cargo load of the Uluburun ship and gives a good performance of the large metal quantities produced in this area. Even the fact that the age of the Karnab mines so far studied , the 1st half of the 2nd millennium B.C., is older than the famous wreck off the coast of Anatolia cannot destroy the hope that more studies possibly may produce dates corresponding to the ship. It remains as a hope that it could be possible to prove analytically the connection between the tin of the ship or other ingots of that time and that one of the mines or to demonstrate the opposite. But no other exploitation of any other tin source in the east in the questionable time is known!

In connection with our mining research it is of special attraction that a settlement of the tin miners not far from the mines of Karnab could be discovered and excavated by Berlin archaeologist Dr. Nikolaus Boroffka in the steppe. Like in the mines the same ceramic of the Andronovo culture was found which belongs to the middle of the 2nd millennium B.C. There were small metal finds in the pit houses and fire places. But primarily it were these hammer stones so typical of Karnab mines with one or two grooves which produced a direct relation between settlement and the mining.[24] This is

24 Parzinger & Boroffka 2001.

Fig. 9: Mushiston. The slope with the prehistoric mining entrances is disturbed by recent geological work in about 3000 m height. The big dump to the left gives the level of the recent gallery from where the prehistoric working ends could be found in the mountain.

only the second time in the West and Central Asian prehistoric mining archaeology that a direct connection between settlement and mines could be established although they were separated spatially at all[25].

Besides Karnab other mining archaeological researches in Uzbekistan have shown that more prehistoric tin ore mining can be proved with safety for the 2nd millennium, e.g. at Lapas and Changali (Fig. 8). In any case the extraction work has had an unexpected large size. Future researches may produce even older mining systems for a safe answer to the question of the tin sources of the Early Bronze Age.

But the Uzbekistan tin mines are not the only ones in Central Asia. At a journey in 1992 and an excursion in 1994 Tajik geologists led our attention to the fact that there were also tin ore bodies in the western foothills of the Pamir-Alai mountains in West Tajikistan[26]. However, from the formerly numerous mining cavities only some had remained at the slope of the mountain Mushiston in 3000 m above sea level since almost every mine had been the victim of the bulldozers of the Soviet prospectors. But finally they also had dug deep galleries in the mountain. Where these galleries approached the slope surface, they had touched or cut prehistoric mines so that these now became accessible by these new galleries. So, before the beginning of the excavations mines of an unknown age were known both at the surface and in 35 m of depth in the mountain. Without digging from the entry at the surface the final cavities and chambers and galleries at the end of the prehistoric mines could be studied. The examinations confirmed that both relics belonged to the same activities and that both were created in the Bronze Age. Confirmation came both by archaeological finds and radiocarbon datings. Logically the mines in the surface of the slopes are older than their endings in the mountain. These mines achieve considerable extents with chambers of up to 10 m of length and 6 m of width. An additional support of the mine's roofs was urgently advisable at such big cavities. On one hand this was achieved by leaving rock remains as mountains pillars on the other wooden props trunks of juniper trees were set. At the always same temperatures of 6° C these props fortunately have rather well survived in the mountain since around 1500 B. C. One of it still stood upright at his place.

Also at Mushiston heavy stone hammers either with grooves or notches served for crushing the rock. Again mining was done by people of the Andronovo culture and started around 2000 BC. The slate forming the mountain wasn't so hard that it would have had to be made crumbly by firesetting, but the veins of quartz containing the tin ore were extremely hard anyway. Here at Mushiston the main mineral of tin is stannite ($CuFeSnS_4$), it occurs together which malachite and azurite in the oxidation zone. More over there is the mineral Mushistonite ($(Cu,Zn,Fe)Sn(OH)_6$) which was discovered here and named after the site. All passable old mining cavities lay in this oxidation zone. At the border of limestone to slate there was a rich ore zone in which the old mining particularly circulated. The ore is trapped here in a white quartz from few centimetres up to half a meter width. The tin ore cassiterite stains it as black spots (<= 1 mm). However, it is the special feature of the Mushiston ore body that the tin ore is mixed with copper ores. Since weathering and oxidation have changed their colour to green and yellowish both ores can and could not been overseen in the quartz. Since it seems difficult to imagine how tin and copper ores, if wanted, could have been separated in the Bronze Age, it easily could have happened that both ores had been smelted together. The copper during the smelting process could have been alloyed in the smelting furnace by the tin also contained in the ore.

[25] The other example is still 2000 years older and dates in the second half of the 4th millennium BC: the site of Madsus is the mining site where the same tools with drilled biconical shafting holes were collected as at the village of miners and smelters at Wadi Fidan 4. Adams & Genz 1995; Hauptmann 2000, 88;91-93.

[26] Cierny 1995.

Fig. 10: Mushiston. Under the dumps of road constructions the entrances of mines could be discovered and cleared.

Fig. 11: Mushiston. Inside the mountain chambers with safety pillars were met and studied.

More or less by chance the copper tin alloy "bronze" then would have been discovered by these unique circumstances. But this alternative concept for the discovery of the bronze in the 3rd millennium B.C. is based on the mixed ores of Mushiston. For more details much more archaeological basic research has to be done in Central Asian tin mines.

Author's address

Jan CIERNY M. A.

Prof. Dr. Gerd WEISGERBER

Deutsches Bergbau-Museum

Herner Str. 45

D-44787 Bochum GERMANY

Email: jan.cierny@bergbaumuseum.de

gerd.weisgerber@bergbaumuseum.de

Bibliography

ADAMS, R.B., GENZ, H., 1995, Excavations at Wadi Fidan 4, a Chalcolithic village complex in the copper ore district of Feinan, southern Jordan. *Palestine Exploration Quaterly* 127, p. 8-20.

ALIMOV, K., BOROFFKA, N., BUBNOVA, M., BURJAKOV, J.F., CIERNY, J., JAKUBOV, J., LUTZ, J., PARZINGER, H., PERNICKA, E., RADILILOVSKYJ, V., RUZANOV, V., SHIRINOV, T., STARŠININ, D. & WEISGERBER, G., 1998, Prähistorischer Zinnbergbau in Mittelasien. *Eurasia Antiqua* 4, p. 137-199.

ALIMOV, K., BOROFFKA, N., BURJAKOV, J.F., CIERNY, J.,LUTZ, J., PARZINGER, H., PERNICKA, E., RUZANOV, V., SHIRINOV, T. & WEISGERBER, G., 1999, Research at Karnab; Uzbekistan. *Istorija Material'noj Kul'tury Uzbekistana* 30, p. 80-87.

BARTELEMY, R. E., 1938, Primitive Tin Metallurgy in Laos. *Mining and Metallurgy* 19, p. 244-246.

BOROFFKA, N., BUBNOVA, M., CIERNY, J., JAKUBOV, J., LUTZ, J., PARZINGER, H., PERNICKA, E., STARSCHININ, D. & WEISGERBER, G., 2000, Issledovanija v rajone drevnich razrabotok olova v severnom Tadschikistane (Muschiston i Takfon). *Archeologitscheskie Raboty v Tadschikistane* 27, p. 71-120.

BOROFFKA, N., CIERNY, J., LUTZ, J., PARZINGER, H., PERNICKA, E. & WEISGERBER, G., Bronze Age Tin from Zentral Asia. In *Ancient interactions east and west in Eurasia* (K. Boyle, C. Renfrero, M. Levine, Eds.) Cambridge 2002.

BASS, G. F., 1967, Cape Gelidonya, A Bronze Age Shipwreck. *Transact. Americ. Philosoph. Society NS* 57, part 8, Philadelphia.

BASS, G. F., 1986, A Bronze Age Shipwreck at Ulu Burun (Kaş), 1984 Campaign. *AJA* 90, p. 269-296.

BASS, G. F, 1989, Das Wrack von Ulu Burun. *Geo* H. 1, p. 84-98.

BELLI, O., 1991, The problem of tin deposits in Anatolia and its need for tin, according to the written sources. In *Anatolian Iron Ages*. The Proceedings of the Second Anatolian Iron Ages Colloquium held at Izmir, 4-8 May 1987, edited by A. Cilingiroglu, & D.H. French, 13, p. 1-9, (= British Institute of Archaeology at Ankara, Monograph).

BERTHOUD, T. et al., 1981, Production, échange et utilisation des métaux, bilan et perspectives des recherches archéologiques récentes dans le domaine oriental. *Paléorient* 7, p. 99-127.

CHAKRABARTI, D. K, 1979, The problem of tin in Early India - a preliminary survey. *Man and Environment* 3, p. 61-74.

CIERNY, J., 1995, Die Gruben von Muschiston in Tadschikistan - Stand die Wiege der Zinnbronze in Mittelasien? *Der Anschnitt* 47, p. 68-69.

CLEUZIOU, S. & BERTHOUD, T., 1982, Early tin in the Near East. A reassessment in the light of new evidence from western Afghanistan. *Expedition* 25, p. 14-19.

CONRAD, H. G. & ROTHENBERG, B. (Eds.), 1980, Antikes Kupfer im Timna-Tal. *Der Anschnitt, Beiheft 1* (Veröff. Dtsch. Bergbau-Museum 20), Bochum.

CRAWFORD, H. E.W., 1974, The problem of tin in Mesopotamian bronzes. *World Archaeology* 5,6, p. 242-247.

DERCKSEN, J. G., 1996, *The Old Assyrian Copper Trade in Anatolia*. Nederlands Historisch Archaeologisch Instituut, Leiden.

DOSSIN, G., 1970, La route de l'étain en Mésopotamie au temps de Zimri-lim. *Revue d'Assyriologie* 64, p. 97-106.

FRANKLIN, A.D., OLIN, J.S. & WERTIME, Th. A. (Eds.), 1978, *The Search for Ancient Tin*. Washington, D.C.

GÜLÇUR, S., 1995, Das bronzezeitliche Wrack von Uluburun bei Kaş. *Antike Welt* 26, p. 453-461.

HAUPTMANN, A., 1985, 5000 Jahre Kupfer in Oman, Bd. 1, Die Entwicklung der Kupfermetallurgie vom 3. Jahrtausend bis zur Neuzeit. *Der Anschnitt, Beiheft 4*, Bochum (Veröff. Dtsch. Bergbau-Museum 33).

HAUPTMANN, A., 1987, Kupfer und Bronzen der südostarabischen Halbinsel. *Der Anschnitt* 39, p. 209-218.

HAUPTMANN, A., 2000, Zur frühen Metallurgie des Kupfers in Fenan, Jordanien. *Der Anschnitt, Beiheft* 11, Bochum (Veröff. Dtsch. Bergbau-Museum 87).

HEIMPEL, W., 1987, Das Untere Meer. *Zeitschrift für Assyriologie u. Vorderasiatische Archäologie* 77, p. 21-91.

HERRMANN, G., 1968, Lapis lazuli, the early phases of its trade. *Iraq* 30, p. 21-57.

KILIAN, I., 1993, Überlegungen zum spätbronzezeitlichen Schiffswrack von Ulun Burun (Kaş). *Jahrb. Röm. Germ. Museum* 49, p. 333-352.

KULKE, H., 1976, Die Lapislazuli-Lagerstätte Sare Sang (Badakhshan). Geologie, Entstehung, Kulturgeschichte und Bergbau. *Afghanistan Journal* 1, p. 43-56.

LITVINSKIJ, B.A., 1950, K istorii dobytschi olova v Uzbekistane. *Trudy Sredneaziatskogo Gosudarstvennogo Universiteta (Archeologija srednej Azii), N.S.* 9, 1950, p. 51-67.

MADDIN, R., STECH WHEELER, T. & MUHLY, J. D., 1977, Tin in the ancient Near East, old questions and new finds. *Expedition* 19, H. 2, p. 35-47.

MADDIN, R., 1989, The copper and tin ingots from the Kaş shipwreck. In *Archäometallurgie der Alten Welt*. Beiträge zum Internationalen Symposium "Old World Archaeometallurgy", Heidelberg 1987, *Der Anschnitt, Beiheft 7*, p. 99-105, Bochum.

MUHLY, J. D., 1973 und Suppl. 1976, *Copper and Tin*. Trans. Connecticut Acad. of Arts and Sciences 43, p. 155-535 und 46, p. 77-136. New Haven. Suppl. Hamden.

MUHLY, J. D., BEGEMANN, F., ÖZTUNALI, Ö., PERNICKA, E., SCHMITT-STRECKER, S. & WAGNER, G.A., 1991, The Bronze Metallurgy in Anatolia and the Question of Local Tin Sources. In *Archaeometry '90*, edited by E. Pernicka & G. A. Wagner. Proceedings of the 27th Symposium on Archaeometry held in Heidelberg April 2 -6, 1990, Basel, p. 209-220.

PARZINGER, H. & BOROFFKA, N., 2001, Eine bronzezeitliche Bergbausiedlung bei Karnab, Usbekistan. Entdeckungen auf der Suche nach dem Zinn der Bronzezeit in Mittelasien. *Archäologie in Deutschland* 17, H. 3, p. 12 ff.

PENHALLURICK, R.D., 1986, Tin in antiquity. London: Institute of Metals.

PERNICKA, E., Die Ausbreitung der Zinnbronze im 3. Jahrtausend. In *Mensch und Umwelt in der Bronzezeit Europas*, edited by H. Hänsel (Abschlusstagung d. Kampagne d. Europarates, Die Bronzezeit, das erste goldene Zeitalter Europas, FU Berlin, 17.-19.3.1997), Kiel 1998, p. 135-147.

PINNOCK, F., 1988, Observations on the Trade of Lapis Lazuli in the 3rd Millennium B.C. In *Wirtschaft und Gesellschaft von Ebla*, edited by H. Hauptmann & H. Waetzoldt. Akten der Internationalen Tagung Heidelberg, 4. - 7. November 1986, Heidelberg, p. 107-110.

PRANGE, M., 1998, Vergleichende Untersuchungen zur Charakterisierung des omanischen Kupfers mittels chemischer und isotopischer Analysenmethoden. *Metalla (Bochum)* 8, 1/2, 2001, p. 1-126.

RODEN, CH., 1985, Montanarchäologische Quellen des ur- und frühgeschichtlichen Zinnerzbergbaus in Europa. *Der Anschnitt* 37,p. 50-80.

RUZANOV, V., CIERNY, J. & WEISGERBER, G., 1999, Ob itogach raboty meshdunarodnoj usbeksko-nemeckoj ekspedicii v 1997-1998 godach po archeologitscheskomu obsledovaniju drevnich istotschnikov olova v Zirabulak-Ziaetdinskich gorach. *Obschtschestvennye nauki v Uzbekistane* 7-8, 1999, p. 68-72.

SABET, A.H., TSOGOEV, V.B., SHIBANIN, S.P., EL-KADI, M.B. & AWAD, S., 1976, The Placer Tin-Deposits of Igla, Abu Dabbab and Nuweibi. *Annals of the Geological Survey of Egypt* 6, p. 169-180.

STECH, T., PIGOTT, V.C., 1986,The Metals Trade in Southwest Asia in the Third Millennium B.C. *Iraq* 48, p. 39-64.

TOSI, M., 1974, The Lapis Lazuli Trade across the Iranian Plateau in the 3rd Millennium B.C. *Gururajamanjarika*, Festschrift Guiseppe Tucci Vol. 1, p. 3-22.

WEISGERBER, G., 1981, Mehr als Kupfer in Oman - Ergebnisse der Expedition 1981. *Der Anschnitt* 33, p. 174-263.

WEISGERBER, G., 1983, Copper Production during the Third Millennium BC in Oman and the Question of Makan. *The Journal of Oman Studies* 6, p. 269-276.

WEISGERBER, G., 1991, Die Suche nach dem altsumerischen Kupferland Makan. *Altertum* 37, 1991, p. 76-90.

WERTIME, TH. A., 1978, The search for ancient Tin, the geographic and historic boundaries. In Franklin, Olin & Wertime 1978, p. 1-6.

YENER, K. A., 2000, *The domestication of metals: the rise of complex metal industries in Anatolia. Leyden.*

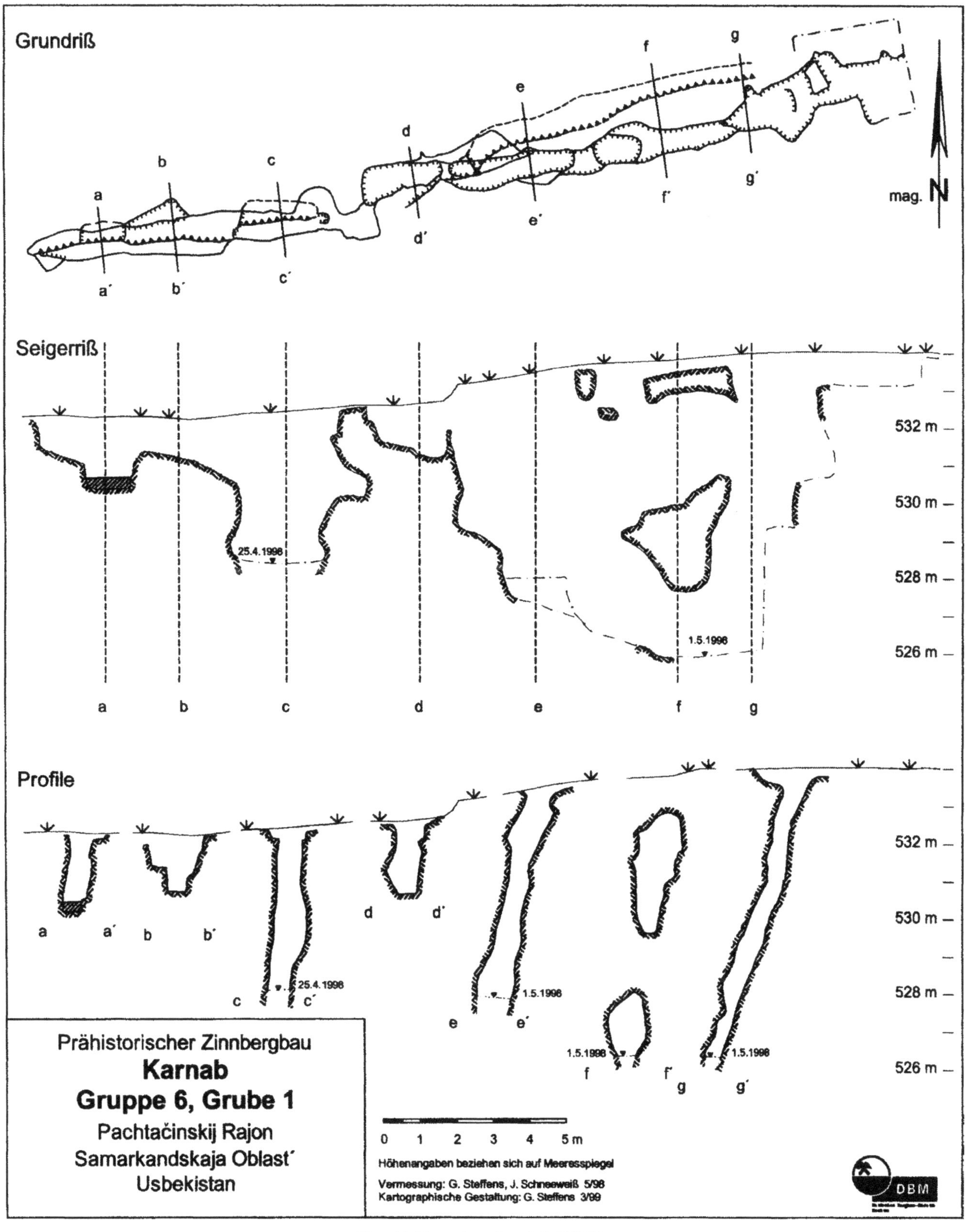

Fig. 12 Karnab, mine 6-1 (1998).

IIème Section – Métallurgie et métallographie - Archéométallurgie en Europe

IInd Section – Metallurgy and metallography - Archaeometallurgy in Europe

CASTING AND METALLOGRAPHY OF TIN BRONZES IN CLAY MOULDS

Q. WANG & B. S. OTTAWAY

Abstract: Compositional data of bronze artefacts of the European Chalcolithic and Bronze Age are now quite numerous. Yet these data give no indication of how the bronzes were made: for instance, in what mould were they cast? With what speed were they cooled? And were they subjected to any cold or hot working? To build up a reference collection for the determination of ancient production methods of cast bronze artefacts, a series of casting experiments with archaeologically relevant alloys has been carried out. Flat axes, typical bronze objects of the European Chalcolithic to Bronze Age period, were cast in moulds of sand, clay and bronze. The compositions of the bronzes, the moulding material and the cooling methods after casting were systematically varied under controlled conditions. The microstructure, dendritic arm spacing/grain size and microhardness of the metals were studied for each casting. This paper presents the results of castings of tin and leaded tin bronzes in clay moulds.

INTRODUCTION

Clay moulds found in the archaeological record were in use in the Middle and Late Bronze Age in Europe. It has been reported that in many cases the clay used for the Bronze Age moulds consisted of two grades: a well-prepared fine inner part and a minimally prepared coarse outer part (Collins 1970, 23; Hodges 1958-1959, 132; Tylecote 1986, 89). The fine inner layer was thought to aid the transfer of decorations from the patterns to the cast metal artefact and the outer coarse layer helped to wrap the two halves together. However, Irish prehistoric casting moulds were found to consist of only one homogenous clay body (Coghlan & Raftery 1961, 243) and the experiments reported here also indicate that two grades of clay for moulds are not necessary to produce a satisfactorily cast artefact.

The majority of the clay moulds found in European Bronze Age sites are bivalve moulds. The registration of the two halves was by means of keys (knobs and pits or dowels and holes) (Coghlan 1960, 156; Tylecote 1986, 89). Indentations could be made using fingertips or small pebbles in the first half of the mould and small lugs were then automatically formed in the second half (Hodges 1958-1959, 132). Nevertheless, keys for registration were not always used in Bronze Age moulds and even in later periods, for instance in the Migration Period, dowels and holes for registration were extremely rare in the moulds used to manufacture jewellery (Lamm 1973, 3).

Analytical data on clay moulds are rare. It has been generally suggested that clay moulds were made of clay with additions of sand and vegetable matter such as straw or dung (Tylecote 1986, 84; Hodges 1958-1959). Clay moulds used in Chinese bronze castings were also of a sand-clay-fiber mixture, which permitted gas to escape during casting (Gettens 1969, 31). The addition of sand reduces contraction and improves refractability. The addition of organic materials, e.g. straw (Tylecote 1986, 84), dung (Tylecote 1986, 84; Shalev 1999) or plant ash (Tan Derui & Huang Long 1996, 67) increases the permeability of the clay mould and thus improves the ability of the molten metal to fill the mould.

Clay moulds were fired at a temperature that reduces the gas content in the clay and results in greater strength and better thermal properties. This ensured the production of high quality bronzes. Estimating the firing temperatures by refiring experiments, Needham (1980, 192) suggests that most European clay moulds were fired to a temperature of about 650°C. The accuracy of these measurement may, however, be doubtful because refiring experiments of ceramics fired to a low temperatures may produce faulty results (Andrews 1993). Assuming that the estimate of the firing temperature of 650°C is correct, the mould would have had to be made of noncalcareous clay as moulds made of calcareous clay had to be fired to 900°C. At this temperature the decomposition of calcite is almost complete and thus complete escape of carbon dioxide from the mould prior to casting is assured.

It has, however, also been reported by Faoláin and Northover (1998, 73) that even firing clay moulds at 650-700°C is not necessarily required, although the failure rate of their experimental castings was higher than with properly fired moulds.

The thickness of the clay mould is also important. If the mould is too thin it is not strong enough to withstand the pouring of molten metals; on the other hand, if it is too thick, it takes a long time to dry and easily breaks during firing if not completely dry. Most of the European Bronze Age clay moulds were about 1-2 cm thick (Coghlan & Raftery 1961, 240; Needham 1980, 197-199).

There are different opinions on preheating of clay moulds prior to teeming, or pouring the hot metal into the mould. Lamm (1973, 6) states that moulds should have a temperature of 700-800°C when the metal is poured in to achieve a complete filling of the mould. However, simulated and actual casting experiments carried out by Ratka (1996, 15) suggest

Table 1. The composition of the 12 different alloys used in the experiments; the balance is copper.

Alloys (wt.%)	Tin bronzes					Leaded tin bronzes						
Sn	2	6	10	15	23	2	2	2	6	6	6	10
Pb	0	0	0	0	0	2	6	10	2	6	10	2

that a mould preheated to a temperature of 350°C could produce a flawless cast and this is supported by the results presented here.

In the experimental work reported here, and to simulate the majority of prehistoric clay moulds as closely as possible, bivalve moulds were made for the bronze casting. A series of 36 castings, including three different cooling rates for each alloy and a total of 12 different alloys, was carried out. The compositions of the alloys are listed in Table 1. The three cooling rates include preheated moulds (preheated/air-cooled), moulds left to cool in room temperature (unpreheated/air-cooled), and quenched in water (unpreheated/water-chilled).

MOULD MAKING

Industrial fireclay was chosen to make the clay moulds. Due to its high content of SiO_2 (65%) in the clay, no sand was added. Sawdust as organic filler was added to increase permeability of the moulds. After some trials, a mixture of clay with 20% sawdust, which had been passed through a 1mm sieve, was chosen to make the clay moulds.

A wooden pattern was used to make the mould. It was found that a good registration could be achieved by making the two pieces of the bivalve mould at the same time. Two wooden frames were used for shaping the clay. Sieved sawdust was spread onto the first half to ease the separation of the two halves of mould. It needs to be mentioned that the addition of sawdust decreased the plasticity of the clay, which would have been hard to shape without the wooden frames. The two halves can be easily separated even without any insulating agents if the clay is dry enough. A runner was made on one side of the cavity at the top part of the mould. Three mini holes were made by drilling to help gas evacuation (Fig 1). The wooden pattern was then removed. The two halves of the mould were located precisely and the whole assemblage was sealed around the sides with a mixture of the same clay and coarse sawdust. A runner bush was added to complete the mould assemblage (Fig 2).

The moulds were dried in air overnight and were then put in a drying oven to dry at 40-50°C for 2 days. Since the clay

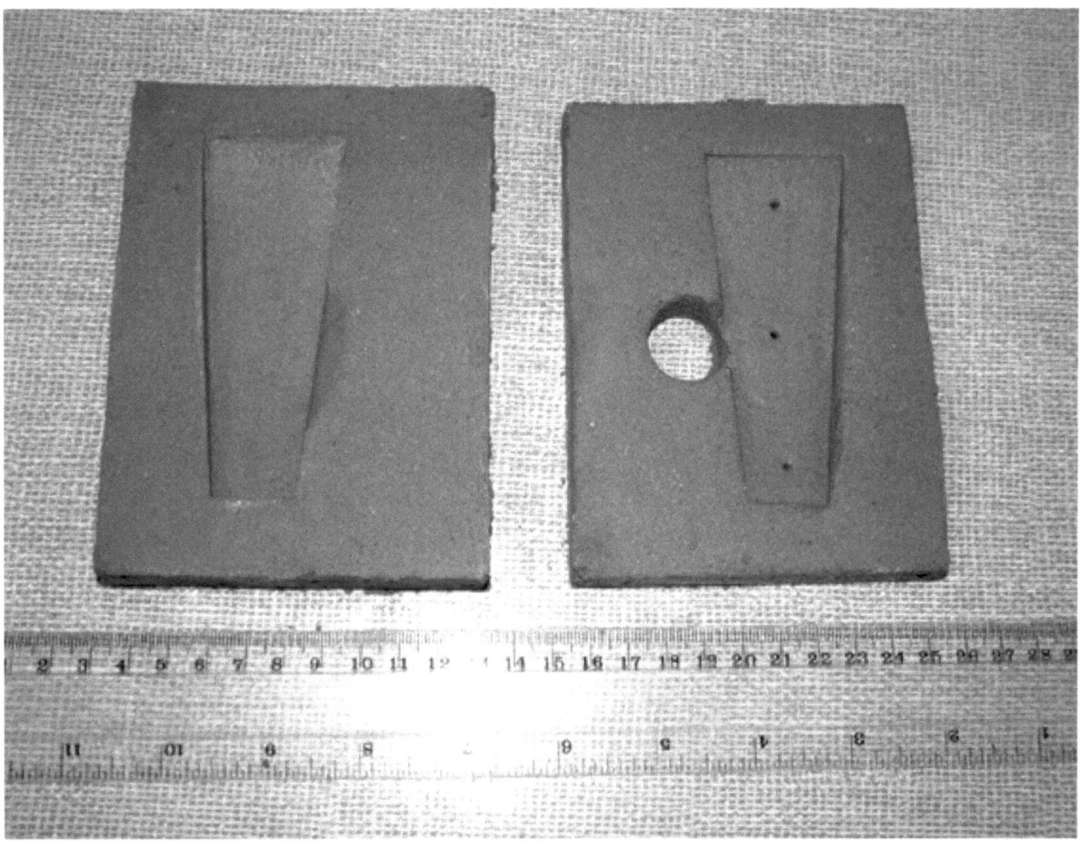

Figure 1. Two completed halves of the clay mould, showing vents (the smaller holes) and runner

Figure 2. A completed mould assemblage

was noncalcareous, the moulds were fired at 700°C. The heat was built up slowly and the firing took approximately10 hours. The sawdust had burnt out, probably around 500°C, and left pores for gas to escape during casting.

EXPERIMENTAL CASTING

Commercial mini copper slugs, tin and lead ingots with 99.9% purity were used for the experiments. The copper was melted in a plumbago crucible using an induction furnace. The capacity of the crucible was 3-6 kg depending on the size of the raw material. Charcoal was put both below and on top of the charge in the crucible to prevent oxidisation of the copper. A thermocouple (NiAl/NiCr) was used to measure the temperature of the metals.

In an induction furnace copper can be heated to a temperature of 1200-1300°C in less than 20 minutes. Such quick heating produces minimum contamination and metal loss (Davis & Oelmann 1985, 73). However, turbulence caused by the induction furnace during melting has a major impact on the casting quality (Campbell 1991, 19). This was confirmed by our earlier experiments with sand mould castings. Therefore, the melt was allowed to rest in the furnace for a short time (30-40 seconds) after the furnace was switched off to allow the gases (oxygen, hydrogen, CO_2, and CO) to escape.

Casting temperature is the temperature of the molten metal just before pouring into the moulds. However, it is difficult to measure the accurate pouring temperature, since the molten metal needs to be poured seconds after removal from the furnace. Here *melting temperature* refers to the temperature of molten metal in the furnace just before the furnace was turned off. The temperature of the molten metals drops dramatically during removal of the crucible, and skimming off the charcoal residue. For example, we measured that a molten charge of 1280°C dropped to 1180°C in a minute. The question is: how high should the casting temperature be? Casting of aluminium alloys (T_{mp} = 660°C), can be successful at casting temperatures of 700-750°C (Campbell 1991, 4), which is about 100°C higher than the melting temperature of aluminium. Experiments carried out by Ratka (1996, 15) suggest that a casting temperature of 1150°C was necessary for a flawless casting of tin bronzes into an unpreheated mould. As expected and supported by our initial experiments, a higher casting temperature provides better quality of castings. To keep all casting parameters as constant as possible, the *melting temperature* of the metal was kept in the range of 1270-1290°C and the *casting temperature* in the range of 1170-1190°C for the whole series of experimental casting.

Flashing was found to be a problem. This could be partially due to shrinkage of the clay and partially due to the high pressure of the molten metal, which pushed the two pieces of the mould apart. To reduce flashing, we put heavy metal ingots on top of the mould at two sides (Fig 3).

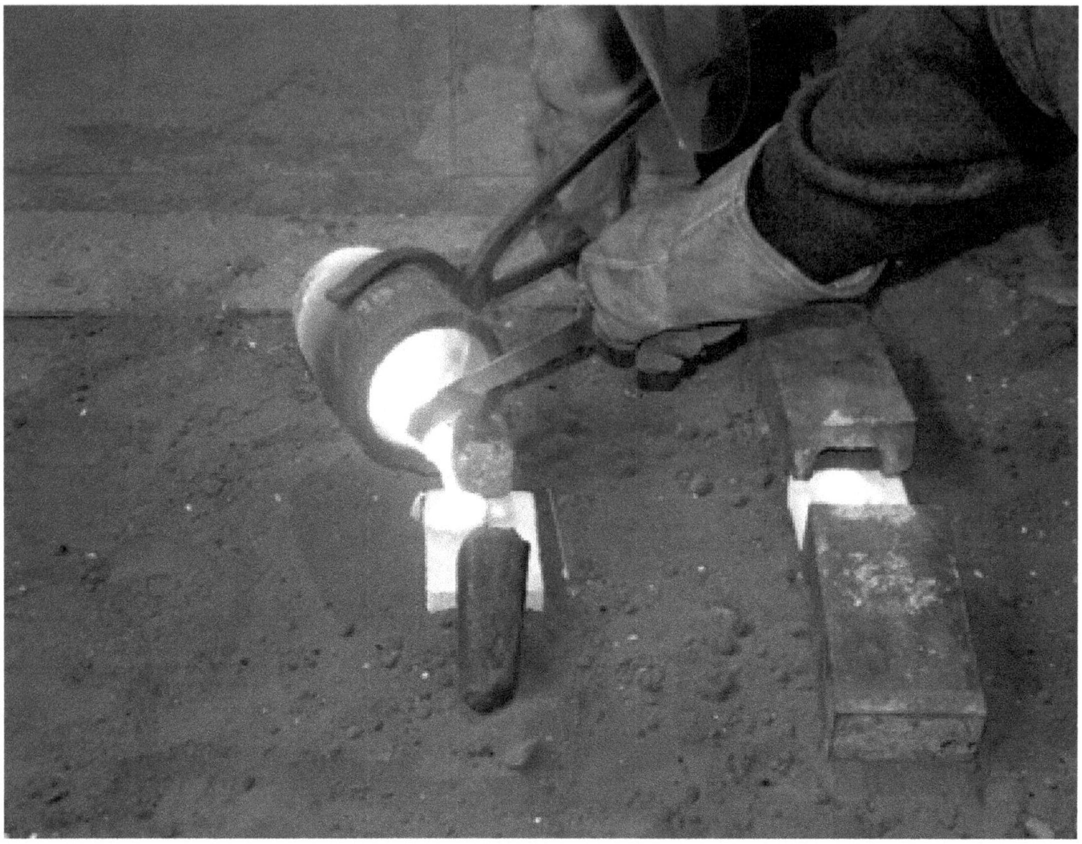

Figure 3. Pouring molten metal into the clay mould, weighed down by heavy metal pieces to prevent excessive flashing

Figure 4. Axe cast in clay mould, showing shrinkage slips on the surface, feeder and some flashing

The three different cooling rates were achieved by pouring the molten metal into three moulds: one preheated (to 350°C) and left to cool at room temperature (air-cooled), one without preheating and air cooled, and one without preheating and water chilled. The preheating temperature was chosen on the basis of simulation and experimental casting conducted by Ratka (1996, 15). Water chilling was achieved by breaking the mould and inserting the cast metal objects into water within a couple of minutes after pouring.

RESULTS AND DISCUSSIONS

Appearances of the casts

Surface shrinkage was noted in some casts (Fig 4). This is probably due to rapid cooling of the surface, where molten metal solidified too quickly to allow the space caused by shrinkage to be filled from the reservoir in the runner bush.

Generally, axes cast in clay moulds have a finer surface than those cast in sand moulds. This can be attributed to the small grain size of clay, which is generally in the range of 5-50 μm. It was found that flashing was almost unavoidable using clay moulds. The actual size of the cast axes was marginally smaller than the pattern due to the shrinkage of the clay. Most casts were free of macropores.

Microstructure of the casts

All cast axes were sectioned across the middle for metallography. The metallographic samples were prepared by the conventional method: polished to 0.25 μm finish and etched with alcoholic ferric chloride. Metallographic pictures were taken using an Olympus microscope equipped with an image grabbing system. A series of photomicrographs are shown in Figures 5-22 and all results are summarised in tabulated form in Appendix I.

Micropores were found to be present in most of the sections. This agrees with simulated and experimental casts conducted by other researches (Bailey 1982, 28; Ratka *et al.* 1999, 302; Staniaszek & Northover 1983; Swiss & Ottaway, in press). The microstructures of the cast bronzes are complex (see Appendix I). Their details are presented and discussed below.

Metallographic examination has shown that some casts have a granular structure and others have dendritic structures, but some casts have a mixed structure of grains and dendrites. In principle, the microstructure of casts with known composition can be predicted on the basis of phase diagrammes. However, phase diagrammes refer to equilibrium situation in which the metals cool very slowly. In our experimental casting and, it is suggested (Scott 1991,11), in metals cast in prehistory, casts were cooled rather faster than equilibrium solidification.

In general, little difference in the microstructures of bronzes cast in a preheated mould and those cast in an unpreheated mould could be found. In the 6% tin bronzes, for instance, both have a dendritic structure and show a few islands of eutectoids (α+δ) (Figs. 5 & 6). On the other hand, water-chilled bronzes are very different from the air-cooled ones. The water-chilled 6% tin bronze, for instance, has, in contrast to the air-cooled one, a very pronounced dendritic structure, smaller grain size and has many interdendritic eutectoids (α+δ) on the grain boundaries in which the δ phase is dark and abnormal in appearance (Figs. 7). Bronzes with 2% tin, cast in preheated mould and in unpreheated moulds and air-cooled, do not have δ phase at all and generally have a granular structure (Fig 8). However, the water-chilled one of the same composition has some δ phase on grain boundaries (Fig 9). The δ phase in water-chilled casts looks dark rather than light blue under the microscope. It's appearance is also different from that in the air-cooled casts. The water-chilled 10% tin bronze has interdendritic needle β phase (Figs 10 & 11). The air-cooled 15% tin bronze has massive interdendritic eutectoid of (α+δ) (Fig 12); while the water-chilled 15% tin has β phase both interdendritically and within a grains (Fig 13).

50μm

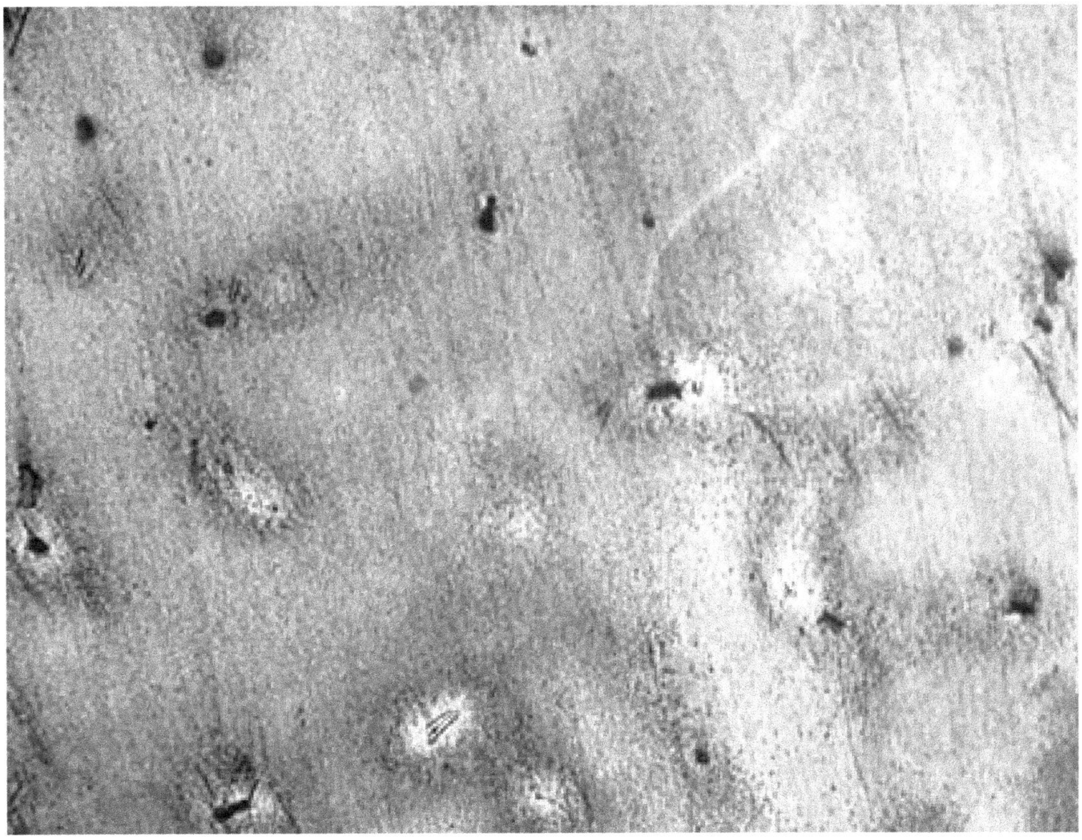

Figure 5. 6% Sn bronze cast in a preheated (350°C) mould and air-cooled, showing very few islands of eutectoid (α+δ)

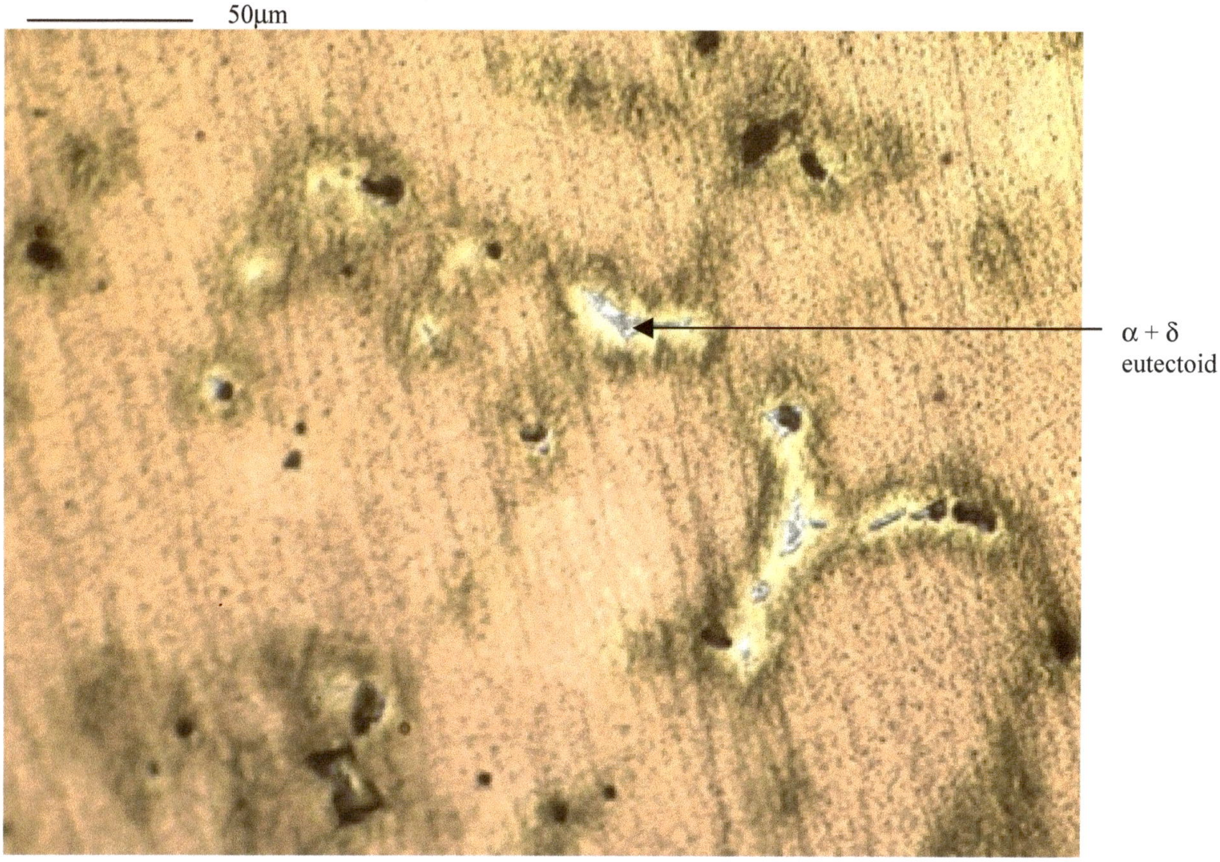

Figure 6. 6% Sn bronze cast in an unpreheated mould and air-cooled, showing presence of eutectoid (α+δ)

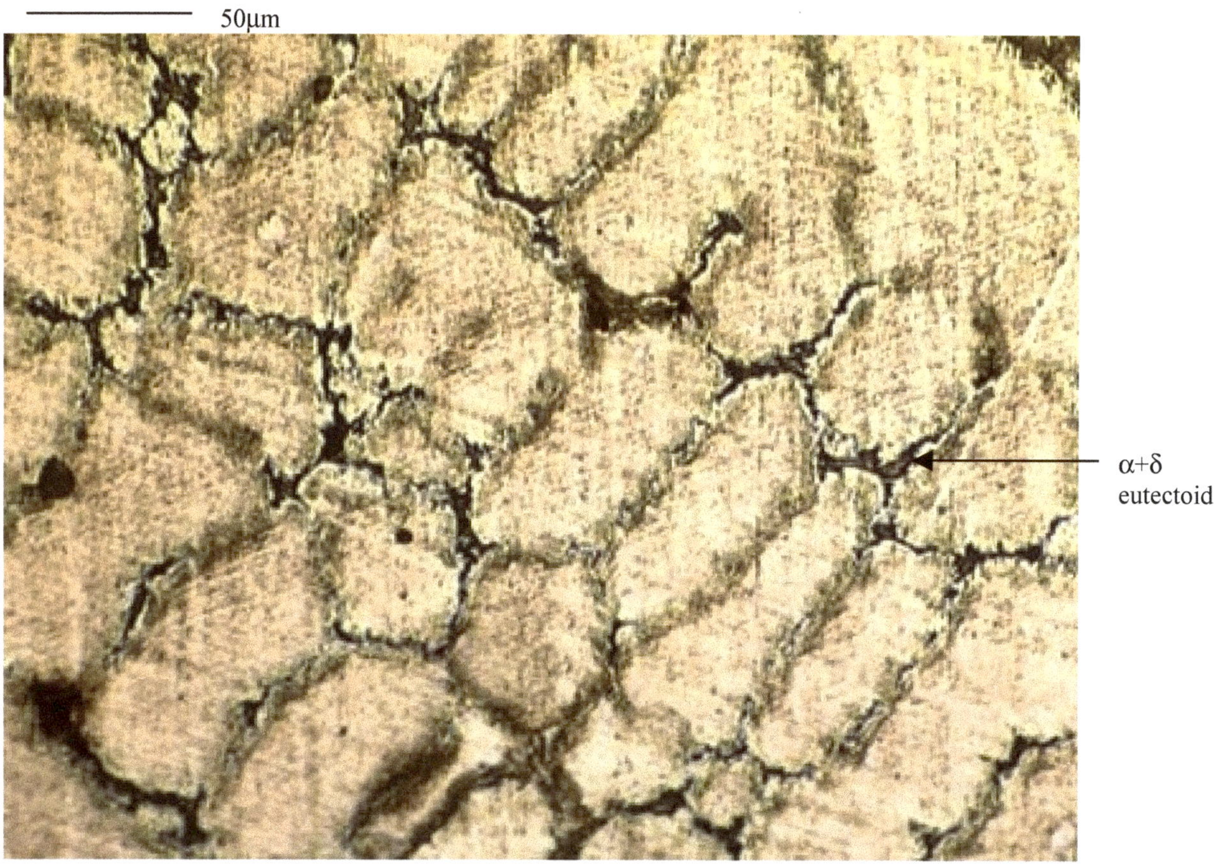

Figure 7. 6% Sn bronze cast in an unpreheated mould and water chilled, showing eutectoid (α+δ) on grain boundaries, in which the δ phase is dark and abnormal in appearance.

Figure 8. 2% Sn bronze cast in an unpreheated mould and air cooled, showing a granular structure

Figure 9. 2% Sn bronze cast in an unpreheated mould and water chilled, showing eutectoid of (α+δ) on grain boundaries

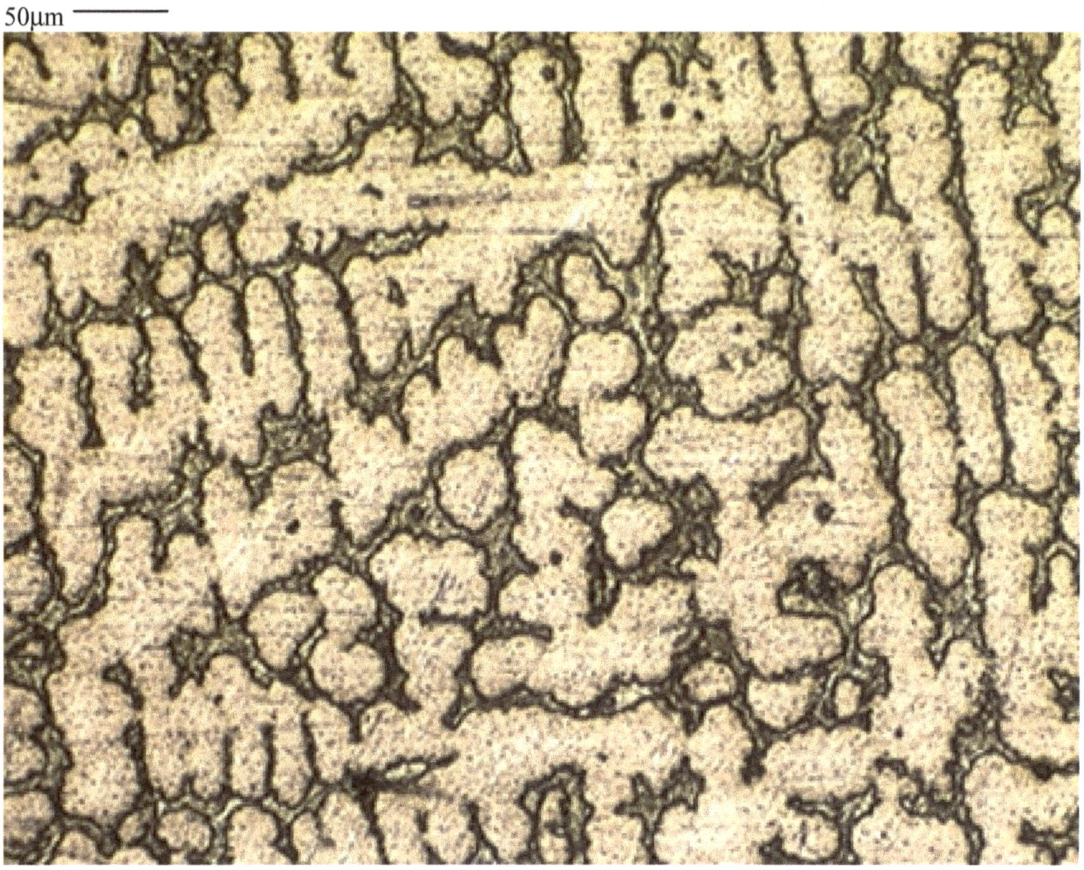

Figure 10. 10% Sn bronze cast in an unpreheated mould and water-chilled, showing pronounced dendritic structure and massive eutectoid ($\alpha+\delta$)

Figure 11. Same sample as Fig. 10 at higher magnification, showing interdentritic needle β phase

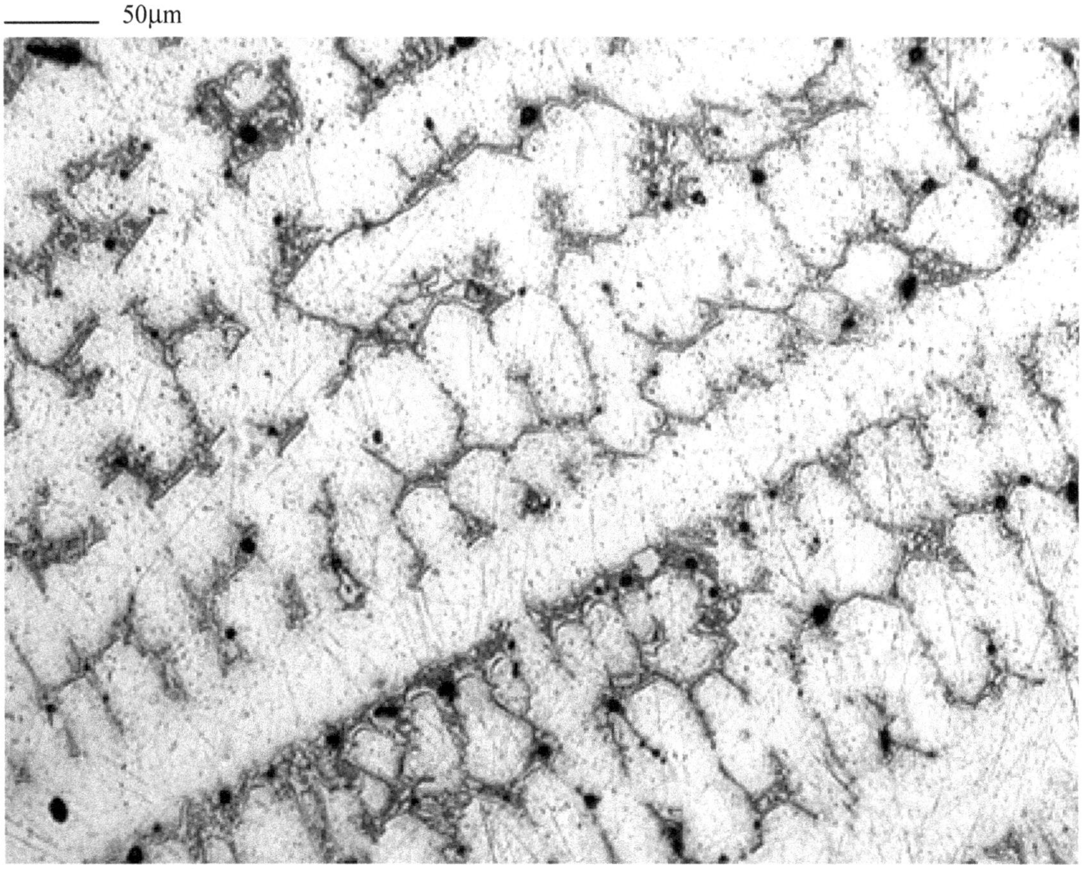

Figure 12. 15% Sn bronze cast in an unpreheated mould and air cooled, showing massive interdendritic eutectoid (α+δ)

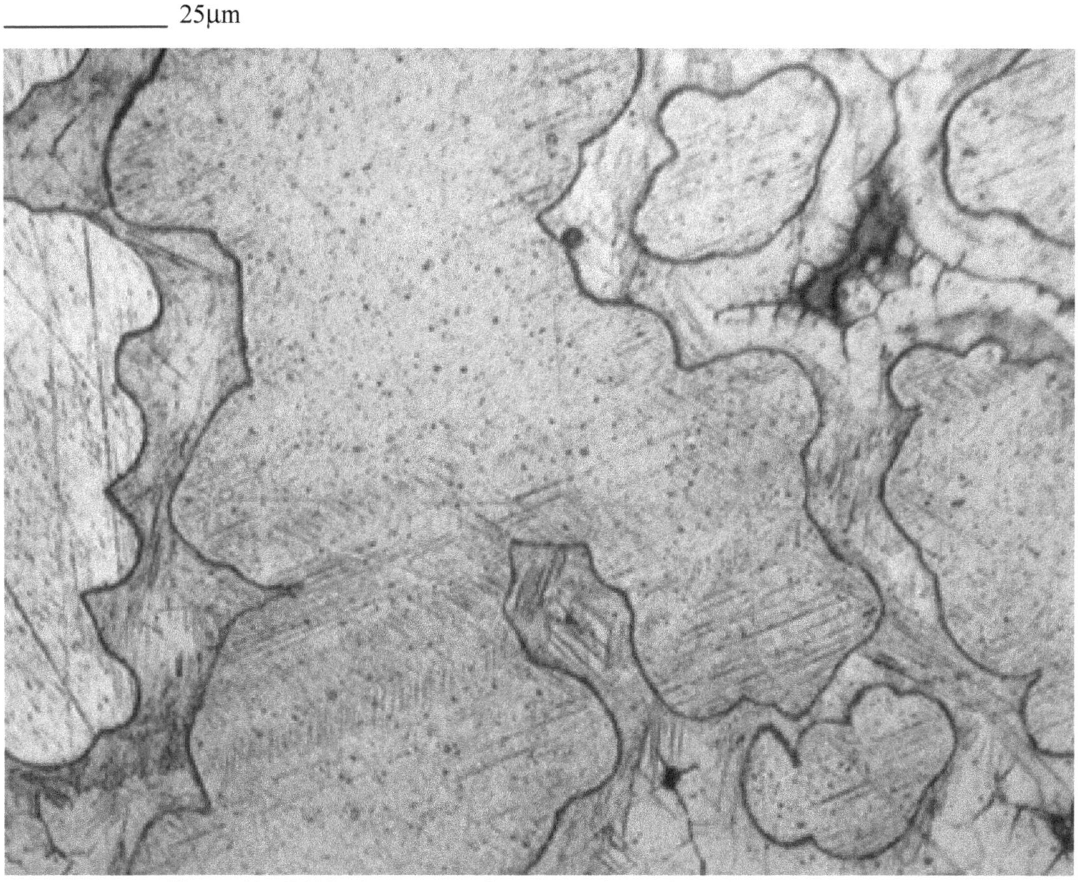

Figure 13. 15%Sn bronze cast in an unpreheated mould and water chilled, showing β phase both interdendritically and within a grains

Bronze containing 23% tin is quite different from the others. The predominant phase of the air-cooled cast bronze of 23% tin is the δ phase (Fig 14). The matrix of the water-chilled one, as SEM analysis confirmed (sample No.125 in Table 2), is needle β phase with islands of α phase and δ phase (Figures 15 & 16). The presence of β rather than γ phase indicates that the quenching temperature was between 586-788°C as γ phase is found in the high-tin bronzes quenched at temperatures between 520-585°C (Chadwick, 1939, 344; Goodway & Conklin, 1987, 10). γ phase, which tends to look a uniform greyish or light brownish colour on etching (Pers. Comm. Scott), does not seem to be revealed in the cast bronzes discussed here. There is a dark phase at grain boundaries (Fig 17). This could be either γ or δ phase. Its morphology suggests that it is more likely to be δ phase (pers. Comm. Scott).

The microstructures of the leaded bronzes are even more complex (see App I). For example, the casts with 2% tin but varying lead contents, e.g. 2, 6, and 10% lead have different microstructures. The cast bronzes with 2% tin + 2% lead all have a dendritic structure regardless of their cooling methods. The bronze with 2% tin + 6% lead cast in a preheated mould, on the other hand, has a granular structure (Fig 18), while the ones cast in moulds at room temperature and air-cooled and water quenched have dendritic structures (Fig 19).

It is well known that lead is immiscible in copper (Scott 1991, 23), and it is usually present as small spherical or dendritric droplets at the grain boundaries (Gettens 1969, 191). This was confirmed in our experimental castings. Figure 20 shows that the lead is almost entirely confined to the grain boundaries. This agrees with experimental casts conducted by Staniaszek and Northover (1983, 263). The lead globules are sometimes difficult to recognise in an etched sample (Fig 21). However, it can clearly be seen in a backscattered electron image (Fig 22).

In principle, lead should not affect the microstructure of the bronzes due to its immiscibility in copper. However, it can improve the fluidity of the alloy in the melt because of its low melting point. This may explain why the microstructures of these leaded bronzes are different. The casting temperature could be the major factor. On one hand, the addition of lead decreased the melting point of the metal; on the other hand, it lowers the casting temperature, because it was added into the molten copper just before pouring into the mould. The combination of these effects may have altered the actual *casting temperature*, although we tried to keep all the *melting temperatures* the same.

Compositional analysis was carried out using a JEOL JSM-6400 Scanning Electron Microscope equipped with EDS for confirmation of metallographic observation. The operating condition was 20KV. The results are listed in Table 2. These data show distribution of alloying elements in different phases; they do not refer to the average composition of the cast object. However, the composition of the matrix (area analysis by SEM/EDS) can provide an estimate of the alloy composition. Accurate compositions will be obtained by ICP-AES in future.

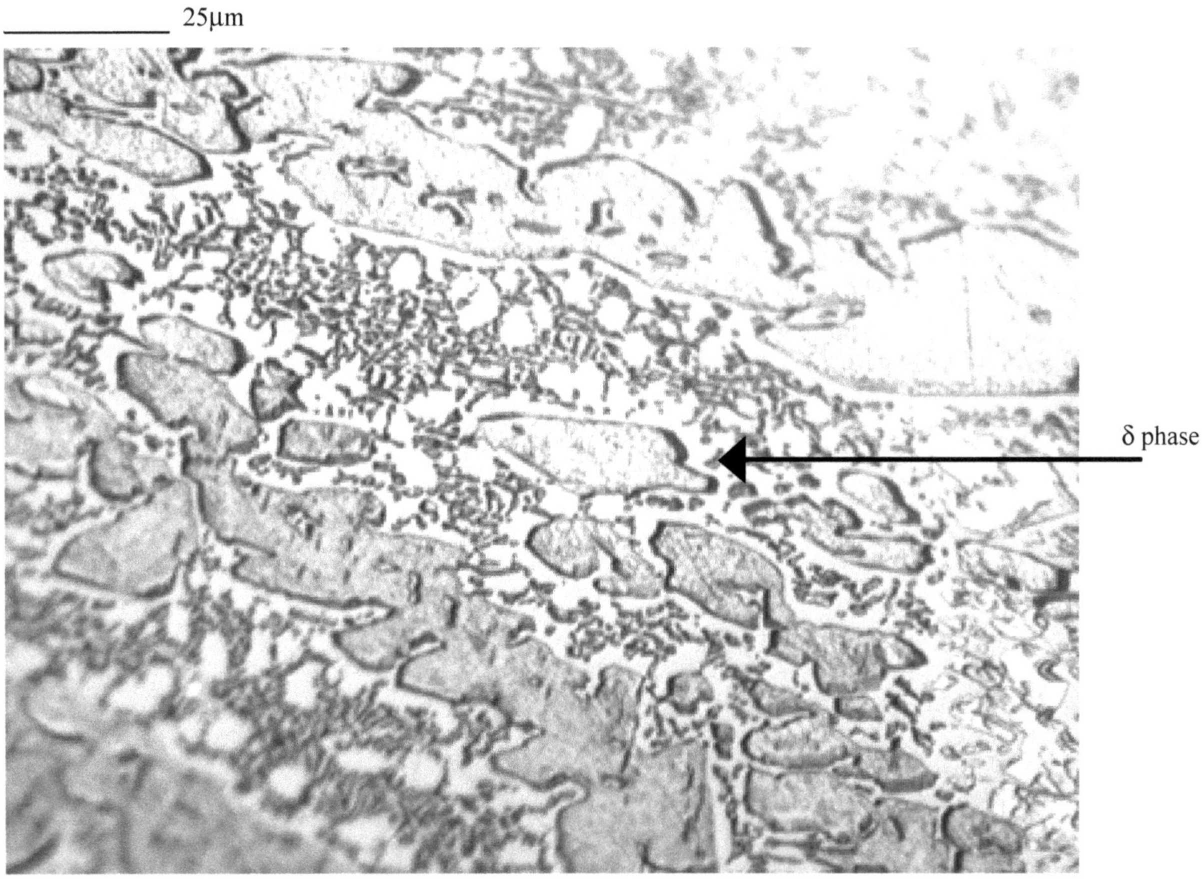

Figure 14. 23% Sn bronze cast in an unpreheated mould and air cooled, showing massive δ phase

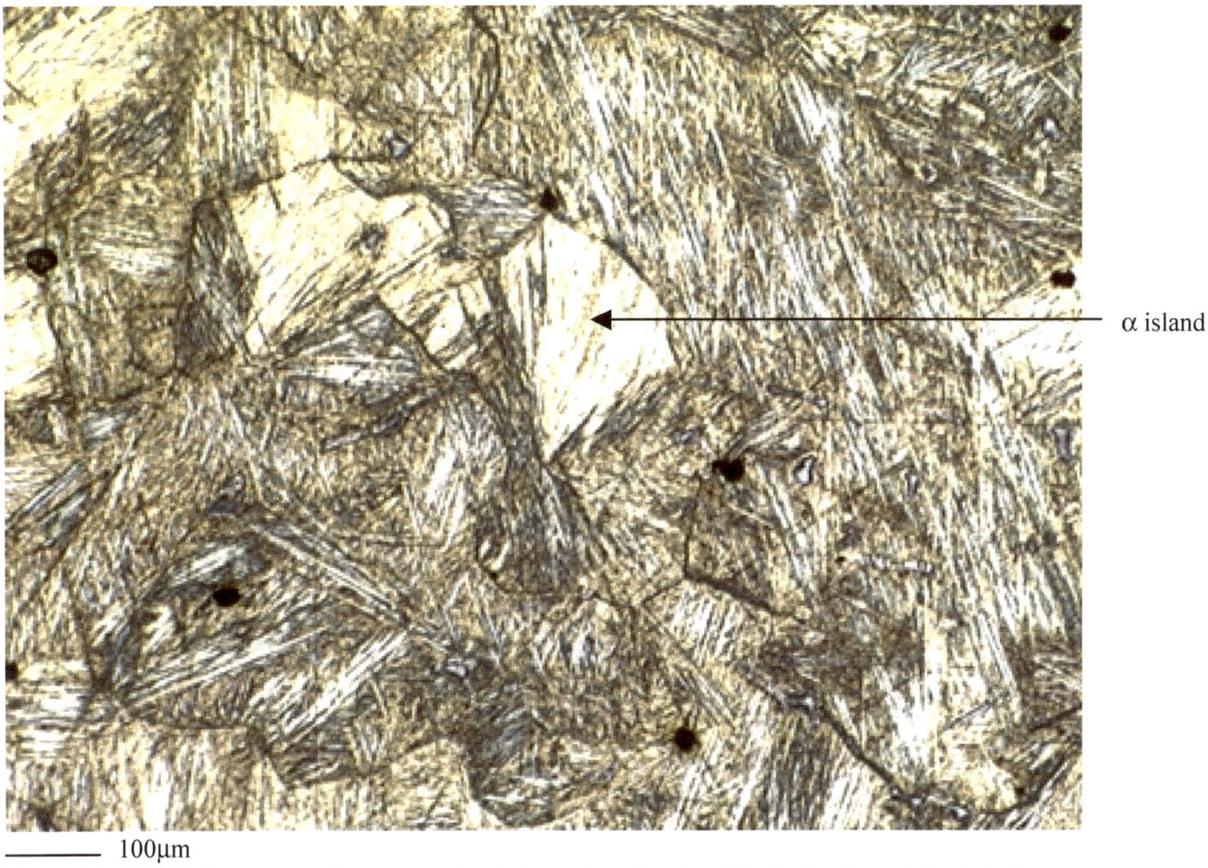

Figure 15. 23% Sn bronze cast in an unpreheated mould and water chilled, showing α islands in the matrix of needle β phase

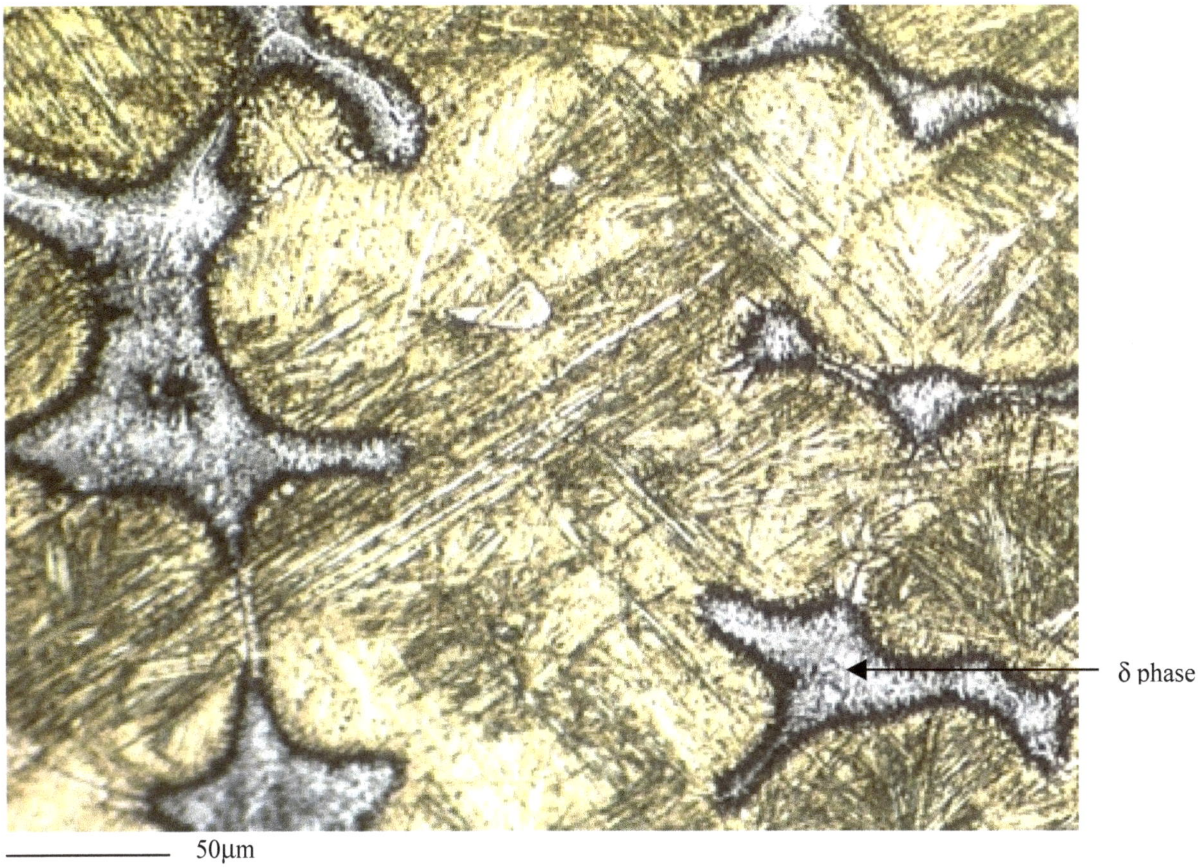

Figure 16. Same sample as figure 15, showing δ phase (gray dotted) in the matrix of β

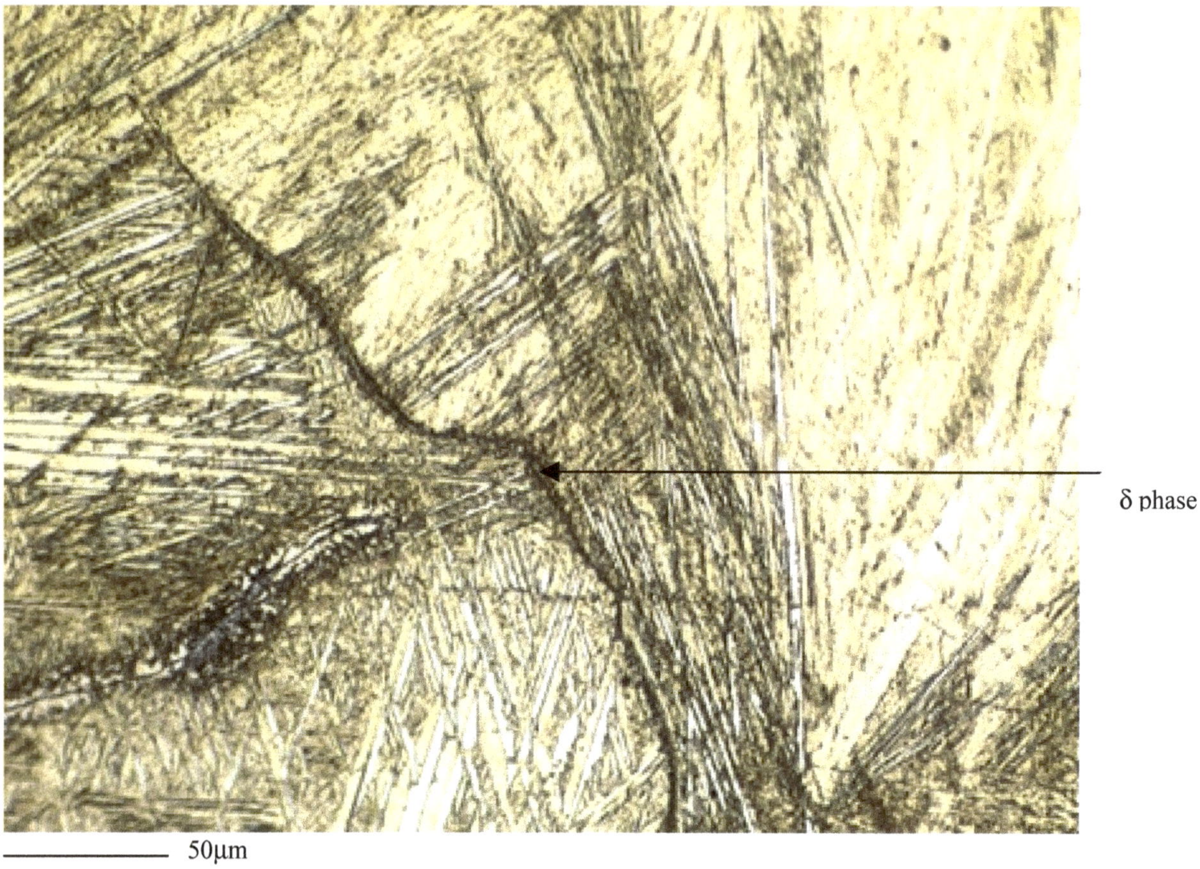

Figure 17. Same sample as figure 15, showing δ phase on the grain boundaries (dark line)

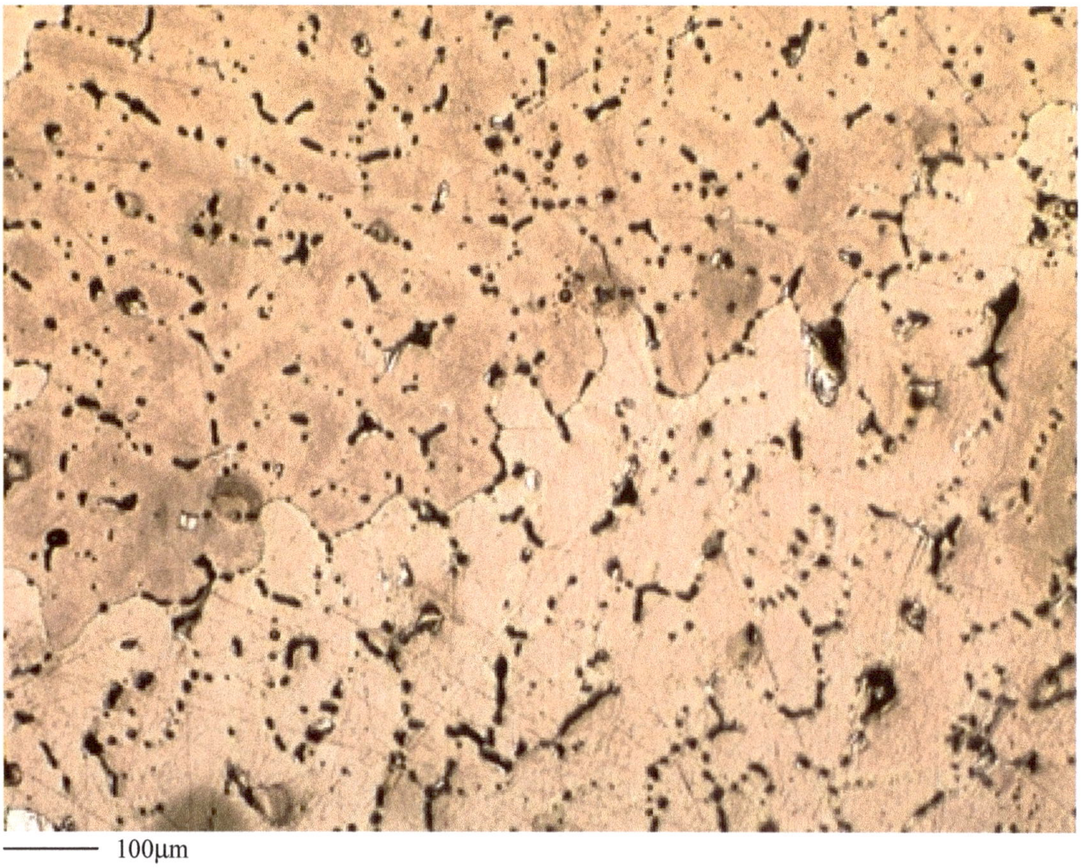

Figure 18. Bronze of 2%Sn + 6%Pb cast in a preheated mould and air cooled, showing a granular structure.

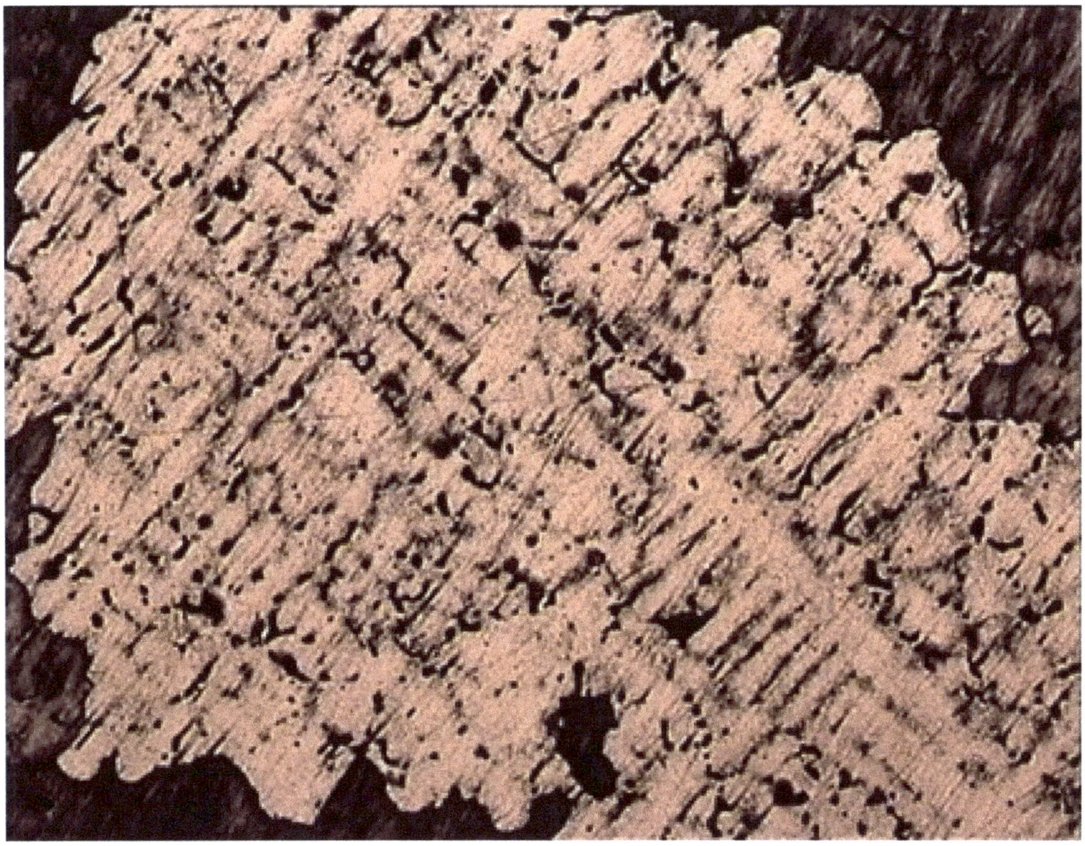

100µm

Figure 19. Bronze of 2%Sn + 6%Pb cast in an unpreheated mould and air cooled, showing a dendritic structure

50µm

Figure 20. Bronze of 2%Sn + 10%Pb cast in an unpreheated mould and air cooled, showing granular structure and Pb on grain boundaries

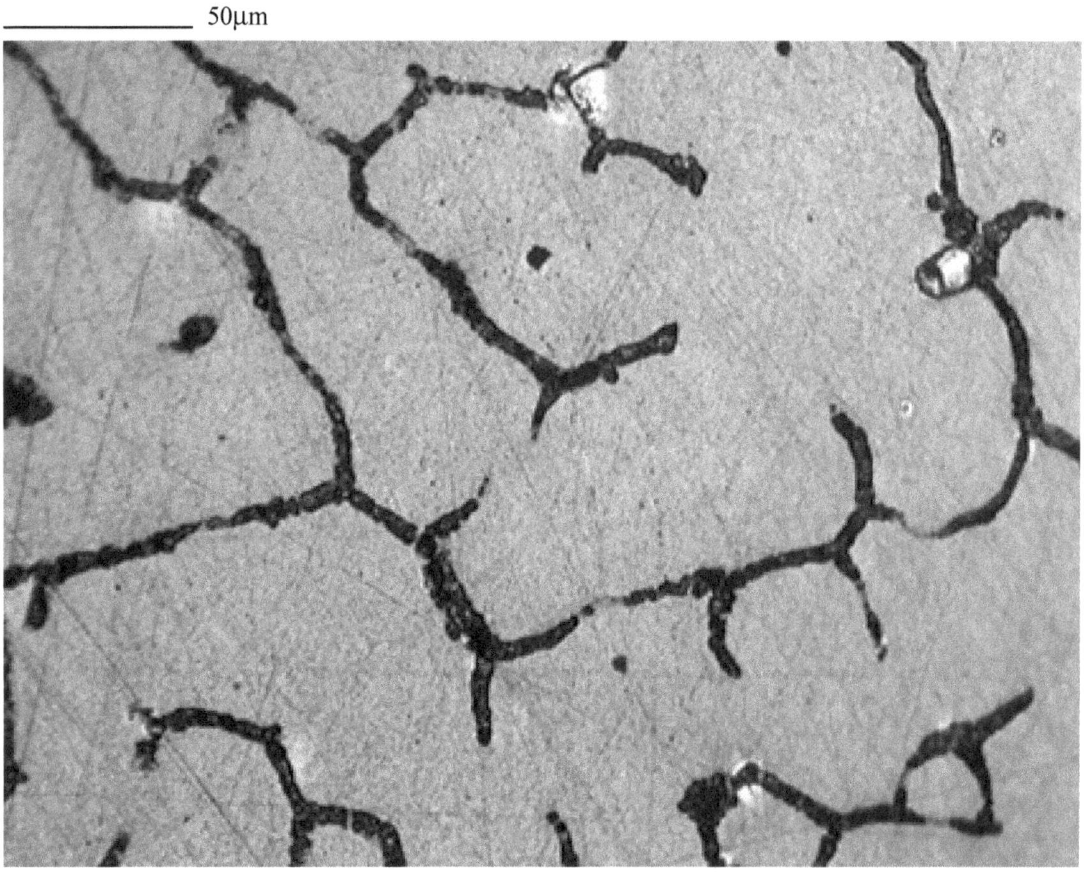

Figure 21. Bronzes of 2%Sn + 2%Pb cast in an unpreheated mould and water chilled, showing a dendritic structure. Pb is almost indistinguishable from the dark phase in the interdendritic regions.

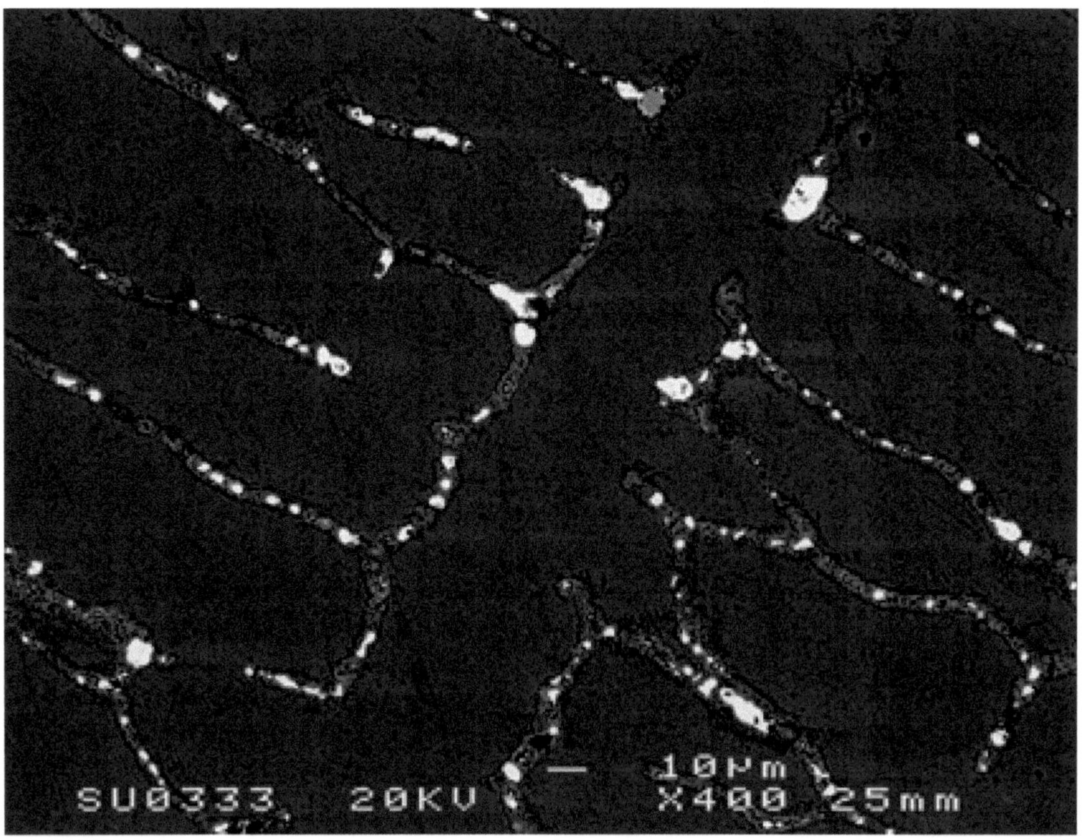

Figure 22. Backscattered electron image of the sample in figure 21, showing lead globules in the interdendritic phase

Table 2. Results of compositional analyses using SEM/EDS;

Sample No.	Original composition (wt.%)			Analysed area	Cu	Sn	Pb
125	Cu	Sn	Pb	Matrix β	76.3	23.7	—
	77	23	0	α	86.9	13.1	—
				δ	68.8	31.2	—
				Grain boundary	77.0	23.0	—
128	96	2	2	Matrix	96.6	1.7	1.7
				α	98.6	1.0	0.4
				Grain boundary without Pb	91.0	8.4	0.6
				Pb on grain boundary	8.4	—	91.6

Grain size and dendrite arm spacing

The grain size of granular structures, or dendrite arm spacing (DAS) of dendritic structure, was measured by the intercept method (Scott 1991, 51). The results of the measurement are listed in Table 3. Here dendrite arm spacing refers to secondary dendrite arm spacing, because it is the most important structural length measurements. The mechanical properties of most alloys are strongly dependent on secondary arm spacing; strength, ductility and elongation increase as DAS decreases (Campbell 1991, 146).

Grain sizes vary dramatically among these casts. There is a general trend to larger grain size in preheated moulds and smaller ones in water-chilled moulds, but there are notable exceptions which do not show any correlation to cooling rates (see Fig. 23 and Table 3). On the other hand, dendrite arm spacing shows a clear correlation to different cooling rates (Fig 23 and Table 3). Water-chilled casts have a smaller dendrite arm spacing than casts from the other two cooling methods.

Grain size cannot be smaller than the dendrite arm spacing. It is independent of solidification time. It is influenced by a large number of important factors such as nucleation events on the sidewalls of the moulds, foreign nuclei, e.g. inclusions, and chance events of damage or fragmentation from a variety of causes. Dendrite arm spacing, on the other hand, is controlled simply by freezing time (Cambell 1991, 146).

Microhardness

The microhardness was measured using a Leco M-400 hardness tester by taking the average of five readings across the metallographic section from one surface to the other with equal intervals. The results are listed in Table 4. The correlation of microhardness of various compositions and different cooling rates is shown in Figure 24. There is not much difference between preheated moulds and unpreheated moulds in microhardness, while most the water-chilled casts have higher microhardness. The graphs also show that addition of low amount of lead, e.g. 6%Sn + 2%Pb and 10%Sn + 2%Pb, can increase hardness in water-chilled casts, although it is generally known that lead decreases hardness of bronzes. It is interesting to note that in their experiments with tin bronzes with varying contents of lead, Staniaszek and Northover (1983, 264) also found that bronzes with 2-4% Pb showed the highest microhardness values. More work needs to be done to investigate this further.

Table 3. Grain size and dendrite arm spacing of the cast bronzes

Composition (wt%)			Dendrite arm spacing (μm)			Grain size (μm)		
Cu	Sn	Pb	Preheated/ air-cooled	Unpreheated/ air-cooled	Unpreheated/ water-chilled	Preheated/ air-cooled	Unpreheated/ air-cooled	Unpreheated/ water-chilled
98	2	0	—	—	—	143	132	116
94	6	0	63	61	36	975	833	2,900
90	10	0	69	58	39	1,167	486	638
85	15	0	67	52	45	1,560	1,140	700
77	23	0	—	—	—	—	—	714
96	2	2	53	61	58	3,800	890	827
92	2	6	—	41	32	1,271	740	780
88	2	10	—	—	—	167	160	109
92	6	2	57	51	34	855	394	488
88	6	6	52	40	28	944	1,300	809
84	6	10	55	52	33	613	840	850
88	10	2	78	79	31	1,100	1,467	353

* The balance is copper

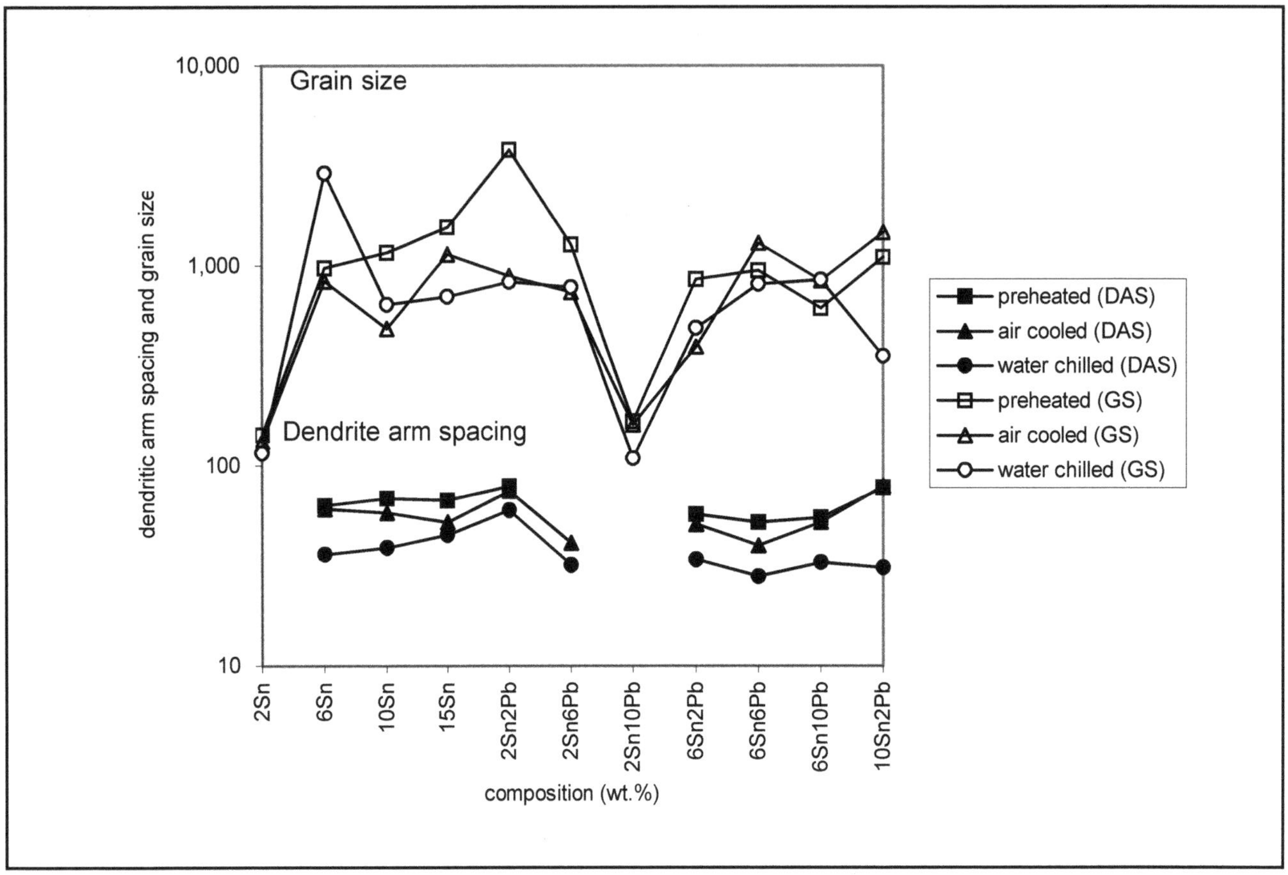

Figure 23. Grain sizes of the cast bronzes in clay mould

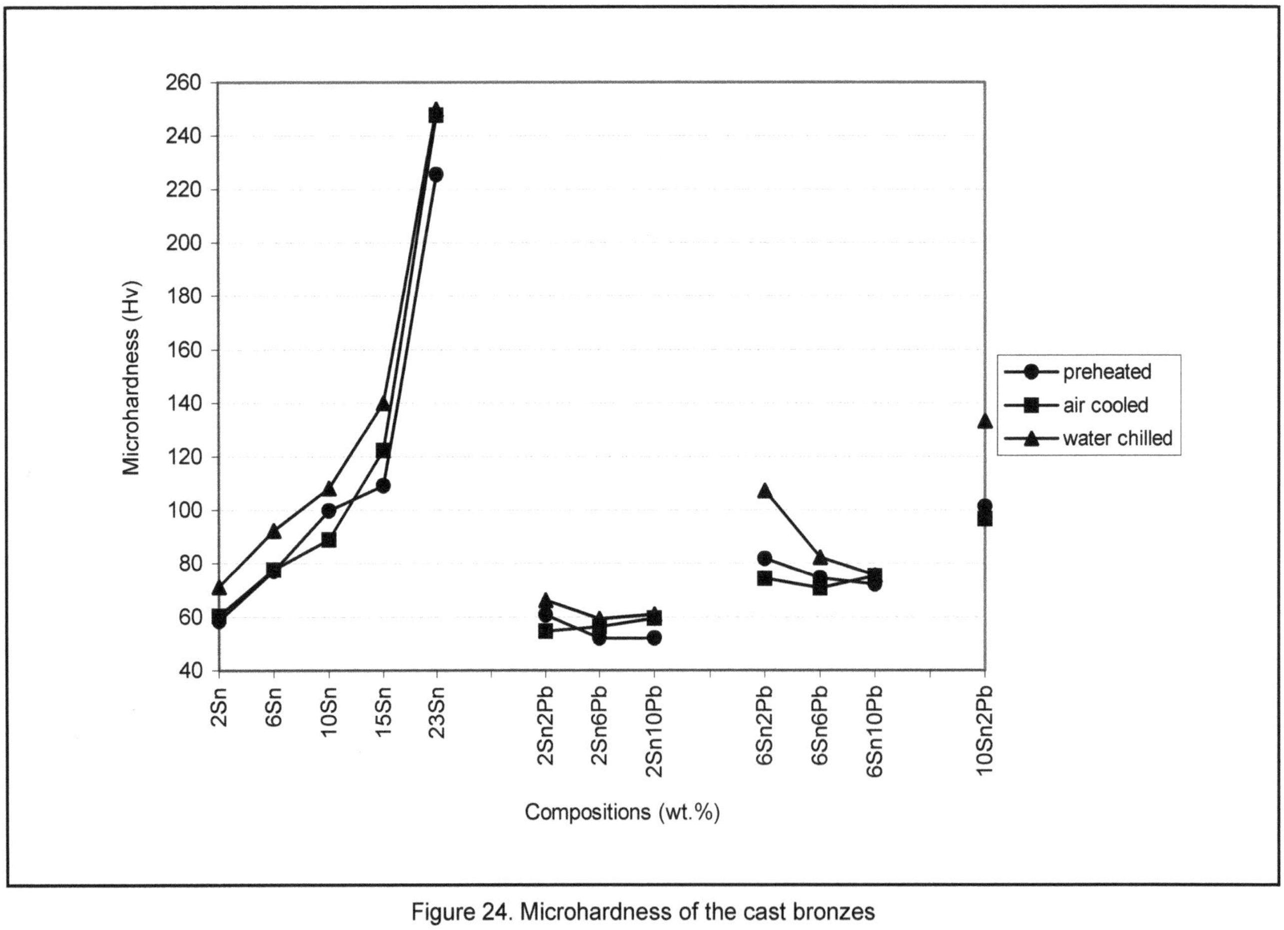

Figure 24. Microhardness of the cast bronzes

Table 4. Microhardness of cast bronzes with different compositions

Compositions (wt.%)*		Preheated/ air-cooled		Unpreheated/ air-cooled		Unpreheated/ water-chilled	
Sn	Pb	Hv	SD	Hv	SD	Hv	SD
2		58.5	2.7	60.4	5.2	71.1	2.5
6		77.1	8.7	77.6	7.1	92.2	8.6
10		99.8	10.9	88.9	3.5	108.1	12.8
15		109.2	11.0	122.3	10.7	140	23.9
23		225.4	44.4	247.8	33.7	250	26.0
2	2	60.7	1.8	54.6	4.7	66.2	4.0
2	6	52.1	4.6	56.2	5.9	59.1	4.3
2	10	51.9	10.5	59.5	3.5	60.9	4.7
6	2	81.7	5.9	74.4	10.5	107.1	20.9
6	6	74.6	1.9	70.7	10.0	82.3	9.5
6	10	72.1	7.9	75.2	7.3	74.4	11.3
10	2	101.2	6.6	96.7	14.7	133.4	19.0

The microhardness values for water quenched bronzes containing 23% tin do, in contrast to the other bronzes investigated here, show little difference to air-cooled bronzes. Yet, most ancient high-tin bronzes were quenched (Goodway & Conklin, 1987, 80; Meeks 1993, 75; Srinivasan 1998, 80). This can most probably be explained by the advantage of the structure of the β phase. A bronze with martensitic needle β phase, such as found in the water-chilled 23% tin bronze (Figs. 15 and 16) is much less brittle than the same bronze slowly cooled to room temperature (Scott 1991, 26). Quenching bronzes at a temperature of 586-798°C prevents formation of brittle δ phase. It has been reported that the Chinese high tin bronze used for the production of the musical instrument *gong*, is ductile and malleable and can be worked easily when quenched, but it is hard, springy, and brittle when cooled slowly in air (Biot 1816, 185 cited in Goodway & Conklin 1987, 24). To investigate the workability of bronzes cast in this series, cold-working experiments are presently being conducted by the authors.

CONCLUSIONS

Thirty-six bronze castings, of 12 alloys and 3 cooling rates for each alloy have been carried out in controlled conditions. The results show that great efforts have to be made in each step, such as the mould making, metal melting and pouring metal into the moulds, to assure a high quality cast. It was found that objects cast in clay moulds have a finer surface than those cast in sand moulds; therefore less effort is needed for finishing the surface of the artefact. However, some flashing seems to be unavoidable using clay moulds.

Our experiments have also shown that preheating a fired clay mould was unnecessary for casting and made little difference to the microstructure or hardness of the cast metal. There was, as expected, an appreciable difference in the microstructure of air-cooled casts compared to the water-chilled casts.

The microstructures of the casts are complex. The low tin (2%Sn) bronzes (sample No.111 - 113) have only granular structures. However, the microstructures of the leaded low tin bronzes (samples 126-134) vary (see App. I); some have a granular structure, and others have a dendritic structure. δ phase was seen in the water-chilled bronzes of 2% Sn, much lower compared to that in the equilibrium phase diagramme. The δ phase in the water-chilled casts looks abnormal in colour and appearance. β phase began to appear in the water-chilled cast bronzes with 10% Sn in our experiments, also much lower than that in the equilibrium phase diagramme. The presence of δ and β phase in such low tin bronzes is peculiar to ancient artefacts.

A clear correlation between cooling rates and dendrite arm spacing was found, whereas grain size does not show such a clear correlation although a general trend towards larger grains in preheated and smaller grains in water-chilled moulds could, as was expected, be observed.

Microhardness values were dependent mainly on the metals' composition, although almost all water-chilled casts in the series were found to have higher values than those cast in preheated or unpreheated and air-cooled moulds. It was interesting to note that whereas addition of lead generally, as is well known, leads to a decrease in hardness, addition of only 2% Pb, seems to increase hardness in water-chilled casts.

Our experiments have shown that far from being able to predict microstructures of copper-based alloys from published phase diagrammes, the behaviour of archaeologically relevant alloys under casting and cooling conditions likely to have prevailed in prehistory needs to be studied in much greater detail. The present study is only a small beginning.

Acknowledgements

This work is a part of the Leverhulme project: "Casting Experiments with Archaeologically Relevant Copper-based

Alloys". All casting experiments were conducted in the Foundry of the Department of Engineering Materials at the University of Sheffield. We would like to thank Alan Harvey, Andy Marshall and Ian Watts for their assistance in castings and other foundry work. Thanks are also due to Phil Staton for his help with image crabbing of the metallography.

Author's address

Authors:

Dr. Quanyu Wang, was Research Associate at the Department of Archaeology & prehistory, the University of Sheffield, while working on the Leverhulme project. She graduated from the Institute of Archaeology, University College London. Her PhD thesis was on "Metalworking technology and deterioration of Jin bronzes from the Tianma-Qucun site, Shanxi, China." Her research interests include technical study of ancient ceramics and metals as well as corrosion and conservation of metalwork.

Professor Barbara Ottaway, taught Archaeological Sciences in the Department of Archaeology and Prehistory, The University of Sheffield. Her research in recent years has concentrated on all aspects of archaeometallurgy, on experimental casting and on use-wear analyses of bronze artefacts. She has now retires and is visiting Professor of the Department of Archaeology, University of Exeter.

Address of both authors:

Quanyu Wang: quanywang@yahoo.com

Barbara Ottaway: b.ottaway@exeter.ac.uk

Bibliography

ANDREWS K C 1993, Slip on something luxurious. Unpublished PhD thesis. University of Sheffield.

BAILEY A R 1982, *The role of Microstructure in Metals* (Surrey).

BIOT J B 1816, Traitéde physique expérimentale et mathématique (Paris).

CAMPBELL J 1991, *Castings* (Oxford).

CHADWICK R 1939, 'The effect of composition and constitution on the working and on some physical properties of the tin bronzes', *Journal of the Institute of Metals* 64(1), 331-346.

COGHLAN H H 1960, 'Prehistoric working of bronze and arsenical copper', *Sibrium* 5, 145-162.

COGHLAN H H and Raftery J 1961, 'Irish Prehistoric casting moulds', *Sibrium* 6, 223-244.

COLLINS A E P 1970, 'Bronze Age Moulds in Ulster', *Ulster Journal of Archaeology* 33, 23-36.

DAVIES D J and OELMANN L A 1985, *Metallurgical Processes and Production Technology* (London).

FAOLÁIN S Ó and NORTHOVER P J 1998, 'The technology of late bronze age sword production in Ireland', *The Journal of Irish Archaeology* IX, 69-88.

GETTENS R J 1969, *The Freer Chinese Bronzes, Vol. II, Technical Studies* (Washington).

GOODWAY M and Conklin H 1987, 'Quenched high-tin bronzes from the Philippines', *Archaeomaterials* 2(1), 21-27.

HODGES H W M 1958-1959, 'The Bronze Age moulds of the British Isles, part 1: Scotland and Northern England – moulds of stone and clay', *Sibrium* 4, 129-137.

HODGES H W M 1960, 'The Bronze Age moulds of the British Isles, part 2: England and Wales – moulds of stone and bronze', *Sibrium* 5, 153-162.

LAMM K 1973, 'The manufacture of jewellery during the Migration Period at Helgö in Sweden', *Bulletin of the Historical Metallurgy Group* 7(2), 1-7.

MEEKS N D 1993, 'Patination Phenomena on Roman and Chinese High-tin Bronze Mirrors and Other Artefacts', in S La Niece and P Craddock (eds.), *Metal Plating and Patination* (Oxford), 63-84.

NEEDHAM S 1980, 'An assemblage of Late Bronze Age metalworking debris from Dainton, Devon', *Proceedings of the Prehistoric Society* 46, 177-215.

RATKA M 1996, Numerical simulation of ancient bronze castings. A lecture given at the 13th International Bronze Congress, Cambridge, USA, May 1996.

RATKA M, SAHM P R, and BUNK W G J 1996, 'Numerical simulation of historical casting processes for bronze', *The Beginnings of Metallurgy* 9, 301-303.

SCOTT D A 1991, *Metallography and microstructure of ancient and historical metals* (Los Angeles).

SHALEV S 1999, 'Recasting the Nahal Mishmar hoard: experimental archaeology and metallurgy', in A Hauptmann, E Pernicka, T Rehren, and U Yalcin (eds.), T*he Beginning of Metallurgy* (Bochum: Der Anschnitt 9, 1999), 295-299.

SRINIVASAN S 1998, 'Highlights of ancient metallurgy in South India: early use of high-tin bronzes, high – carbon steel, metallic zinc, bronze smelting and solid cast images', in *Proceedings of the Fourth International Conference on the Beginning of the use on Metals and Alloys, May 25-27, 1998, Kunibiki Messe, Matsue, Shimane, Japan*, 79-84.

STANIASZEK B E P and NORTHOVER J P 1983, 'The properties of leaded bronze alloys', *Proceedings of the 22nd Symposium on Archaeometry, University of Bradford, Bradford, UK. March 30th – April 3rd 1982*, 262-272.

SWISS A J and OTTAWAY B S in press, ' Investment or investment? Casting and using a bronze mould', in J P Northover and C Salter (eds.), *Proceedings of Founders, Smiths and Platers, an International conference on metal forming and finishing of the earliest times* (Oxford).

TAN Derui and HUANG Long 1996, 'Shang Zhou Qing Tong Qi Tao Fan Cai Liao Yan Jiu (A study on treatment techniques about ancient clay mould materials)', in Ma Chengyuan (ed.), *WEN Wu Bao Hu Ke Xue Lun Wen Ji* (A collection of the Shanghai Museum essays on science of conservation) (Shanghai), 47-70.

TYLECOTE R F 1986, *The Prehistory of Metallurgy in the British Isles* (London).

Appendix I. Casting conditions and microstructures of the bronze axes cast in the clay mould

Sample No.	Composition* (wt.%)		Cooling method	Microstructure	Figure No
111	2	—	Preheated/air-cooled	Granular, α+δ eutectoid is absent	—
112			Unpreheated/air-cooled	Granular, α+δ eutectoid is absent	8
113			Unpreheated/water-chilled	Granular, α+δ eutectoid on grain boundaries	9
114	6	—	Preheated/air-cooled	Dendritic, very few α+δ eutectoids	5
115			Unpreheated/air-cooled	Dendritic, very few α+δ eutectoids	6
116			Unpreheated/water-chilled	Dendritic, α+δ eutectoids in the interdendritic regions, no β	7
117	10	—	Preheated/air-cooled	Dendritic, α+δ eutectoids can clearly be seen	—
118			Unpreheated/air-cooled	Dendritic, α+δ eutectoids can clearly be seen	—
119			Unpreheated/water-chilled	Dendritic, β infilled between dendrites	10 & 11
120	15	—	Preheated/air-cooled	Dendritic, massive α+δ eutectoids	—
121			Unpreheated/air-cooled	Dendritic, massive α+δ eutectoids	12
122			Unpreheated/water-chilled	Dendritic, β infilled between dendrites and within α grains	13
123	23	—	Preheated/air-cooled	Dendritic, massive α+δ eutectoids	—
124			Unpreheated/air-cooled	Dendritic, massive α+δ eutectoids	14
125			Unpreheated/water-chilled	β matrix with islands of α and δ	15-17
126	2	2	Preheated/air-cooled	Dendritic, α+δ eutectoids is absent	—
127			Unpreheated/air-cooled	Dendritic, α+δ eutectoids is absent	—
128			Unpreheated/water-chilled	Dendritic, dark phase in the interdendritic regions (α or δ? See Table 2 sample No.128)	21-22
129	2	6	Preheated/air-cooled	Granular, α+δ eutectoids is absent, Pb is on grain boundaries	18
130			Unpreheated/air-cooled	Dendritic, α+δ eutectoids is absent	19
131			Unpreheated/water-chilled	Dendritic, α+δ eutectoids is absent	—
132	2	10	Preheated/air-cooled	Granular, α+δ eutectoids is absent, Pb is both on grain boundaries and within the grains	—
133			Unpreheated/air-cooled	Granular, α+δ eutectoids is absent, Pb is both on grain boundaries and within the grains	20
134			Unpreheated/water-chilled	Granular, α+δ eutectoids is absent, Pb is both on grain boundaries and within the grains	—
135	6	2	Preheated/air-cooled	Dendritic, very few α+δ eutectoids	—
136			Unpreheated/air-cooled	Dendritic, very few α+δ eutectoids	—
137			Unpreheated/water-chilled	Dendritic, eutectoid with dark δ phase in the interdendritic regions	
138	6	6	Preheated/air-cooled	Dendritic, very few α+δ eutectoids	—
139			Unpreheated/air-cooled	Dendritic, very few α+δ eutectoids	—
140			Unpreheated/water-chilled	Dendritic, eutectoid with dark δ phase in the interdendritic regions	—
141	6	10	Preheated/air-cooled	Dendritic, very few α+δ eutectoids	—
142			Unpreheated/air-cooled	Dendritic, very few α+δ eutectoids	—
143			Unpreheated/water-chilled	Dendritic, eutectoid with dark δ phase in the interdendritic regions	
144	10	2	Preheated/air-cooled	Dendritic, α+δ eutectoids in the interdendritic regions	—
145			Unpreheated/air-cooled	Dendritic, α+δ eutectoids in the interdendritic regions	—
146			Unpreheated/water-chilled	Dendritic, eutectoid with dark δ phase in the interdendritic regions; β might be present within the eutectoid	—

* The balance is copper.

THE "ETRUSCAN TIN": A PRELIMINARY CONTRIBUTION FROM RESEARCHES AT MONTE VALERIO AND BARATTI-POPULONIA (SOUTHERN TUSCANY, ITALY)

M. BENVENUTI, L. CHIARANTINI, L. NORFINI, A. CASINI, S. GUIDERI & G. TANELLI

Abstract: Etruria Mineraria has been the seat of relevant metal production in the 1st millennium BC, especially for iron, copper, lead and precious metals. Several authors have repeatedly suggested that the Etruscans also exploited tin from the Monte Valerio area (Campiglia Marittima polymetallic district). However, unquestionable archaeological evidences of tin exploitation at Monte Valerio in pre-Roman times are still lacking. In this paper we present the preliminary results of a "geo-archaeological" survey in the Monte Valerio area. Four main types of tin ores have been identified based on direct field observations and analysis of museum's specimens. Only the now exhausted ore type called "massello" by past century miners, constituted by massive concentrations of cassiterite, could have been suitable for exploitation by ancient (Etruscan?) miners for a number of concurrent factors (occurrence at, or near, the surface; easily separable gangue minerals; medium to very high tin grade). At Monte Valerio, however, cassiterite preferentially occurs in disseminated-type ores (our type-A), which show definitely lower tin grades, probably too low for ancient mining technologies. The textural features and trace element composition of a "bronze" scrap recently found at the nearby Etruscan site of Baratti-Populonia (dating no later than the 3rd cent. BC) have been also analysed. It probably represents an unsuccessfull bronze-making process, as evidenced by the occurrence of unreacted, euhedral cassiterite crystals. Minor and trace element (Fe, Ti, W, Nb, Ta, Cu) composition of cassiterites from Monte Valerio and other ancient tin districts (Cornwall, Portugal, Sardinia) have been compared with those shown by cassiterite crystals within the Baratti "bronze" fragment. Although some distinctive compositional trends have been observed (distinctive enrichment in W of C-type ores from Monte Valerio, and in Nb-Ta of Portoguese samples), the results so far obtained do not permit to ascertain whether or not Monte Valerio tin has been used by the Etruscans for bronze making industry at Baratti-Populonia. Provenance studies could benefit from multi-tracer approach, including ongoing analysis of lead isotope signature.

1. INTRODUCTION

Quoting Northover & Gillis (1999) "tin holds a special fascination for those concerned with the evolution of metallurgy from the earliest time. Here is a metal whose ores have a very restricted geographical distribution but which had a very widespread impact on the development of copper alloys", namely (tin) bronze. The present archaeological evidence indicates that (tin) bronze appeared in southwest Asia in the third millennium BCE and production centres of bronze artefacts spread rapidly over the whole ancient world, from the Mediterranean region to the Persian Gulf (Pernicka, 2002). Given the limited number of tin deposits (at least those which could be profitably exploited with ancient mining technologies), this would imply that tin was widely traded, and over long distances, and bronzes were manufactured in places far removed from where tin (or its ores) originated. A more precise definition of such a trade network, connecting tin mining areas and metalworking centres, represents a major challenge for future archaeometallurgical researches.

A large number of bronze items have been produced in Etruria Mineraria since the first Iron Age up to Roman times (IX-III cent. BC). Although extremely variable in shape, use and metal technology, all the Etruscan bronze alloys show a remarkably constant composition, with about 8 to 10% tin contents (Craddock, 1986). This implies a significant consumption of tin. Nowadays we don't know where the tin worked by Etruscans came from. The hypothesis that "Etruscan tin" was (at least partly) exploited from the the Monte Valerio mine (located within the polymetallic district of Campiglia Marittima) has been put forward by Simonin (1858) and many others after him, but it has not received so far strong support from archaeological investigations (cf. Tizzoni, 1999).

In this paper we report the first results of a "geo-archaeological survey" carried out at the Monte Valerio mine and the nearby Etruscan metalworking site of Baratti with the purpose to evaluate on a more quantitative ground the Simonin's hypothesis.

2. THE ARCHAEOLOGICAL BACKGROUND

2.1 Tin sources in Western Europe

The main mineral in which tin occurs is cassiterite, SnO_2, although it may also appear as stannite, Cu_2FeSnS_4. Cassiterite is commonly found in hydrothermal deposits related to intermediate to acid magmatic rocks (porphyry tin, skarns, veins and sheeted veins around granites, greisens) as well as in pegmatites. When released from parent rocks by weathering processes, cassiterite may be readily concentrated in placer deposits due to its high density and chemical stability. Tin placer deposits were surely the favourite target of ancient miners, because placer mining was (and is) much easier and inexpensive compared to hard-rock mining.

Main ancient tin mines in western Europe are confined to outcrops of Hercynian granites, namely in south-west Britain (Devon and Cornwall), north-west Iberia and the Erzgebirge, on the borders of Saxony and Czech Republic. Minor tin production in the antiquity may have arisen in France (Brittany, Massif Central). As to Italy, the only known localities where tin has been exploited in the more or less recent past include Sardinia (Monte Mannu, Iglesiente) and southern Tuscany, namely the M. Valerio mine. Although minor, the latter deposit – if ever exploited in pre-Roman times! – would have played an important role in the complex metallurgical framework of ancient Europe.

3. THE MONTE VALERIO MINE

3.1 Geology

As schematically shown in Figure 1, the terrains outcropping in the M. Valerio area belong to three main structural Complexes (from bottom to top): the Tuscan Domain Complex, the Austroalpine Domain Complex, and the Neoautochtonous Complex (Costantini et al., 1993). The Tuscan Domain Complex is made of a sequence of calcareous formations (Massive Limestone, Red Ammonitic Limestone and Cherty Limestone of early Jurassic age), which are overlain by Jaspers (Late Jurassic) and Tuscan Scaglia (Cretaceous-Oligocene). The Austroalpine Domain is here only represented by the S. Barbara Formation, constituted

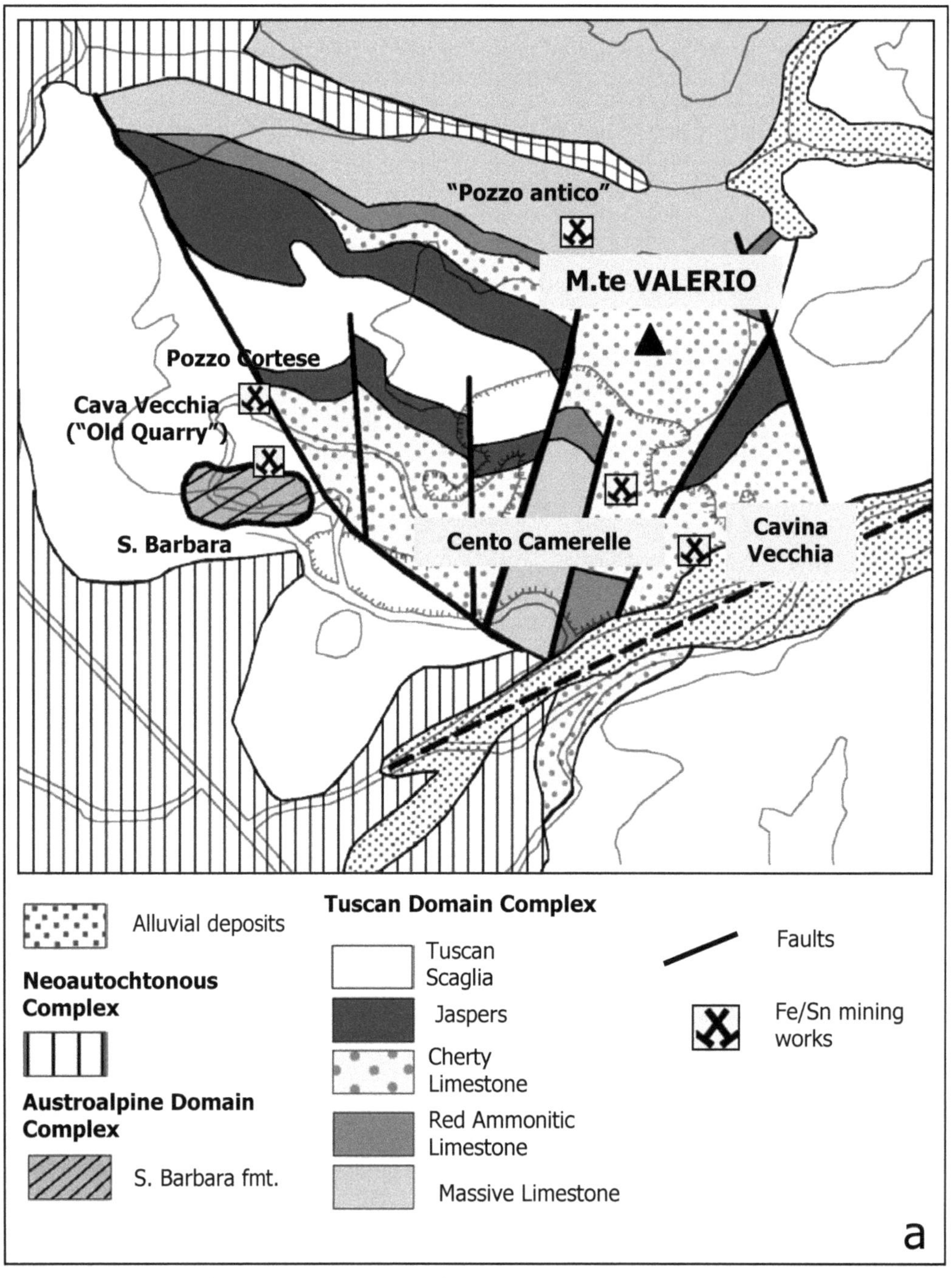

Figure 1a. Geological sketch map of M. Valerio area, with Fe/Sn mining works

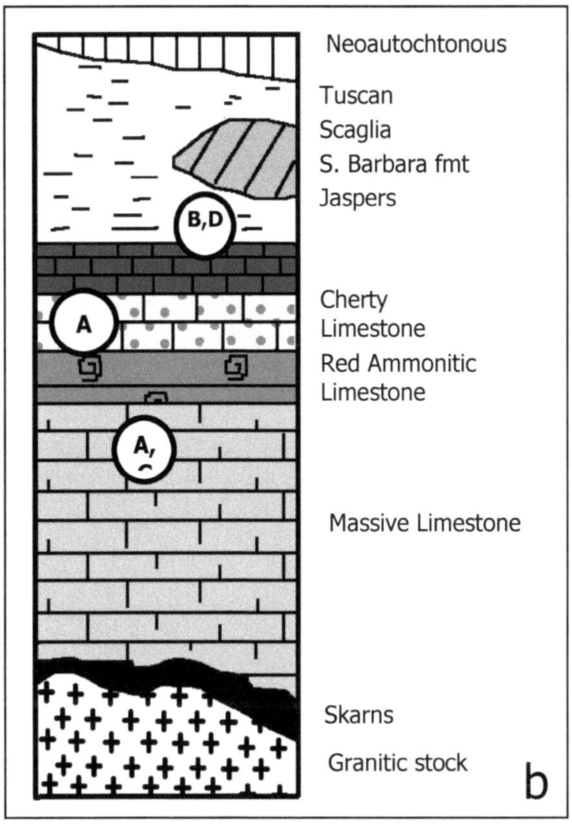

Figure 1b. Stratigraphic column with location of the four types of tin ore (see text).

by fine-layered grey limestone and dolomites. The most recent geological formations in the M. Valerio area are the upper Pleistocenic terrains of the Neoautochtonous Complex, namely the Val di Gori Sands Fmt. and the Donoratico red-orange Sands Formation.

A borehole drilled in the Monte Valerio area encountered a quartz-monzonitic stock (the "Mt. Valerio pluton") at 1148 metres below the surface (Rodolico, 1945). This stock belongs to the Tuscan Magmatic Province, which is constituted by intrusive (7.0-4.0 Ma) and subordinate volcanic products (5.0-0.3 Ma) emplaced in Mio-Pliocene during the post-deformation, extensional stage of the Apenninic orogeny (Peccerillo et al., 1987, and references therein). The intrusion of the plutonic bodies and dikes into the Mesozoic carbonate rocks of the Campiglia M.ma district caused widespread thermo-metamorphism and hydrothermal alteration, with formation of several skarn bodies. As a result of this intense hydrothermal circulation and later supergene processes, different kinds of Fe, Cu, Pb, Zn and Sn mineralization occur in the district. According to Tanelli et al. (1993), they include the Cu, Fe±Zn, Sn skarn deposits of Monte Rombolo-Monte Spinosa, the Cu, Pb, Zn±Ag skarn deposits of Valle del Temperino-Valle Lanzi, the residual Fe hydroxide±calcite deposits of Campo alle Buche and Colle S.Antonio, and, finally, the Fe-Sn mineralization of Monte Valerio-Cento Camerelle.

As thoroughly described by Bertolani (1958) and Venerandi-Pirri & Zuffardi (1981) – to whom the reader is referred for further details - Sn ore types at Monte Valerio include both strata-bound, massive concentrations of cassiterite + tourmaline (± quartz, calcite, goethite, hematite, pyrolusite, siderite, pyrite, chalcopyrite, galena, sphalerite, rutile) and crosscutting deposits developed along fractures and/or karst structures. According to Venerandi-Pirri & Zuffardi (1981) the highest Sn grades (>0.3 wt%) occur in strata-bound and vein-type orebodies hosted in the upper portion of Massive Limestone Fmt. and within the Jasper Fmt. The richest ore shoots (up to 70 wt% Sn), called by miners "massello" (="small mass"), are reported to occur within strata-bound lenses or at their intersection with vein-type ores. The "massello" oebodies were distinguished in two types - "calcareo" (calcareous) and "pietra" (stony, i.e., quartzous) – on the basis of associated gangue minerals. The observed geologic setting and minero-textural features of tin mineralization would be in agreement with a two-stage genetic model: 1) syn-sedimentary deposition of cassiterite+tourmaline (Trias-Dogger); 2) post-Pliocene supergene remobilization of syn-sedimentary ores with deposition of vein- and/or karst-type orebodies (Venerandi-Pirri & Zuffardi, 1981).

3.2 Mining history: the state of the art

Although commonly regarded in the archaeological literature as an "ancient tin mine", the first documentary evidence of mining activity at Monte Valerio dates back only to XVII century AD, and it is referred to iron, not tin! It is a Magona del Ferro's internal report (dating 1638), where Monte Valerio mine is said to yield a good quality limonitic ore. Up to the last quarter of XIX century Monte Valerio was thus exploited for iron alone (cf. D'Achiardi, 1872). In 1875, however, the mining engineer F. Blanchard discovered the presence of cassiterite intimately admixed with iron oxides/hydroxides: since that time up to 1948 Monte Valerio became a tin mine, yielding no less than 1500 tonnes tin. After the Second World War, however, metal exploitation was replaced by limestone quarrying for concretes, which is still active nowadays. One regrettable consequence of such an intense mining and quarrying activity over the last two centuries is that almost all the traces of ancient mine workings have gone lost! These "ancient" workings were described as «*excavations anciennes peu profondes communiquant entre elles par de chemineés tres étroites*» (Simonin, 1858), or as «*une succession de chambres et galeries les reliant entre-elles, d'où le nom Cento Camerelle*» (Blanchard, 1876). Other descriptions have been provided by Church (1879) and, more recently, by Bertolani (1958). Taking into account the geologic setting of ore bodies and the distribution of these adits, tunnels and rooms, both Blanchard (1876) and Church (1879) maintain that these workings were undoubtedly excavated to exploit tin ore, and not merely iron oxyhydroxides. If we accept this interpretation, the Cento Camerelle workings should predate the XVII century (the Magona's document attests that iron was exploited at that time). But to say that these workings date back to Etruscan times seems to be very hazardous. Little help come from archaeological findings, which are scanty and of dubious attribution. In particular, the copper and bronze mine lamps and picks found at the Cava Vecchia stope, which were considered Etruscan in age by Simonin (1858), could instead be younger, i.e. Roman (Tizzoni, 1999).

4. THE GEO-ARCHAEOLOGICAL SURVEY

In order to reconstruct – as far as possible - the evolution of metal exploitation at Monte Valerio on the basis of surviving archaeological evidences, i.e., on a strictly quantitative basis, a detailed field mapping was undertaken in the Monte Valerio area a few years ago by G.S.A.L. (Gruppo Speleologico Archeologico Livornese). After a careful review of historical documents, mining reports, and all available written documents, a detailed survey of the mine area has been undertaken, including also underground (spelaeological) exploration (*cf.* Cascone and Casini, 1998; GSAL, 2000). The main results will be briefly summarized hereafter. Although limestone quarrying has destroyed the greatest part of mine workings, it has been still possible to establish the location of different stopes, adits, etc., mentioned in the old documents and reconstruct the evolution of mining activity at least since the XIX century (i.e., after Blanchard's discovery of tin ores). As far as pre-Industrial mining at Monte Valerio is concerned, the archaeological evidences are very scanty and uncertain. As already said, the "Cento Camerelle" workings (together with those at "Cavina vecchia"), surely predate the XIX century. Although their original location has been reconstructed (Fig.1), they are no more observable, having been completely destroyed by the limestone quarry. GSAL (2000) mentions few additional sites where evidences of "ancient" mining are still observable. The most important is "Pozzo Antico ("old shaft"), located to the NW of Monte Valerio (Fig.1): it is a sub-vertical pit with an internal diameter of 1-1.5 metres, extending to depth for about 23 metres. The shaft was apparently excavated to follow very closely the ore, and looks very similar to those found in the nearby Valle del Temperino (dating from Etruscan times to Middle Age: cf. Casini and Francovich, 1993). Unfortunately no traces of tin minerals have been found within it! In summary, no clear-cut evidences of Etruscan mining activity has been detected at Monte Valerio; in addition, even admitting that some workings (e.g., Pozzo Antico) are Etruscan, there is no proof that tin was the exploited metal! It seems, therefore, that different tools are needed to test the hypothesis that the Etruscans exploited or not tin mineralization from Monte Valerio. In this respect, we decided to:

a) re-examine the geologic setting, mineralogy, and textural features of tin mineralization at Monte Valerio (chapter 6.1);

b) carry out a survey in the nearby metalworking centre of Baratti-Populonia looking for any waste material (slags, partially smelted mineral charge) referable to tin smelting in Etruscan times; in particular we have analysed a "bronze" scrap recently found during archaeological excavations (chapter 6.2);

c) compare the trace element composition of Monte Valerio Sn mineralization and other potential tin sources in the antiquity (Cornwall, Iberia, Sardinia) with the Cu-Sn metal scrap found at Baratti (chapter 6.3).

5. SAMPLING AND ANALYTICAL PROCEDURES

Many of the analysed samples have been taken directly in the field during the "geoarchaeological survey", both from surface (Cava Vecchia, Pozzo Cortese, Santa Barbara stopes) and underground outcrops (XX century "Dux" adit, at Cento Camerelle). We have also analysed a few museum's specimens (kindly supplied by the Florence Museum of Mineralogy), consisting of cassiterite-rich samples from both Monte Valerio mine and other European tin mines elsewhere (Sardinia, Portugal, and Cornwall).

Mineralogy of the samples was determined by optical (reflected light) and X-Ray powder diffraction (Philips PW 1830, Cu kα/ Ni rad., 40 kV, 20 mA; Philips PW 1130/90, Co kα/ Fe rad., 40 kV, 20 mA).

Major elements were determined by X-ray fluorescence (Philips PW 1480 equipped with Mo anticathode).

Microchemical analyses were performed by scanning electron microscopy (SEM Philips 515, equipped with EDS Edax 9800) and electron microprobe (JEOL JXA 8600 operated at 15 kV accelerating potential and 30 nA sample current). Standard used for quantitative micro-analyses of cassiterite included cassiterite (Sn), ilmenite (Ti, Fe), rodocrosite (Mn), cuprite (Cu), galena (Pb, S), tugtupite (Cl,Si) and metallic elements for W, Nb, and Ta, whereas pure metals (Sn, Cu, Fe) were used as standards for microanalyses of the metal scrap from Baratti.

6. ANALYTICAL RESULTS

6.1 Sn mineralization at Monte Valerio

Based on both previous data (Bertolani, 1958; Venerandi-Pirri & Zuffardi, 1981), and our own field and laboratory work, we can distinguish four main types of tin mineralization at Monte Valerio:

a) lenses and veins of cassiterite + Fe-oxides and hydroxides

b) small cassiterite veins with actinolite and tourmaline

c) "massello calcareo" (massive cassiterite hosted in calcareous rocks)

d) "massello pietra" (massive cassiterite hosted in siliceous rocks).

Our goal was not only to describe in detail the mineralogy, textural features and mineral chemistry of the different Sn ore types, but also to evaluate which of them could have been profitably exploited (if ever) by ancient (Etruscan?) miners.

A-type Sn ores occur in lenses, veins or karstic structures hosted by either carbonatic (Massive Limestone; Red Ammonitic Limestone) or siliciclastic rocks (Tuscan Scaglia).

We could stll observe outcrops of type-A mineralization (samples DuxL5A, DuxL5B, 5959) in the central-eastern portion of M. Valerio mine, i.e., where the Cento Camerelle stope was approximately located. Mineralogy includes anhedral, often rounded crystals of cassiterite (about 100 mm in size), disseminated through a groundmass of goethite (Fig. 2). The latter mineral is commonly pseudomorph after pyrite. A-type tin ore bodies yielded average grades of 0.4 to 1 wt% Sn. Much of tin output from Monte Valerio since Blanchard's times has come from this ore type.

B-type ore bodies occur in limited outcrops only in the northern part of S. Barbara stope (Pozzo Cortese). The sample we examined (SB4/2) consists of millimetric veins of cassiterite cutting through the siliciclastic matrix of the Tuscan Scaglia. Cassiterite is associated with tourmaline and abundant actinolite crystals of millimetric size (Fig. 3). Only minor tin production may have arisen from this kind of mineralization.

Samples of C-type ore bodies coming from Cento Camerelle were kindly provided by the Florence Museum of Mineralogy (samples 5431, 15069, 15064). They resemble very closely type-A Sn ores, but show much higher cassiterite contents (up to 60-70 wt% SnO_2). Mineralogy typically includes anhedral, often rounded crystals of cassiterite which tend to cluster into decimetric nodules surrounded by a groundmass of calcite ± limonite (Fig. 4). C-type massive cassiterite ore bodies cannot be more observed, neither at the surface nor in underground adits. Mining reports from XIX-XX cent. indicate that this ore type, although extraordinarily rich in tin, did not yield significant production.

D-type ore bodies are very similar to B-type, except for distinctly higher cassiterite contents. The sample 15060 is representative of this ore type. It consists of decimetric nodules made up of microscopic crystals of cassiterite with quartz, locally associated with anhedral, millimetric masses of Fe-hydroxides (Fig. 5). Low tin tonnages and limited outcrops characterise D-type ore bodies.

In summary, C-type tin ores, thanks to their favourable geological setting (orebodies outcropping at or near the surface), minero-textural features (easy beneficiation cassiterite from carbonatic gangue), and high tin grades, would have represented a preferred target for ancient miners.

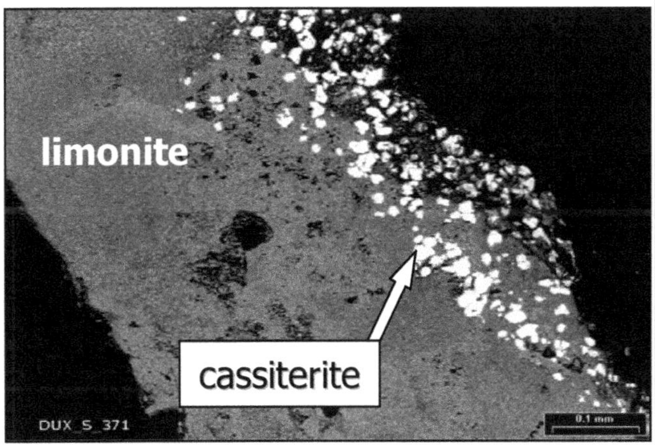

Figure 2. SEM Backscattered Electron image of A-type ore. Fe oxides (grey) with cassiterite crystals (white).

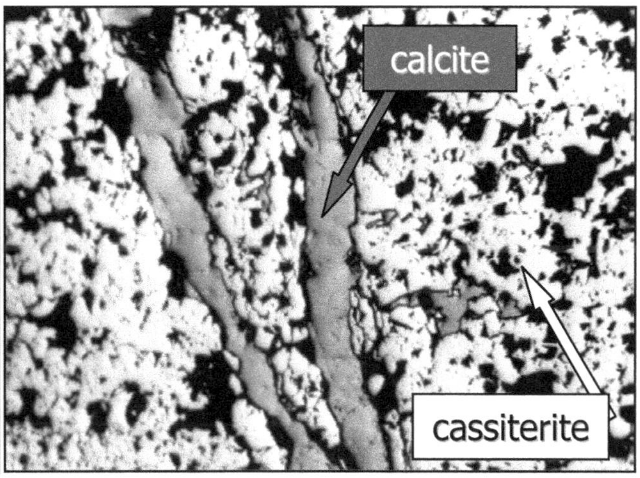

Figure 4. Reflected light photo of C-type ore. Massive cassiterite (white) crosscut by calcite veins (grey).

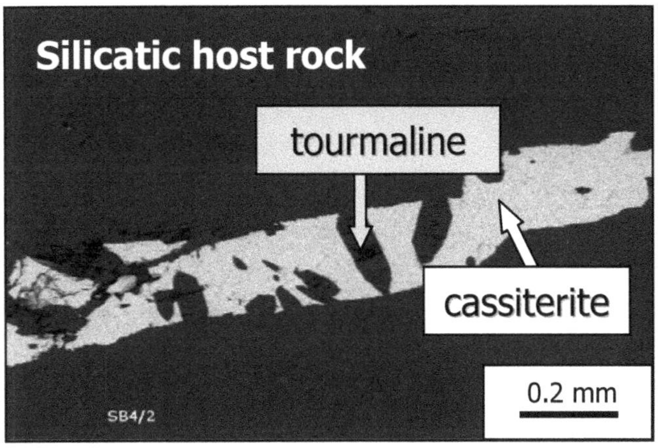

Figure 3. SEM Backscattered Electron image of B-type ore. Vein of cassiterite (white) + tourmaline crosscutting a siliceous rock (dark grey)

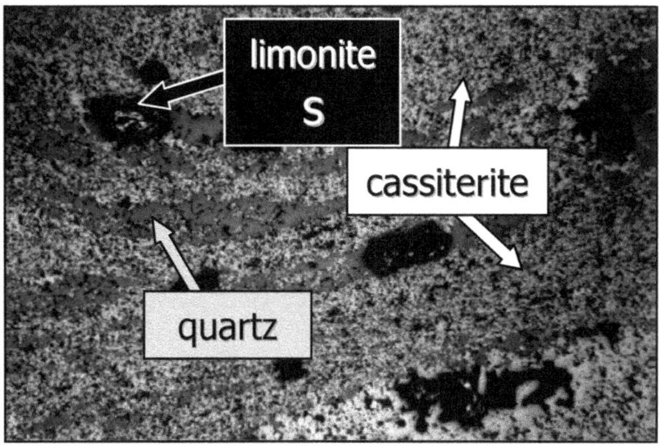

Figure 5. Reflected light image of D-type ore. Cassiterite (bright grey) mixed with quartz (dark grey) and iron oxides/hydroxides (black).

On the contrary, the low tonnages, low tin grade and, especially, the unfavourable minero-textural features (intimate intergrowth of cassiterite and gangue silicates) of B- and D-type orebodies lead us to postulate that they could have not been exploited in the antiquity. As to A-type Sn ores, they were presumably the most abundant and easily accessible Sn ore type to ancient miners, although it is highly dubitative that the latter were able to distinguish cassiterite from the very similar earthy iron hydroxides embedding it.

6.2 Tin metallurgy at Baratti (Populonia)

We could try to tackle the problem of "Etruscan tin" by a different approach: are there any evidences of the utilization of Monte Valerio tin in (nearby) Etruscan metallurgical sites? In order to fully answer this question one should ideally find relics of cassiterite-bearing charge at some metallurgical sites and compare them with Monte Valerio tin ores by means of distinguishing features ("source markers"). We followed this approach and chose to investigate the Baratti-Populonia archaeometallurgical site searching for any surviving evidence of tin working.

The role of Baratti-Populonia as one of the most important iron-working centres of Europe during Etruscan (7th-4th century BC) and (Etrusco-) Roman (3rd century BC-1st century AD) times is well documented (cf. Corretti & Benvenuti, 2001, and references therein). Although local production of bronze artefacts is well documented at Populonia, especially between the end of the VI cent. BC and the first half of the V cent. BC (Fedeli et al., 1993), information about non ferrous metal production in the area is still scarce.

In the course of an extensive archaeometallurgical research on several excavation sites in the Baratti industrial zone (P. Porcareccia, Casone, Campo VI, S. Cerbone slag beach deposit: see Fig. 6), significant tin anomalies (ranging from 200 ppm to 0.4 wt% tin) have been detected in many iron slags dating earlier than the 3rd cent. BC (Benvenuti et. al., 2000). In these samples tin is present as micrometric ($\leq 5\mu m$) globules of FeSn and $FeSn_2$ intermetallic compounds. Benvenuti et al. (2000) suggested that such anomalous tin contents could be due to the utilisation of iron ores with disseminated (undetected?) cassiterite, very likely exploited from the nearby Campiglia Marittima district (i.e., the A-type tin ores from Monte Valerio). By smelting iron, trace tin present in the mineral charge would bind with some iron forming the observed iron-tin alloys.

An additional evidence of tin working at Baratti-Populonia is provided by one little Cu-Sn metal scrap, less than 1 cm large, recently found in the Campo VI area and dating no later than the 3rd cent. BC (Mariani, 2000). It is characterised by a copper-rich groundmass embedding crystals of SnO_2. The latter are unevenly distributed, being preferentially concentrated in one portion of the fragment (Fig. 7). The SnO_2 crystals may reach up to 200 μm in length, and range from euhedral (with various habits: low pyramidal, long prismatic, pseudo cubic) to anhedral, partly resorbed by the copper-rich groundmass. To be noticed that the elongated habit (Fig. 8) resembles very closely the so-called "needle tin" textures of natural cassiterite (cf. Mantell, 1949; Taylor, 1979). Electron Microprobe analyses show that the SnO_2 crystals have very homogeneous composition, with high tin contents (> 99 weight %) and low amounts of iron (< 0,5 wt%, see next chapter) (Tab. 1). The groundmass composition changes significantly from almost pure copper in the "cassiterite"-rich portion, with Sn constantly below instrumental detection limits (zone A Fig. 7 and Tab. 2), to a Cu-Sn alloy with variable Sn content (max. 2 wt%) (zone B Fig. 7, Tab. 2). We think that this metallic item represents a

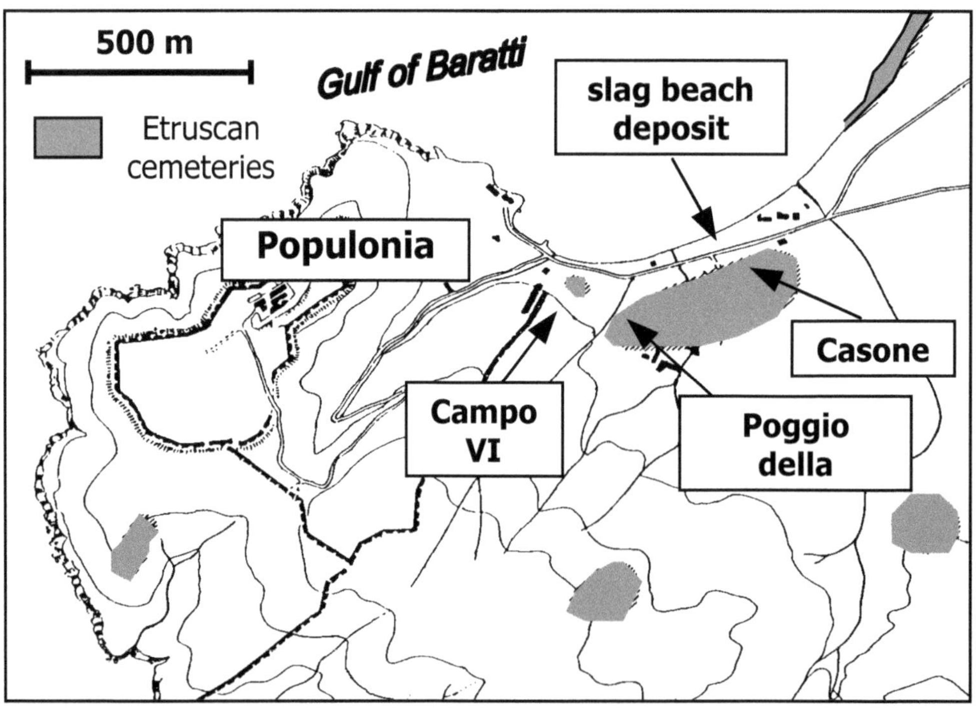

Figure 6. Map of the Populonia-Baratti area showing the investigated sites.

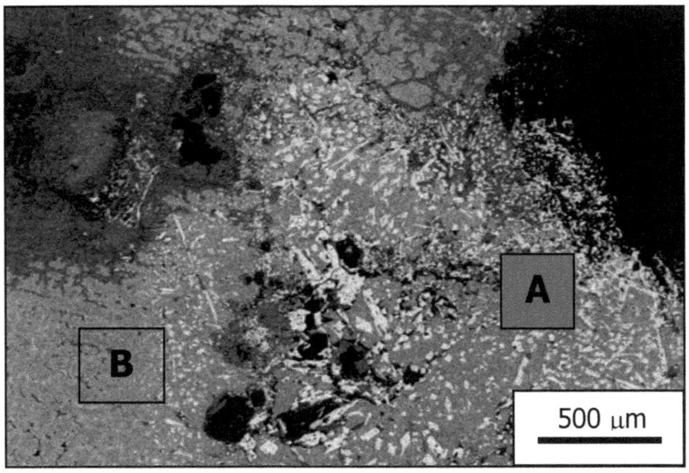

Figure 7. SEM Backscattered Electron image of copper-tin metal scrap from Baratti. SnO_2 crystals (bright grey) are disseminated within a groundmass of copper and copper oxides (dark grey); A = "cassiterite"-rich portion, B = "cassiterite"-poor portion (see Table 2 for analyses).

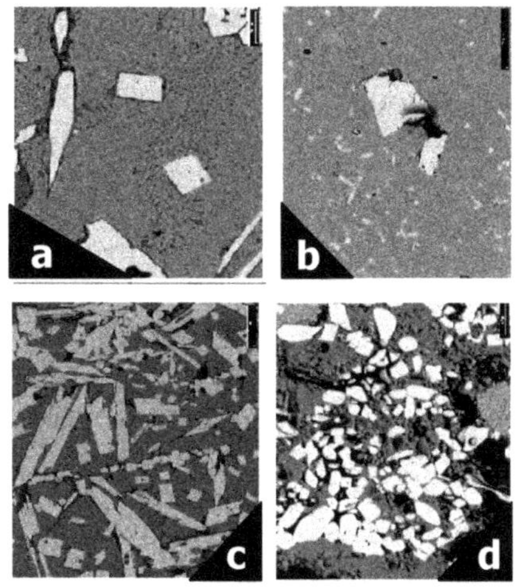

Figure 8. Close up views of SnO_2 with different habits: a) rhombohedral and rectangular habits; b) crystal showing partial resorption by the groundmass; c) "needle tin" texture; d) anhedral, partly resorbed crystals.

Table 1. Microprobe analysis of SnO_2 phase from the metal scrap found at Populonia. The average (*ave*), minimum and maximum values, and the standard deviations (*sd*) of 8 analytical spots are reported as wt %. *<dl* = below detection limits (FeO = 0.04%±0.01; TiO_2 = 0.05±0.01; WO_3 = 0.19±0.04; Nb_2O_5 = 0.06±0.01; Ta_2O_5 = 0.14±0.03; MnO = 0.02±0.005; Cu_2O = 0.05±0.01; SnO_2 = 0.04±0.01)

	SnO_2	FeO	TiO_2	WO_3	Nb_2O_5	Ta_2O_5	MnO	Cu_2O
ave	99,72	0,16	*<dl*	*<dl*	*<dl*	*<dl*	*<dl*	0,26
sd	0,55	0,12	*<dl*	*<dl*	*<dl*	*<dl*	*<dl*	0,32
max	100,8	0,48	*<dl*	*<dl*	*<dl*	*<dl*	*<dl*	0,75
min	98,99	0,04	*<dl*	*<dl*	*<dl*	*<dl*	*<dl*	*<dl*

Table 2. Microprobe analysis of copper-rich groundmass of Cu-Sn artefact found at Populonia (reported as % atoms). *<dl* = below detection limit.

	Cu	Sn	Fe	total
SnO_2–poor area	98.38	1.69	*<dl*	100.1
	98.86	1.94	*<dl*	100.8
⇑	99.39	0.33	*<dl*	99.72
	96.87	0.15	*<dl*	97.03
	98.38	0.04	*<dl*	98.42
SnO_2–rich area	100	0.05	*<dl*	100.01

scrap from an unsuccessful bronze-making process: the occurrence of partially unreacted cassiterite crystals could be due to operating conditions not enough reducing in the furnace (or crucible). Our sample provides many textural analogies with the Raquela Vilaca crucible (Meredith, 1998, p. 155-160), where unsmelted, crushed cassiterite crystals are embedded in a copper-rich, Sn-free matrix. If our interpretation is valid, then the bronze scrap could have been produced by cementation, i.e., by adding unsmelted cassiterite to refined copper to produce bronze, although this hitherto unique evidence is not enough to make any final judgement about bronze production by the Etruscans at Baratti-Populonia.

6.3 Mineral chemistry of cassiterite

Cassiterite is largely a non-stoichiometric mineral phase. Sn is frequently replaced by Fe^{3+} (up to about Fe/Sn = 1:6); other common substitutions may involve Nb, Ta (up to 8% Ta_2O_5 in pegmatitic cassiterite), Mn, Ti, W, Zr, Al, In, Hf and many other elements (cf. Gaines et al., 1997; Plimer at al., 1991;

Neiva, 1996). In principle, such large compositional variations of cassiterite may be used to distinguish among different Sn deposits from a certain district and, eventually, to trace back the provenance of cassiterite-bearing charge found in ancient metallurgical sites (cf. Rapp, 1978; Rapp et al., 2000).

Just as a preliminary approach to provenance studies, we have investigated the chemical composition of Monte Valerio cassiterite from the four ore types above described, and made a comparison with museum's specimens of cassiterite coming from Beira Alta province (Portugal), Cornwall and S. Vittoria - Perdu Cara (Sardinia). The first two are major tin districts surely known and exploited since the antiquity (cf. Meredith, 1998), the third one could have been exploited by the Phoenicians (Ingo et al., 1997).

Based on literature data (cf. Rapp et al., 2000) we have selected for the analytical run at the electron microprobe the following elements: Fe, Ti, W, Nb, Ta, Mn, and Cu. The main results of microprobe analyses are reported in Table 3 and Figs. 9 and 10. From a rough, qualitative inspection of analytical results it seems that the four types of tin mineralization from Monte Valerio are not much different, with the relevant exception of higher W contents of C-type ore (Fig. 9). Among European cassiterites the Beira Alta samples are characterised by distinctively higher Ta-Nb contents, with respect to all other samples (Fig 10). The SnO_2 crystals from for the Baratti metal scrap show only minor contents of Fe and Cu, the latter possibly due to contamination from copper rich groundmass, while Ti, W, Nb, and Ta are constantly below detection limits. The Baratti sample falls at

Table 3. Microprobe analyses of cassiterites (reported as wt. %) from M. Valerio, Cornwall, Portugal and Sardinia. *m* = average of N analytical spots; *s* = standard deviation; ... = below detection limits (see Table 1).

Provenance	Type of ore	Sample	N		SnO_2	FeO	TiO_2	WO_3	Nb_2O_5	Ta_2O_5	MnO	total
M.Valerio	A	sample 5959	8	m	98,98	1,13	0,09	...	...	...	...	100,27
				s	2,00	1,16	0,08	...	...	...	...	0,93
				max value	100,60	3,66	0,21	...	...	...	...	
				min value	94,60	0,15	0,06	...	...	...	...	
	B	sample SB4/2	11	m	98,26	0,80	0,20	...	...	...	...	99,25
				s	0,77	0,22	0,44	...	...	...	...	0,36
				max value	99,85	1,06	1,38	...	...	...	...	
				min value	96,86	0,28	...	...	...	...	...	
	C	sample 5431	10	m	99,23	0,08	0,14	0,51	...	...	...	100,41
				s	0,39	0,08	0,20	0,37	...	...	...	0,90
				max value	99,86	0,25	0,57	0,96	...	...	...	
				min value	98,69	...	...	...	...	...	...	
		sample 15069	12	m	98,59	0,04	0,10	0,80	...	...	...	99,40
				s	0,58	0,05	0,15	0,42	...	...	...	0,32
				max value	99,43	0,13	0,51	1,42	...	...	...	
				min value	67,81	...	...	...	...	...	...	
		sample 15064	9	m	98,90	0,07	...	0,66	...	...	...	99,66
				s	0,57	0,05	...	0,36	...	...	...	0,38
				max value	99,49	0,13	...	1,26	...	...	...	
				min value	97,81	...	...	0,30	...	...	...	
	D	sample 15060	12	m	99,53	0,12	0,11	0,09	...	...	...	100,43
				s	0,16	0,12	0,14	0,2	...	...	...	0,56
				max value	99,76	0,32	0,46	0,54	...	...	...	
				min value	99,23	...	...	...	...	...	...	
Cornwall		sample 5382	8	m	99,02	0,07	0,18	0,02	0,10	0,06	0,003	99,40
				s	0,53	0,03	0,09	0,07	0,11	0,11	0,01	0,43
				max value	99,66	0,14	0,30	0,19	0,30	0,28	0,02	
				min value	98,13	0,04	...	...	...	...	...	
		sample 5375	7	m	98,99	0,01	0,43	0,03	...	...	...	99,49
				s	0,40	0,02	0,21	0,08	...	...	...	0,29
				max value	99,63	0,04	0,66	0,22	...	...	...	
				min value	98,29	...	0,18	...	...	...	...	
Portugal		sample 5372	11	m	98,38	0,12	0,37	...	0,32	0,32	...	99,57
				s	0,78	0,12	0,13	...	0,16	0,53	...	0,46
				max value	99,23	0,43	0,64	...	0,60	1,82	...	
				min value	97,60	...	0,23	...	...	...	...	
		sample 5373	11	m	99,29	...	0,64	...	0,01	...	0,0001	100,52
				s	0,22	...	0,18	...	0,03	...	0,0004	0,30
				max value	99,56	...	1,02	...	0,09	...	0,02	
				min value	98,85	...	0,37	...	...	...	...	
Sardinia		Perdu Cara sample SPC1	6	m	99,33	0,54	0,05	0,10	...	...	...	100,38
				s	0,27	0,24	0,09	0,25	...	...	...	0,35
				max value	99,79	0,69	0,21	0,61	...	...	...	
				min value	99,01	0,11	...	...	...	...	...	
		Punta S. Vittoria sample SSV2	6	m	99,19	0,36	0,14	0,25	...	...	...	100,42
				s	0,27	0,29	0,12	0,39	...	...	...	0,48
				max value	99,51	0,84	0,32	0,76	...	...	...	
				min value	98,85	0,05	...	...	...	...	...	
		Punta S. Vittoria sample SSV3	6	m	99,28	0,61	...	0,03	...	...	...	100,63
				s	0,27	0,33	...	0,08	...	...	...	0,33
				max value	99,87	1,00	...	0,19	...	...	...	
				min value	99,21	...	...	...	...	...	...	

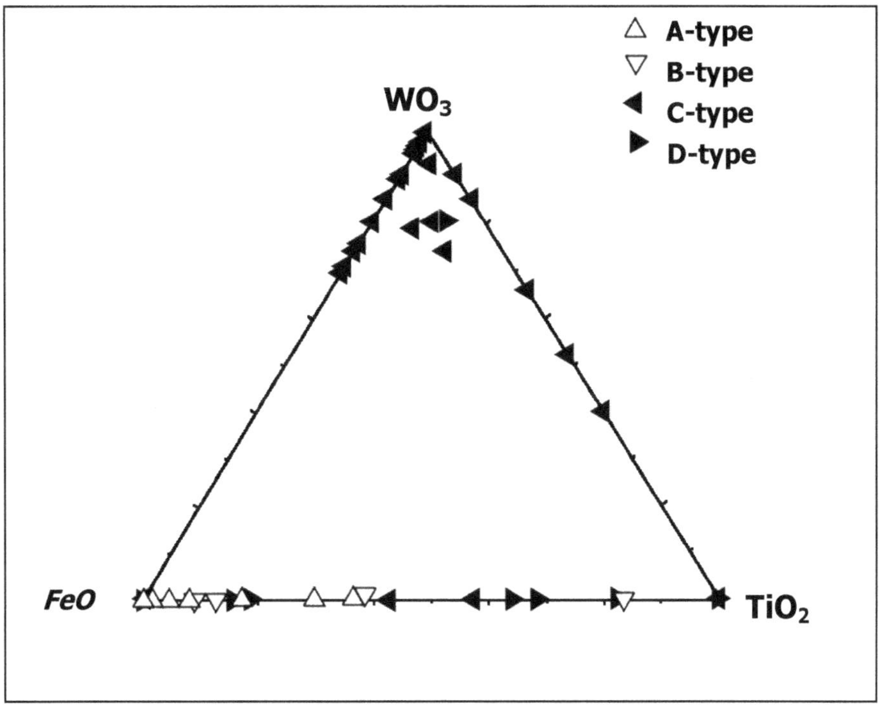

Figure 9. Trace elements content of cassiterite ores from M. Valerio in terms of ternary diagram FeO-TiO_2-WO_3.

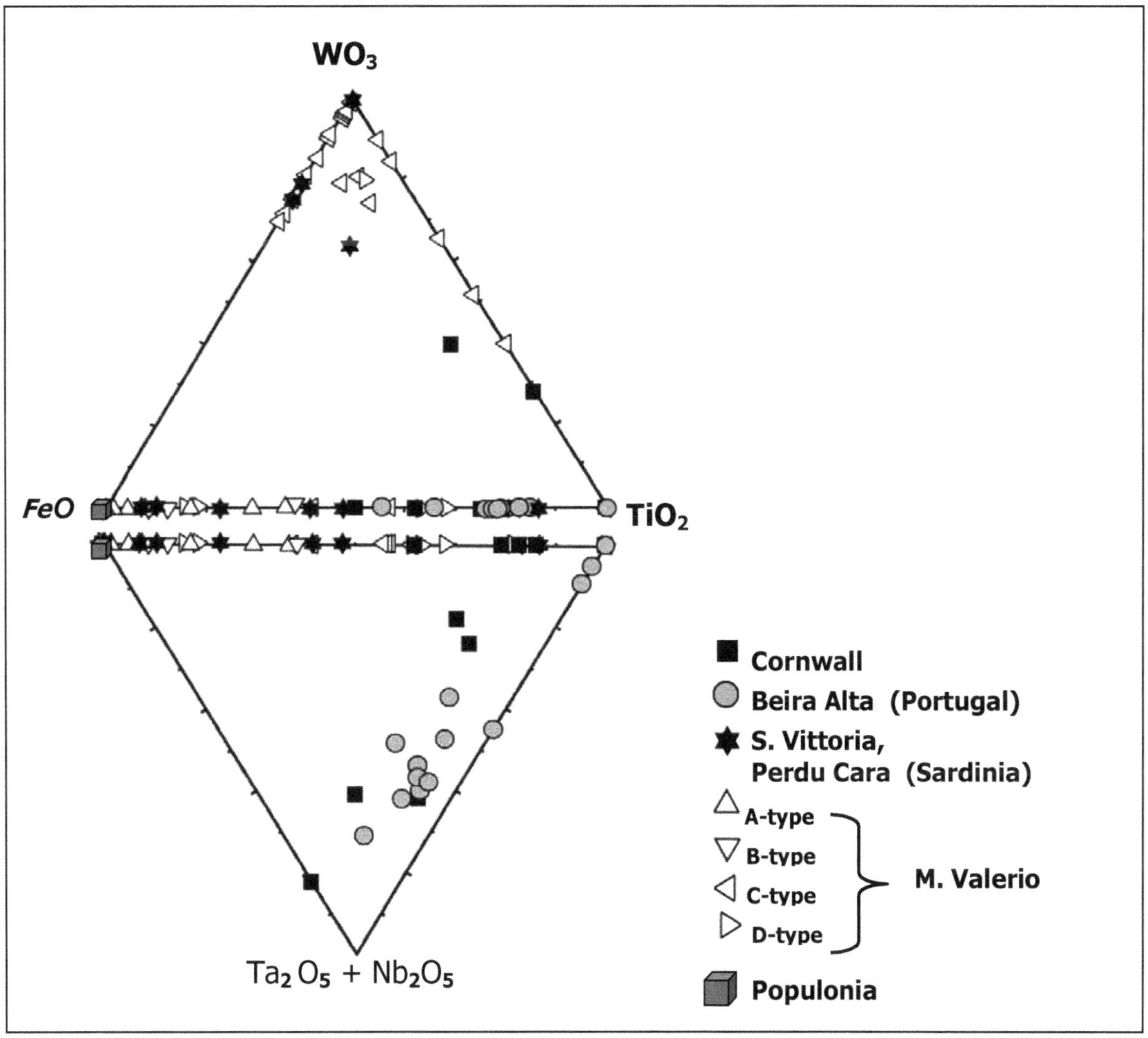

Figure 10. Trace element compositions of cassiterite from various provenances are reported in the ternary diagrams FeO-TiO_2-WO_3 and FeO-TiO_2-(Ta_2O_5+Nb_2O_5). The average trace element composition of SnO_2 phases (interpreted as relics of natural cassiterite) from the Baratti "bronze" scrap is also illustrated.

the Fe corner in the ternary diagram of Fig. 10, very close to a number of samples of various provenance (A- and B- type ores from Monte Valerio; Perdu Cara, Sardinia). Thus it is impossible to gather information (even qualitative) on tin sources for the Baratti metal scrap based exclusively on the selected elements.

7. CONCLUSIONS

We can summarize the state of the art of our knowledge on the old vexed question of "Etruscan tin" in the following points:

a) up to nowadays there are not "direct" archaeological evidences of tin exploitation from Monte Valerio ore bodies in pre-Roman (Etruscan) times;

b) the "geo-archaeological survey" carried out at Monte Valerio led to the identification of four main types of tin mineralization at Monte Valerio, of which only type-C (massive cassiterite) would have been exploitable with profit by ancient miners;

c) there are however "indirect" evidences" that cassiterite-bearing iron ores from Monte Valerio (probably the A-type limonitic orebodies with disseminated cassiterite) were employed at Baratti by the Etruscans, although for iron reduction, and not tin smelting;

d) the Cu-Sn metal scrap recently discovered at Baratti (dating the 3rd cent. BC or earlier) was produced by cementation, i.e., by adding unsmelted cassiterite to refined copper to produce bronze

e) preliminary analyses of minor and trace element (Fe, Ti, W, Nb, Ta, Cu) composition of cassiterite from different provenances (metal scrap from Baratti; four Sn ore types from Monte Valerio; ancient tin districts such as Cornwall, Beira Alta, Sardinia) has been performed in the attempt to verify the "Etruscan tin" hypothesis;

f) a few compositional trends of analysed cassiterites become evident even at this preliminary stage of the research (e.g., distinctive enrichment in W of C-type ores from Monte Valerio, and in Nb-Ta of Portoguese samples);

g) it is impossible to trace back the source deposits of Baratti's cassiterites (showing almost stoichiometric composition with only minor Fe contents) uniquely from the selected minor and trace element signature.

Future research will include an enlargement of the analytical database to include other "chemical tracers" by using a high sensitivity analysis technique like PIXE. More sophisticated discriminant analysis will also be of help in unravelling possible provenance of smelted "Etruscan tin". Finally, we have started to investigate the lead isotope composition of both tin sources and different kinds of Sn-bearing metallurgical wastes (slags, metal artefacts) from Baratti (*cf.* Chiarantini et al., 2002). This "multi-tracer" approach will hopefully contribute new answers to the vexed and fascinating question of "Etruscan tin".

Acknowledgements

This work was financed by CNR through the 'Progetto Finalizzato Beni Culturali' (G.Tanelli) and MIUR ex-60% grants (M. Benvenuti). We acknowledge the assistance of Mario Paolieri with SEM/EDS and Electron Microprobe analyses. The Florence Museum of Mineralogy is thanked for providing us with specimens of cassiterite from various localities.

Authors' addresses

M. BENVENUTI, L. CHIARANTINI, L. NORFINI, G. TANELLI
Dipartimento Scienze della Terra - Univ. Firenze (Italy).
Via La Pira 4
J - 50121 Firenze
E-mail: m.benv@geo.unifi.it

S. GUIDERI
Soc. Parchi Val di Cornia, Campiglia M.ma.
Via Lerario 90
I - 57025 Piombino (LI)
E-mail: guideris@parchivaldicornia.it

A. CASINI
Gruppo Speleologico Archeologico Livornese
Via Roma, 234
57127 Livorno

Bibliography

BENVENUTI M., MASCARO I., COSTAGLIOLA P., TANELLI G., ROMUALDI A., 2000 - *Iron, Copper and Tin at Baratti (Populonia): smelting processes and metal provenances*, Historical Metallurgy, vol. 34 (2), pp. 67-76.

BERTOLANI M., 1958 – *Osservazioni sulle mineralizzazioni metallifere del Campigliese (Livorno)*, Periodico di Mineralogia, vol. 27, pp. 311-344.

BLANCHARD F., 1876 - *Sulla scoperta della Cassiterite a Campiglia Marittima*, Boll. R. Com. Geol. d'Italia, vol. 7, pp. 52-54

CASCONE G., CASINI A., 1998 - *La Miniera di Stagno di Monte Valerio (Campiglia Marittima)*, 18° Congresso Nazionale di Speleologia, Chiusa '98.

CASINI A., FRANCOVICH R., 1993 - *Problemi di archeologia mineraria nella Toscana medievale: il caso di Rocca San Silvestro.* In: Les techniques Minéres de l'Antiquité au XVIII siècle, Actes 113e Congr. Nat. Soc. Savantes, Paris, Editions du CHTHS, pp. 249-265.

CHIARANTINI L., BENVENUTI M., COSTAGLIOLA P., MASCARO I., VILLA I. M., 2002 - *Copper smelting at Populonia-Baratti (Southern Tuscany, Italy) in the 1st Millennium BC.* 33rd Int. Symp. on Archaeometry, 22-26 April 2002, Amsterdam.

CHURCH A. H., 1879 – *La scoperta del minerale di stagno in Italia e sua relazione con la lavorazione del bronzo presso gli antichi.* Boll. R. Com. Geol. d'Italia, vol. 10, pp. 382-394; 545-556.

CORRETTI A., BENVENUTI M., 2001 - The beginning of iron metallurgy in Tuscany with special reference to *Etruria Mineraria*, Mediterranean Archaeology, vol. 14, pp. 127-145.

COSTANTINI A., LAZZAROTTO A., MACCANTELLI M., MAZZANTI M., SANDRELLI F., TAVARNELLI E., ELTER F. M., 1993 - *Geologia della provincia di Livorno a sud del fiume Cecina.* In: I quaderni del Museo di Storia Naturale di Livorno, vol. 13 (2), pp. 1-164.

CRADDOCK P.T., 1986 – *The metallurgy and composition of Etruscan bronze*, Studi Etruschi, vol. 52, pp. 211-271.

D'ACHIARDI A., 1872 – *Mineralogia della Toscana*, Ed. Nistri, Pisa, vol. 1, pp. 276.

GAINES, R.V., SKINNER H.C.W., FOORD E.E., MASON B., ROSENZWEIG A. (Eds.), 1997 - *Dana's New Mineralogy: The System of Mineralogy of James Dwight Dana and Edward Salisbury Dana*, New York: Wiley, 1872 pp.

FEDELI F., GALIMBERTI A., ROMUALDI A., 1993 - *Populonia e il suo territorio profilo storico-archeologico*, Ediz. All'insegna del Giglio, Firenze, 149 pp.

G.S.A.L. (GRUPPO ARCHEOLOGICO SPELEOLOGICO LIVORNESE), 2000 - *Studio e rilevamento del sistema minerario di Monte Valerio (Campiglia Marittima, Toscana).* Unpublished Report, 83 pp.

INGO G.M., ACQUARO E., MANFREDI L.I., BULTRINI G., CHIOZZINI G., 1997 - *La pirometallurgia*, In: *Progetto Tharros*, E. Acquaro , M.T. Francisi, G,M, Ingo, L.I. Manfredi (Eds.), Agorà Edizioni, La Spezia, pp. 29-46.

MANTELL C.L., 1949 - *Tin: its mining, production, technology and applications*, Reinhold Publishing Corporation, 573 pp.

MARIANI A., 2000 – *Studio archeometallurgico di resti dei forni siderurgici etruschi provenienti dai quartieri industriali di Baratti (Populonia)*, Unpublished thesis, Università di Firenze, 152 pp.

MEREDITH C., 1998 – *An archaeometallurgical survey for ancient tin mines and smelting sites in Spain and Portugal*, BAR Int. Series, B714, BAR Publishing, Oxford, 205 pp.

NEIVA A.M.R., 1996 - *Geochemistry of cassiterite and its inclusions and exsolution products from tin and tungsten deposits in Portugal*, Canadian Mineralogist, vol. 34 (4), pp. 745-768.

NORTHOVER J.P., GILLIS C., 1999 - *Questions in the analysis of ancient tin.* In: Metals in antiquity, S.M.M. Young, A.M. Pollard, P. Budd, R.A.Ixer (Eds.), BAR Publishing, BAR S792, pp. 78-85.

PECCERILLO A., CONTICELLI A., MANETTI P., 1987 - *Petrological characteristics and the genesis of the recent magmatism of southern Tuscany and northern Latium,* Periodico di Mineralogia, vol. 56, 157-172

PERNICKA E., 2002 – *Origins of tin – What is new?* Proceedings of the 33rd Int. Symp. on Archaeometry, 22-26 April 2002, Amsterdam, 117-118.

PLIMER I.R., Lu J., KLEEMAN J.D., 1991 - *Trace and rare earth elements in cassiterite; sources and components for the tin deposits of the Mole Granite, Australia*, Mineralium Deposita, vol. 26 (4), p. 267-274.

RAPP G.,Jr., 1978 - *Trace elements as a guide to the geographical source of tin ore; smelting experiments,* In: A.D. Franklin, J.S. Olin, T.A. Wertime (Eds.), *The search for ancient tin*, U. S. Gov. Print. Off., Washington, D.C., pp. 59-63.

RAPP G.,Jr., RUSSELL R., ZHICHUN J., 2000 - *Using neutron activation analysis to source ancient tin*, in: Metals in antiquity, BAR S792, BAR Publishing, Oxford, pp. 153-162.

RODOLICO F., 1945 – *Ragguagli sul granito del Campigliese.* Mem. Soc. Tosc. Scienze Nat., vol. 52, pp. 125-132.

SIMONIN M. L., 1858 – *De l'exploitation des mines et de la metallurgie en Toscane pendant l'antiquité et le Moyen Age.* Annales des mines, 5° serie, vol. XIV, pp. 557-615

TANELLI G., MORELLI F., BENVENUTI M., 1993 - *I minerali del Campigliese: "beni ambientali, culturali e industriali".* Boll. Soc. Geol. It., vol. 112, pp. 715-728.

TAYLOR R.G., 1979 - *Geology of tin deposits,* Elsevier Scientific Publishing Company, Amsterdam, pp. 543 pp.

TIZZONI M., 1999 - *Etruscan tin: a legend?* Pallas, vol. 50, pp. 89-100.

VENERANDI PIRRI I., ZUFFARDI P., 1982 - *The tin of Monte Valerio (Tuscany): pneumatolytic-hydrothermal or sedimentary-remobilization processes?* In: Ore genesis, the state of the art, G.C. Amstutz (Ed.), Berlin, Springer, pp. 57-65.

ANALYSE ELEMENTAIRE DES DEPOTS LORRAINS : ESSAI DE CARACTERISATION D'UNE PRODUCTION METALLIQUE DE LA FIN DE L'AGE DU BRONZE

Cécile VEBER, Benoît MILLE & David BOURGARIT

Résumé : Le Bronze Final IIIb (fin du IXe siècle - début du VIIIe siècle avant J.C.) est marqué en Lorraine et en Sarre par un phénomène de déposition que l'on rencontre très peu au cours des périodes précédentes de l'âge du Bronze de ces régions. Ces dépôts contiennent un mobilier très homogène et caractéristique, et ont de ce fait donné lieu à plusieurs types d'objets éponymes. Trois cents analyses de composition ont été réalisées sur ces objets par ICP-AES, pour explorer la possibilité d'individualiser un «métal lorrain» .

Si les résultats obtenus mettent en évidence une homogénéité certaine à l'échelle de la Sarre-Lorraine - qu'il est nécessaire de pondérer quelque peu par des données comme la catégorie des objets, et au-delà par leur mode de fabrication - la comparaison des dépôts de Lorraine et de Sarre avec des ensembles contemporains des Alpes françaises du Nord et de Suisse ne permet pas d'isoler une métallurgie lorraine autonome. Cette confrontation laisse au contraire apparaître de grandes similitudes dans la composition de l'alliage : la proportion de plomb est partout comprise entre 1 et 3 % pour une teneur moyenne en étain de 8 %. La composition en impuretés est également très semblable, tant au niveau du schéma de composition que des teneurs cumulées qui sont comprises entre 1,5 et 2,4 %. Tous ces ensembles semblent donc se rattacher à une même tradition métallurgique.

Par contre, les ensembles contemporains de l'Ouest - pour lesquels une synthèse des analyses est pour la première fois esquissée - montrent d'importantes différences avec les ensembles de l'Est. Ainsi, d'un côté les taux d'alliage sont systématiquement supérieurs à 16 %, avec des teneurs en plomb qui comptent au minimum pour 1/3 de l'alliage. De l'autre côté, le taux cumulé des impuretés est cette fois-ci beaucoup plus bas, il ne dépasse pas 0,5 %. Cette partition des ensembles de l'Est et de l'Ouest confirme bien celle qui est observée, par le biais de la typologie des objets, entre un domaine continental et un domaine atlantique. L'étude de la composition élémentaire des objets métalliques révèle ainsi que des traditions métallurgiques différentes ont été en usage, à l'extrême fin de l'âge du Bronze, dans ces deux grands complexes techno-culturels de l'espace Ouest-européen.

Abstract: A deposition phenomenon occurred in Lorraine and Sarre at the IIIb period of the Final Bronze Age (end 9th cent. - beginning 8th cent. BC.), whereas it is rarely observed in the previous periods of the Bronze Age. The artefacts within the deposits constitute homogeneous groups and exhibit characteristic typologies, so that several deposits have led to eponymous types. Three hundred chemical analyses were performed by ICP-AES, in order to try identify a local metal from Lorraine.

The results show a clear homogeneity in the metal compositions at the regional level of Sarre-Lorraine, but also at a larger level, as shown by the comparison with deposits from the French Northern Alps and Switzerland. Everywhere the lead content ranges from 1 to 3 wt. %, and the tin mean content is around 8 %. Furthermore the impurity pattern shows very few variations, both in composition and total concentration which ranges from 1.5 to 2.4 %. Consequently this homogeneity does not allow us to identify an autonomous metallurgy. On the contrary, all these groups seem to belong to the same metallurgical tradition.

Nevertheless, when looking at the contemporaneous groups from Western France –for which a synthesis of the analytical data is drawn for the first time – big differences appear with the Eastern groups. On the one hand, the alloying elements contents are always higher than 16 %, with lead representing at least one third of the total amount. On the other hand, the impurity contents are quite lower and do not exceed 0.5 %. This partition West / East reflects the separation Atlantic area / Continental area previously put into light by means of typological observations. Thus, this analytical study reveals that different metallurgical traditions were in use in these two important techno-cultural entities of Western Europe.

INTRODUCTION

La majorité des dépôts de Lorraine et de Sarre ont été découverts à la fin du XIXe et au début du XXe siècle de notre ère, ces ensembles sont presque tous attribués à la fin du IXe / début du VIIIe siècle avant notre ère (Bronze Final IIIb en chronologie française ou Hallstatt B2 en chronologie allemande). Les spécialistes de l'époque notent immédiatement leur grande similitude typologique qui est mise en parallèle avec l'existence d'anciennes mines de cuivre dans la région du Warndt, au nord de la Moselle (Beauprès 1902, Linkenheld 1930, Millotte 1965). La plus ancienne exploitation reconnue date des premiers siècles de notre ère, il s'agit de la mine de Sainte Barbara sur la commune de Vaudrevange en Sarre (Jacquot 1857) ; de plus une occupation de la fin du Bronze Final est découverte à Falk sur un affleurement de cuivre (Hamon, Blouet 1993). Les auteurs parlent alors d'une production originale et locale qui se développerait durant cette dernière période de l'Age du Bronze Final (Beauprès 1902, Linkenheld 1930, Millotte 1965). Cependant, aucune trace d'exploitation n'a pu être attribuée à cette période, de plus, aucune étude n'a été menée sur ces objets particuliers et sur leur lien avec le cuivre local. On est donc en droit de se demander s'il existe réellement une production métallurgique autonome en «Sarre-Lorraine» à l'extrême fin de l'âge du Bronze. La découverte en 1991 du dépôt de Farébersviller a permis de relancer les recherches et d'apporter des premiers éléments de réponse à cette question d'une production locale[1].

[1] Ce travail a été effectué dans le cadre de la thèse de doctorat de Cécile Veber, soutenue à l'université de Dijon sous la direction de C. Mordant.

L'hypothèse de la production d'objets à partir des ressources locales en minerais de cuivre peut être testée de deux manières, d'une part en confrontant l'analyse de la composition des objets de Sarre et de Lorraine avec celle des productions des régions voisines, et d'autre part en comparant la composition des objets avec celle des minerais locaux pour établir la filiation minerai / métal. Le présent article fait état du premier volet de ce travail : tout comme on a pu isoler un «répertoire lorrain» à partir de l'étude typologique, nous explorerons la possibilité d'individualiser un «métal lorrain» à partir de l'analyse élémentaire des objets. Nous présenterons dans un premier temps les dépôts de «Sarre-Lorraine», leur répartition et la description des objets par catégories fonctionnelles. La deuxième partie sera consacrée à la caractérisation du métal des objets. Nous examinerons enfin dans la troisième partie le mode d'intégration de la composition du métal des objets de «Sarre-Lorraine» au sein du domaine continental, et nous mettrons en perspective le résultat obtenu en le comparant avec l'analyse élémentaire de quelques dépôts du domaine atlantique.

I. LES DEPOTS DE SARRE-LORRAINE

I. 1. Définition et répartition des dépôts de Sarre-Lorraine

Les dépôts dits de «Sarre-Lorraine» correspondent à des ensemble d'objets en bronze plus ou moins fournis[2] découverts en Lorraine, en Sarre et pour certains dans le sud du Palatinat. Le corpus traité ici rassemble les dépôts de Farébersviller (Moselle), Bouzonville (Moselle), Basse-Yutz 1 et 2 (Moselle), Frouard (Meurthe-et-Moselle), Champigneulles (Meurthe-et-Moselle), Lay-Saint-Rémy (Meurthe-et-Moselle) et Rosières-aux-Salines (Meurthe-et-Moselle), voir Véber 1999 et 2002 pour Farébersviller, Reboul et Millotte 1975 pour les autres dépôts. Vient s'ajouter le célèbre dépôt de Vaudrevange (Sarre) conservé au Musée des Antiquités Nationales à Saint-Germain-en-Laye (Kölling 1968). Tous ces dépôts sont attribués à la dernière période de l'âge du Bronze correspondant au Bronze Final IIIb en chronologie française ou Hallsttat B2 en chronologie allemande, c'est-à-dire en chronologie absolue de la fin du IXe au début du VIIIe siècle avant notre ère.

Les dépôts sont principalement répartis tout au long de la vallée de la Moselle, et dans une moindre mesure sur le cours de la Meurthe, de la Sarre, et dans la région du Warndt située au nord du département de la Moselle et au sud de la Sarre (figure 1). Dans plusieurs cas, les découvertes de dépôts ont été multiples, le second dépôt de Basse-Yutz par exemple, se trouvait à une centaine de mètres du premier. De la même façon, quatre dépôts ont été exhumés dans les alentours immédiats de Vaudrevange. Cette répartition aux abords des fleuves et rivières est un phénomène connu et plusieurs fois mentionné, auquel s'ajoute souvent de nombreuses trouvailles faites directement dans le lit du cours d'eau en question. Dans notre cas, il faut avoir à l'esprit que si les vallées de la Moselle et de la Meurthe ont pu avoir une importance tout à fait légitime à l'âge du Bronze du point de vue de leur position géographique, elles sont restées des axes de circulation privilégiés jusqu'à nos jours. La vallée de la Moselle demeure d'ailleurs l'axe structurant de la région Lorraine, et elle a fait l'objet d'aménagements et de travaux plus nombreux que dans le reste de la région, multipliant ainsi les occasions de découvertes archéologiques.

I. 2. Une identité typologique très forte

Les objets contenus dans ces ensembles appartiennent à toutes les catégories fonctionnelles (tableau 1 et figure 2). Les armes et outils comprennent pour l'essentiel des haches, on note aussi une épée du type de Mörigen, des gouges, des ciseaux, des couteaux à douille et des faucilles à languette. La parure se résume à des bracelets de différents types, souvent proches des types de Homburg et de Balingen ; les épingles, à tête vasiforme, ne sont que rarement présentes et toujours en petit nombre. D'autres pièces sont attribuées aux objets de prestige (*tintinabulum*, tubuccins, phalères) et à l'harnachement des chevaux (mors et canon de mors). Seule la vaisselle semble manquer.

Les dépôts de Sarre et de Lorraine se distinguent clairement des autres ensembles contemporains en raison d'une part de la typologie particulière de certains objets, et d'autre part à cause de l'association récurrente de ces pièces à l'intérieur même des dépôts. Plusieurs des dépôts de Sarre-Lorraine sont ainsi devenus site éponyme d'un type particulier d'objet. Tel est par exemple le cas pour les larges bracelets dits du type de Vaudrevange, ou encore pour les haches à douille ronde et lame en méplat dites du type de Frouard. Les haches à ailerons du type de Homburg tirent quant à elles leur nom d'un dépôt du sud du Palatinat, tout comme les bracelets décorés du type de Homburg et de Balingen. Enfin, quelques types de pièces sont simplement qualifiées de «lorraines» à l'image des *tintinabulum* et des tubuccins. On ajoutera que le répertoire des objets constitue un ensemble non seulement original, mais également typologiquement très homogène.

II. CARACTERISATION DU METAL DES OBJETS LORRAINS

Pour chacun des dépôts l'ensemble des pièces a été analysé, ce qui conduit à un nombre total de 300 analyses. L'objectif est de tenter d'isoler un «métal lorrain» : retrouve-t-on dans la composition chimique élémentaire du métal l'homogénéité et l'originalité du répertoire typologique ?

II. 1. Conditions d'analyse

L'ensemble des analyses a été réalisé par ICP-AES (Inductive Coupled Plasma - Atomic Emission Spectroscopy), hormis les analyses des objets des dépôts de Bouzonville et Basse-

[2] Un des plus petits dépôts de Lorraine ne contient que quatre pièces (Bouzonville, Moselle : Reboul et Millotte 1975), mais sa découverte est ancienne et il est possible qu'il soit incomplet, le plus gros en contient plus de cent trente (Farébersviller, Moselle, Veber, 2002).

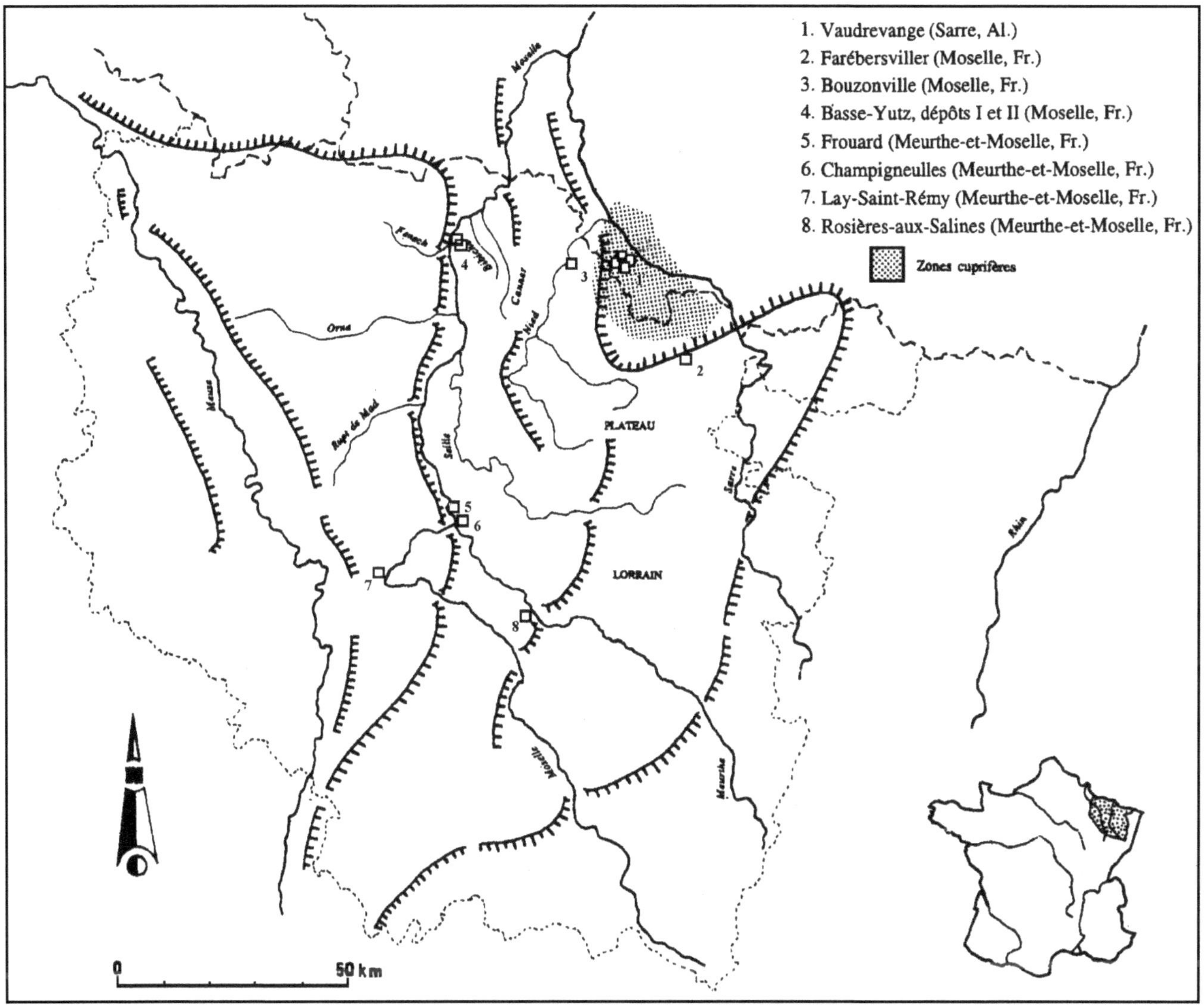

Figure 1 : Carte de localisation des dépôts de Sarre et de Lorraine étudiés et de la zone cuprifère du Warndt (Moselle).

Yutz 1 et 2 qui ont été faites antérieurement par DCP-AES (Direct Current Plasma). Des copeaux métalliques sont prélevés à l'aide d'une micro-perceuse en prenant soin d'éliminer les produits de corrosion. Chaque prélèvement a fait l'objet d'une vérification sous la loupe binoculaire, et d'un tri le cas échéant afin de ne conserver que du métal sain. Les échantillons sont mis en solution dans de l'eau régale. Un nouveau protocole d'analyse, mis en place récemment a permis d'élargir la gamme d'éléments mesurés, vingt-neuf sont actuellement dosés (voir Mille et Bourgarit 2000).

Il est important de souligner que ce nouveau protocole d'analyse ne nécessite pas plus de 10 mg de métal, ce qui autorise l'emploi de forets de faible diamètre (1 mm, voire 0,6 mm si nécessaire), donnant ainsi accès aux objets de faible épaisseur qui habituellement ne sont pas analysés. Dans le cas présent, ce mode de prélèvement a par exemple permis l'analyse des boucles filiformes du dépôt de Vaudrevange.

II. 2. Les alliages (figure 3)

D'une manière générale, la distribution des teneurs en étain est relativement peu étalée sur l'ensemble des dépôts lorrains[3], avec une moyenne légèrement inférieure à 8 % et un écart type de 2,5 %. Les teneurs moyennes en plomb se situent autour de 3 % avec une variation relative plus importante que pour l'étain. La composition des alliages ne semble pas réellement dépendre des grandes catégories fonctionnelles précédemment définies (tableau 1). On notera principalement que le groupe des objets de prestige montre les teneurs les plus élevées en étain, et une dispersion plus importante des teneurs en plomb. Le mode de fabrication, par fonderie ou déformation plastique a vraisemblablement une influence. Ainsi certains tubuccins (ceux de Frouard), objets coulés, ont un taux de plomb très élevé, alors que les phalères formées essentiellement par martelage ne contiennent quasiment pas de plomb. Cependant, il existe aussi des types d'objets tout à fait similaires en apparence dont certains individus présentent des compositions qui semblent aberrantes. C'est le cas de certains bracelets du type de Vaudrevange qui contiennent plus de 9 % de plomb, alors que la moyenne pour ce type est de 2 % si l'on ne prend pas en compte ces individus (Véber, Pernot, 2003). Si l'on s'intéresse maintenant à la répartition par proximité

3 Sur les 300 analyses réalisées 11 ne figurent pas dans les calculs. Elles correspondent à la série de culots de fonderie du dépôt de Farébersviller.

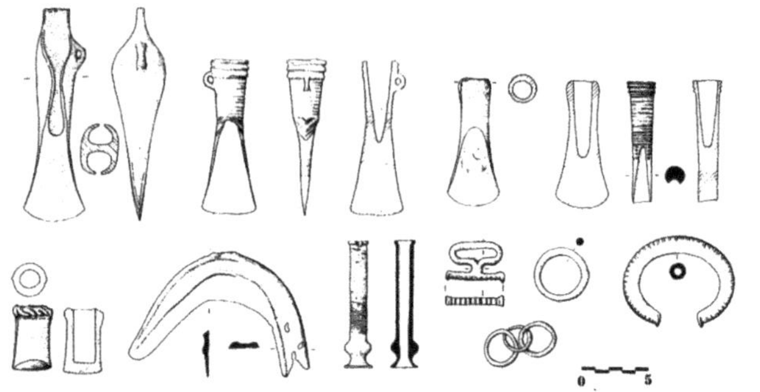
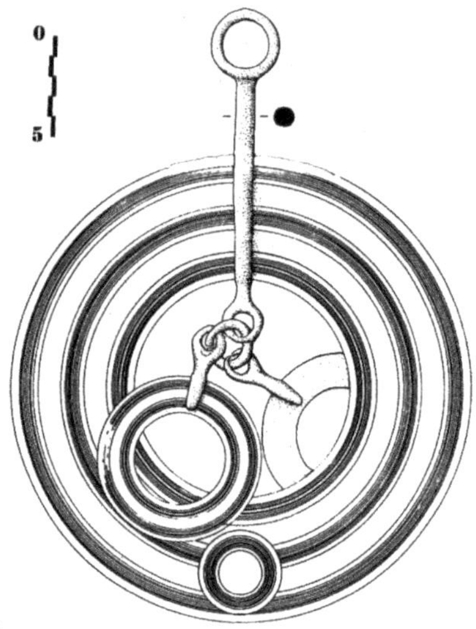

A: Extrait du mobilier du dépôt de Frouard (54) d'après Reboul et Millotte 1975.

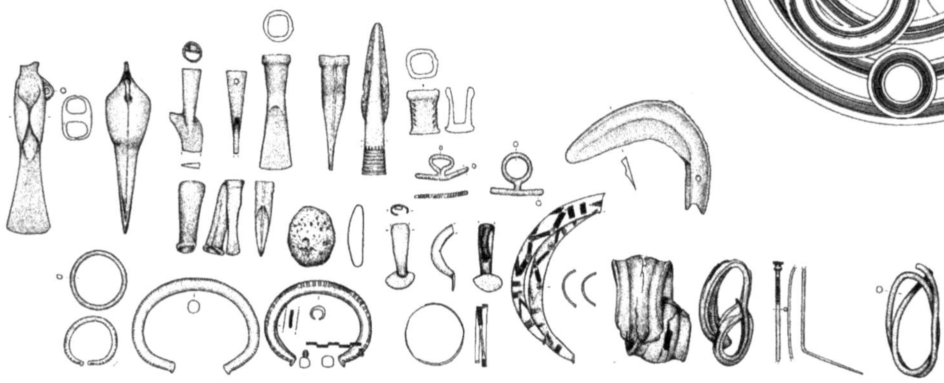

B: Extrait du mobilier du dépôt de Farébervillier (57) d'après Véber à paraître, dessins : T. Logel.

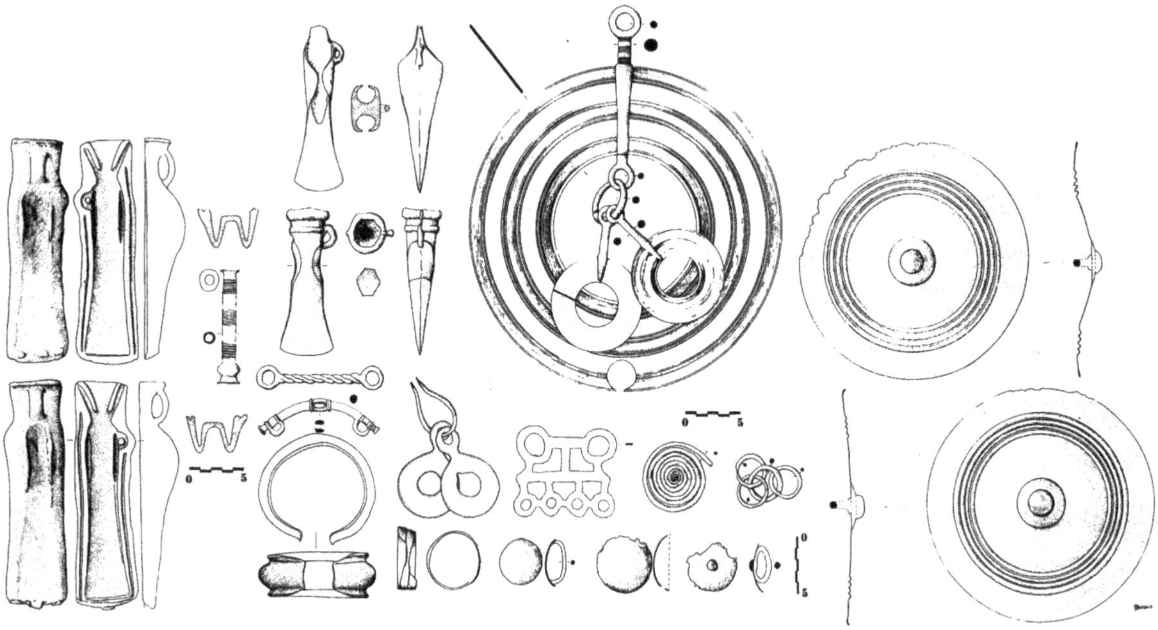

C: Extrait du mobilier du dépôt de Vqndrevqnge, dépôt 1 (Sarre) d'après Reboul et Millotte 1975.

Figure 2 : Mobilier des dépôts de Frouard (Meurthe-et-Moselle) (A), Farébersviller (Moselle) (B) et Vaudrevange (Sarre) (C). Les extraits présentés contiennent pour chaque ensemble un seul exemplaire de chacun des types d'objets représentés. Ces trois dépôts, considérés par leur composition en objets comme représentatifs du corpus de Sarre-Lorraine, ont livré respectivement 41, 130 et 69 pièces.

Tableau 1 : Répartition des objets en fonction des catégories fonctionnelles dans les dépôts de Sarre-Lorraine étudiés

	Catégories	Types	Nb de pièces analysées
Armes/Outils	Haches	À ailerons subterminaux, type Homburg	21
		À douille ronde, type frouard	8
		À douille, autre type	1
	Couteaux	À douille	3
	Gouges	À douille	2
	Ciseaux	À douille	2
	Marteaux	À douille	2
	Faucilles		29
	Lance		3
	Épée	Indéterminé	1
		type Mörigen	2
Parures	Bracelets	Type Homburg/Balingen et en oméga	60
		Type Vaudrevange	33
		Autres bracelets	4
	Épingles	À tête vasiforme	1
	Boucles filiformes		8
Prestige	Tintinabulum (tous éléments confondus)		12
	Tubuccins		11
	Canons de mors		2
	Montants de mors		3
	Anneaux	Triple/quadruple	29
	Boucles de suspension		2
	Pendentifs (pendentifs et tiges)		8
	Plaques ajourées		2
	Phalères (tous éléments confondus)		8
Divers	Fil		1
	Crochet		1
	Cloche		1
	Moule	De hache à ailerons	2
	Anneaux	simple	7
	Spirales		2
	Boutons	À bélière	9
		À pied	2
	Goutte/culot		7
		Total	**289**

géographique, le schéma correspondant au taux de l'alliage est très similaire pour les quatre groupes de dépôts. Les légères différences constatées s'expliquent à chaque fois par la composition très atypique de quelques objets. Ainsi, la présence d'une série de huit faucilles en cuivre non allié est la cause principale de la teneur moyenne plus faible en étain du dépôt de Farebersviller (7,0 %). De même, le très fort taux de plomb des quatre tubuccins de Frouard (de 14 à 21 %) élève de plus d'un point la valeur moyenne en plomb du groupe de dépôts de la région de Nancy (4,4 % au lieu de 3,3 %).

II. 3. Les impuretés (figure 4)

Les impuretés principales, par ordre de teneurs décroissantes sont l'antimoine, l'arsenic, le nickel, l'argent, le cobalt et le fer. La teneur moyenne cumulée des impuretés s'élève à 1,5 % pour l'ensemble des dépôts lorrains. Comme pour les éléments d'alliage, la distribution des teneurs en impuretés est très similaire pour les quatre groupes de dépôts. En particulier, le schéma de composition reste toujours le même, quand bien même la teneur cumulée varie (de 1,2 % pour les dépôts de la région de Nancy à 1,8 % pour le groupe de Basse-Yutz/Bouzonville) : l'antimoine domine toujours, suivi par l'arsenic, puis viennent argent et nickel.

III. DES TRADITIONS MÉTALLURGIQUES DIFFÉRENTES

Les compositions chimiques montrent relativement peu de variation au sein du corpus étudié, aussi bien pour les éléments d'alliage que pour les impuretés. Ceci va nous permettre de considérer, dans le cadre de cet article, la production lorraine comme un ensemble homogène du point de vue de la composition du métal. Reste à savoir si cette production peut se différencier des productions voisines, à l'échelle d'une

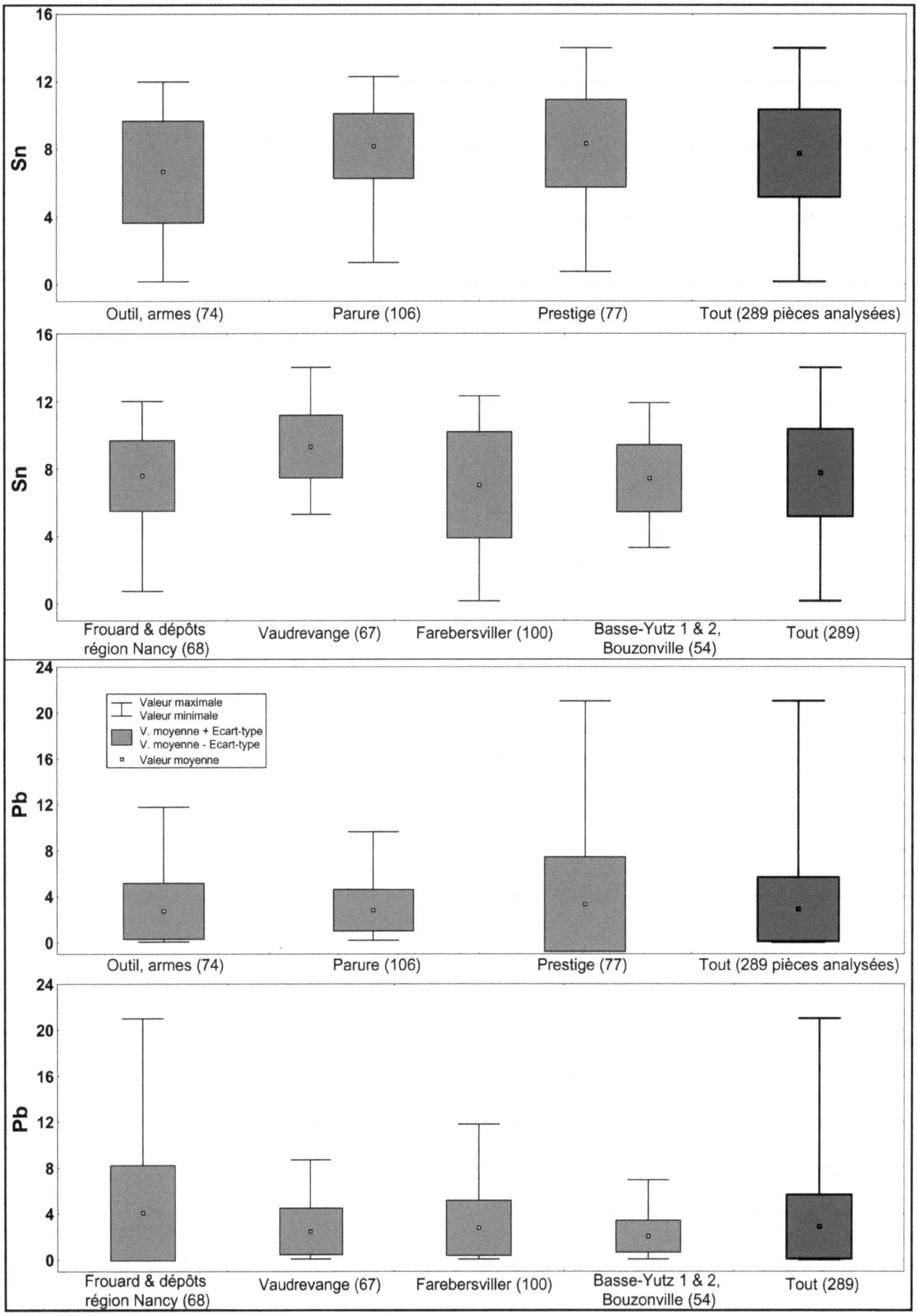

Figure 3 : Composition en éléments d'alliage des objets des dépôts de Sarre-Lorraine (en % massique). Les teneurs en étain et plomb sont comparées en répartissant l'ensemble des objets analysés soit par catégories fonctionnelles (armes/outils, parures et objets de prestige, cf. tableau 1), soit par groupes de proximité géographique (i) Frouard, Lay-Saint-Rémy, Rosières-aux-Salines et Champigneulles (Meurthe-et-Moselle), (ii) Vaudrevange (Sarre), (iii) Farébersviller (Moselle), (iv) Basse-Yutz 1 et 2 et Bouzonville (Moselle). La valeur moyenne, l'écart-type et les teneurs extrêmes sont indiqués sur le graphique pour chaque groupe d'objets ainsi défini, le nombre d'objets pris en compte est indiqué entre parenthèses. On notera que la distribution des taux d'alliage est très similaire, quel que soit le mode de regroupement choisi.

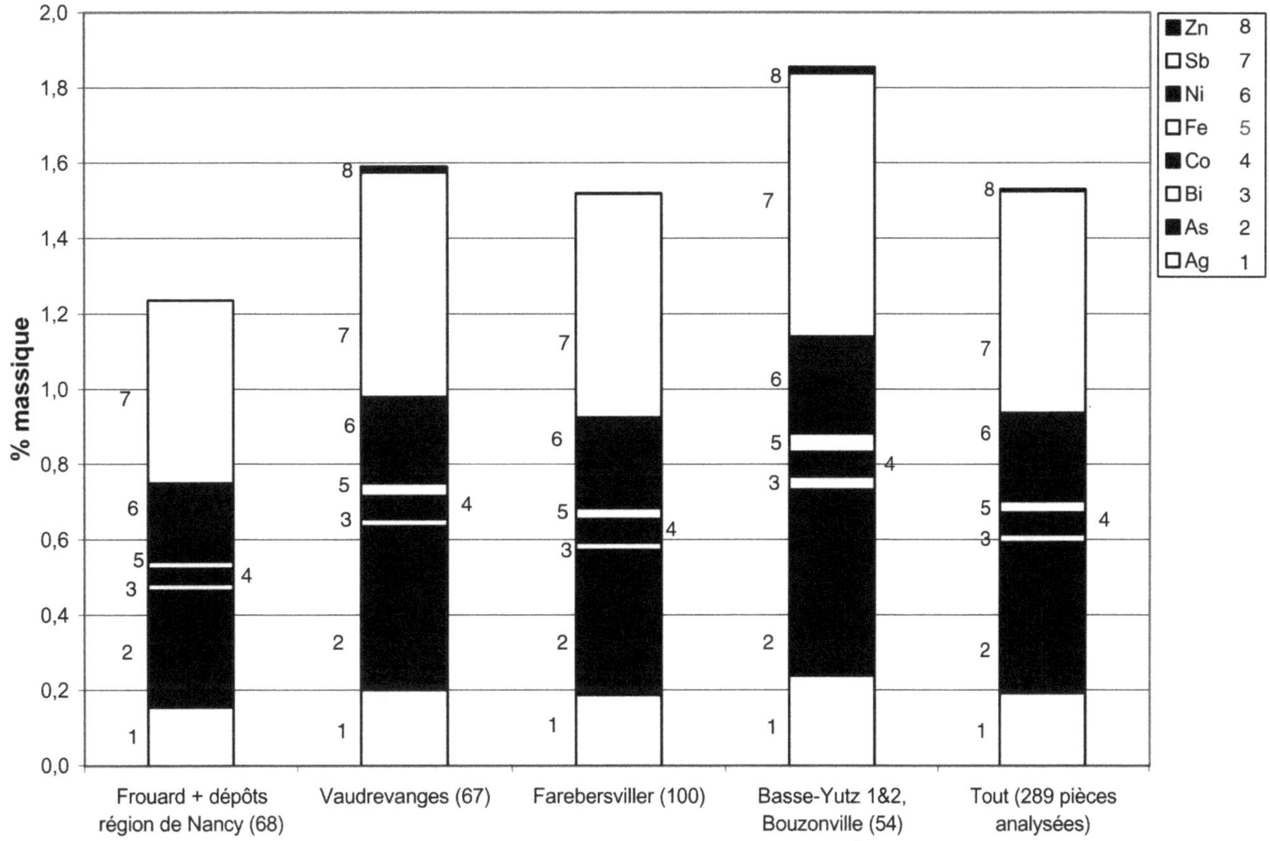

Figure 4 : Cumul des teneurs moyennes en impuretés des objets des dépôts de Sarre-Lorraine. Les analyses ont été regroupées en fonction des mêmes groupes de proximité géographique que ceux précédemment définis (figure 3). Remarquer que le classement relatif des impuretés est le même pour les quatre groupes : Sb> As > (Ag = Ni) > Co, ...

part du domaine continental auquel elle appartient, d'autre part vis à vis du domaine atlantique voisin[4].

Outre les ensembles lorrains, le domaine continental sera représenté par deux autres études de synthèse, la première concerne la Suisse (de l'Est, du Centre-Ouest et de l'Ouest, Rychner et Kläntschi, 1995), la seconde est consacrée au Nord des Alpes françaises (Verney et Bocquet, 1998) ; nous y ajouterons l'analyse de cinq objets du dépôt de Choisy-Le-Roi dans le bassin parisien (Mohen, 1977). Ces corpus comprennent tous les objets attribués au BF IIIb / Ha. B2 par leurs auteurs, issus ou non de dépôts (lingots non compris). Concernant le domaine atlantique, il n'existe à l'heure actuelle aucune synthèse sur la composition du métal à l'âge du Bronze final. Nous avons donc dû choisir parmi les analyses existantes quelques dépôts devant servir de référence, guidés en cela par les travaux de définition du Bronze Final IIIb atlantique effectués par J. Briard puis P. Brun (Briard, 1965 et Brun, 1991). Nous avons ainsi retenu l'analyse des objets provenant des dépôts de Planguenoual «le Crapon» dans les Côtes d'Armor (Bourhis *et al.*, 1975, p. 46), de Vénat à Saint-Yrieix en Charente (Coffyn et al., 1981) et d'Amiens «le Plainseau» dans la Somme (Blanchet, 1984). Nous y avons ajouté l'analyse de deux dépôts récemment découverts dans la vallée de l'Oise par J.-C. Blanchet : Compiègne «les Rossignols» et Lacroix-Saint-Ouen «La Haute Queue»[5].

La lecture des histogrammes qui figurent l'ensemble des corpus de la Suisse à la Bretagne permet de distinguer nettement deux groupes de composition homogène (figures 5 et 6) :

- à l'Ouest les teneurs de l'alliage sont nettement plus élevées, les valeurs cumulées dépassent dans tous les cas 16 %. Le plomb est bien sûr présent en teneurs beaucoup plus importantes, comme cela a déjà été mis en évidence à de multiples reprises pour le Bronze Final atlantique. Mais la surprise vient surtout de la teneur en étain, qui est elle aussi plus importante dans le domaine atlantique que dans le domaine continental (moyenne toujours supérieure à 10 %).

- les valeurs cumulées en impuretés présentent un gradient inverse : les taux sont très faibles à l'Ouest, dépassant à peine 0,5 % alors que pour les sites de l'Est, Choisy-le-Roi compris, toutes les teneurs sont supérieures à 1,2 % avec pour les Alpes françaises du Nord une valeur proche de 2,5 %. Les variations apparaissent plus importantes à l'Est qu'à l'Ouest, où le taux d'impuretés est très stable.

[4] On entend par «domaine», continental et atlantique, des entités techno-culturelles recouvrant un domaine géographique particulier pour une période déterminée. Le domaine continental ou Nord-alpin recouvre l'Allemagne, la Suisse et la France de l'Est jusqu'à une limite qui traverse le Nord de la France, le bassin parisien et se prolonge vers la Loire. Le domaine dit atlantique se trouve à l'ouest de cette limite. La partition entre ces deux domaines est directement issue des partitions observées aux périodes précédentes du Bronze final à partir de la répartition du mobilier métallique (Brun 1988).

[5] Ces deux dépôts sont en cours d'étude au C2RMF par Benoît Mille.

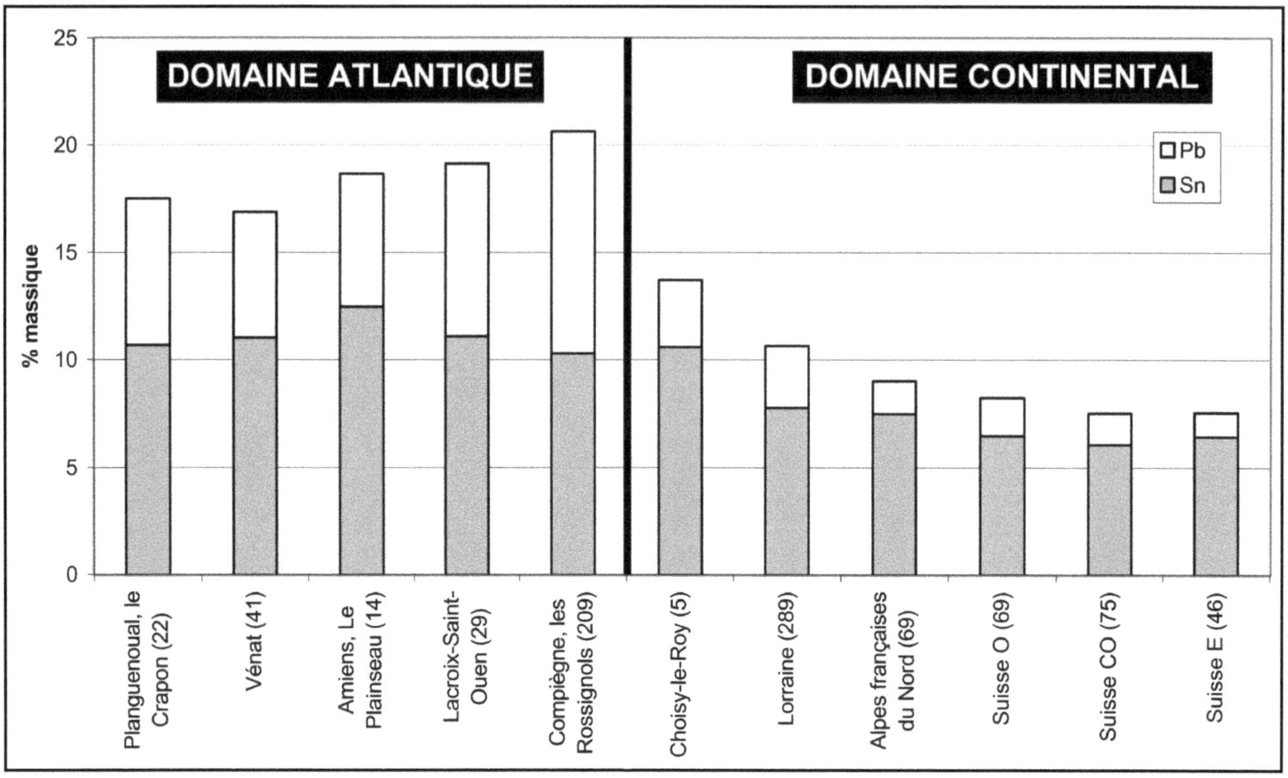

Figure 5 : Comparaison de la teneur moyenne cumulée en éléments d'alliage d'objets issus des deux grands domaines d'influence techno-culturelle Ouest-européen à l'extrême fin de l'âge du Bronze (BF IIIb/Ha. B2). Les recettes d'alliage sont très distinctement différenciées : l'ajout d'étain comme de plomb est systématiquement plus important au sein du domaine atlantique, conduisant à un total supérieur à 16 % (dont plus de 10 % d'étain), tandis que pour le domaine continental la teneur cumulée ne dépasse généralement pas 10 %. On remarquera de plus l'accroissement progressif des teneurs à mesure que l'on s'approche du domaine atlantique (surtout pour le plomb).

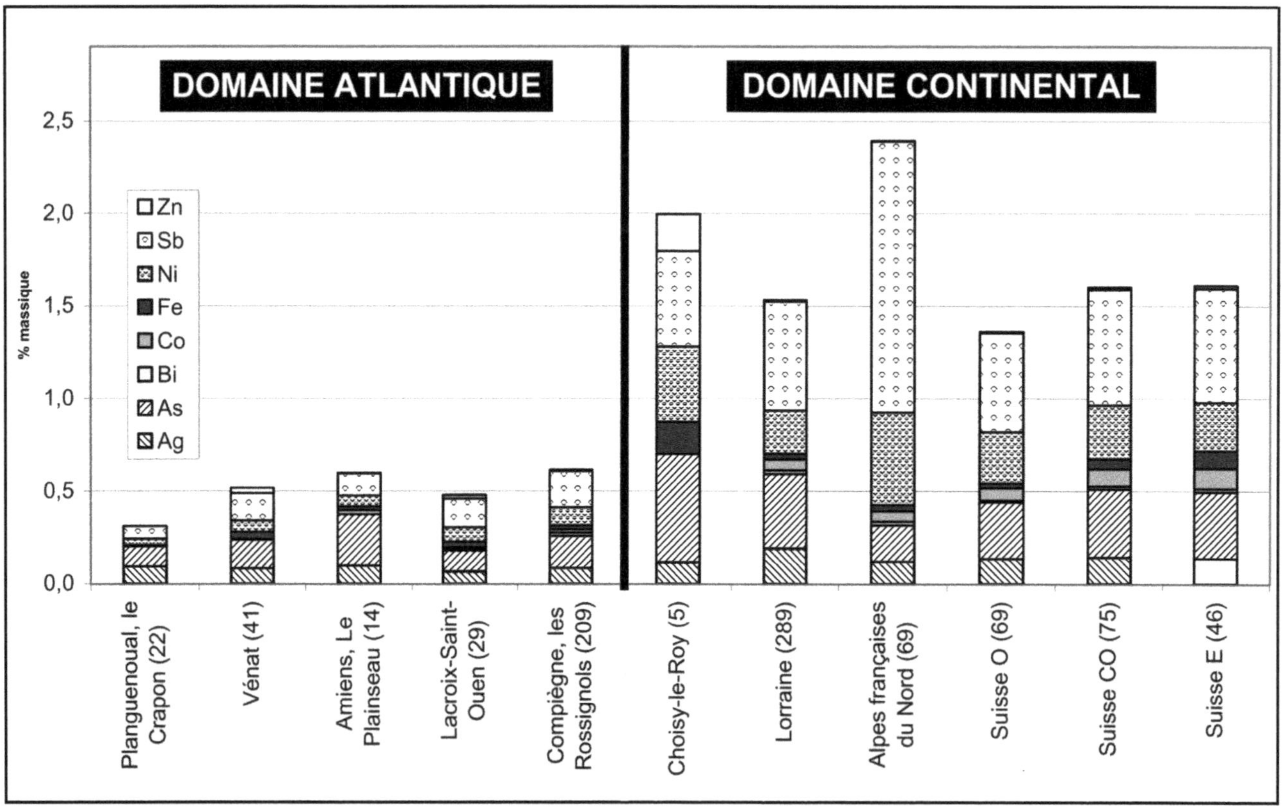

Figure 6 : Comparaison de la teneur moyenne cumulée en impuretés d'objets issus des deux grands domaines d'influence techno-culturelle Ouest-européen à l'extrême fin de l'âge du Bronze (BF IIIb/Ha. B2). La partition des deux domaines par le spectre d'impuretés est flagrante, le métal est nettement moins pur dans le domaine continental (la moyenne n'est jamais inférieure à 1,3 %) que dans le domaine atlantique (moins de 0,5 %).

Cette mise en perspective montre que la composition du métal lorrain s'intègre parfaitement dans celle du domaine continental. On observe ainsi une gradation progressive des teneurs en plomb et en étain à mesure que l'on s'approche du domaine atlantique, et la Lorraine obéit à cette tendance. De même, avec une teneur cumulée en impuretés de 1,5 %, la Lorraine est tout à fait dans la moyenne observée. Par conséquent la production lorraine ne peut pas se distinguer, par la composition du métal, des productions voisines du domaine continental.

En revanche la production lorraine, et avec elle les productions voisines, se distinguent nettement de ce que l'on peut observer au sein du domaine atlantique, à la fois par les éléments d'alliage et par les impuretés : le métal de l'Ouest est plus allié que celui de l'Est, et nettement plus pur. Ces résultats mettent clairement en évidence des différences dans les traditions d'alliage entre les deux domaines. Ainsi, il ne saurait être question du remplacement de l'étain par le plomb dans le domaine atlantique, mais de recettes d'alliage différentes : à l'Est, le bronze est préparé en ajoutant 8 % d'étain et 1 à 3 % de plomb, tandis qu'à l'Ouest on ajoute au moins 6 % de plomb et plus de 10 % d'étain. Concernant la plus grande pureté du cuivre utilisé dans le domaine atlantique, nous ne sommes pas pour l'instant en mesure de fournir d'explication : cette pureté est-elle le fait de cuivres de provenances différentes, ou celui d'un procédé métallurgique différent qui conduirait à un meilleur affinage du cuivre ?

CONCLUSION

Les analyses élémentaires des objets lorrains confirment dans une certaine mesure l'homogénéité de la production lorraine, pour la période du BF IIIb / Ha. B2, constatée auparavant par la typologie, tant au niveau de la composition de l'alliage que des taux d'impuretés. En particulier la recette d'alliage montre très peu de variations (8 % d'étain, 3 % de plomb), les contraintes liées à la mise en forme des objets (martelage ou moulage) semblent le plus souvent responsables des variations de teneurs parfois observées.

Cette homogénéité de composition des objets en alliage à base de cuivre se retrouve à l'échelle de tout le domaine continental, plaçant le mobilier étudié au cœur d'une tradition métallurgique qui dépasse les frontières de la Lorraine. L'inter-comparaison de la composition du métal des objets ne permet donc pas à elle seule d'isoler une production proprement lorraine, et la question de l'utilisation d'un cuivre lorrain dans les objets reste entière. Pour y répondre, il nous faudra faire appel à des marqueurs plus discriminants pour établir ou non un éventuel approvisionnement en cuivre à partir des ressources en minerai de cuivre local. Quelques pistes peuvent être proposées : il s'agira avant tout de confronter la composition élémentaire des objets à celle des minerais, en prêtant une attention toute particulière aux éventuels marqueurs géochimiques spécifiques des minerais de cuivre lorrains. Une analyse comparée de la composition isotopique du plomb sera probablement un complément indispensable.

Par ailleurs, la mise en parallèle des résultats d'analyse obtenus pour les dépôts lorrains avec d'autres ensembles du BF IIIb / Ha. B2 met en évidence, pour la première fois, l'existence de traditions métallurgiques différentes entre domaine atlantique et domaine continental. Ainsi, le métal des dépôts de l'Ouest est plus allié, tant en étain qu'en plomb, et d'une plus grande pureté que celui de l'Est. Ces premiers résultats d'analyse demandent à être complétés, plusieurs régions n'ayant pas pu être représentées faute d'analyses disponibles. Toutefois cette étude permet d'ores et déjà de confirmer, pour l'extrême fin de l'âge du Bronze, une vision bipartite de l'espace Ouest-européen entre domaine atlantique et domaine nord alpin, reconnue jusqu'alors uniquement par la typologie.

Addresses des auteurs

Cécile VEBER
Institut National de Recherche en Archéologie Préventive, Base archéologique des chais, route de Moulins, 57160 Scy-Chazelles / Université de Dijon

Benoît MILLE
David BOURGARIT
Centre de Recherche et de Restauration des Musées de France, UMR 171 du CNRS, 6, rue des pyramides,
75041 Paris cedex 01

Bibliographie

BEAUPRÈ, J., 1902, *les études préhistoriques en Lorraine de 1889 à 1902*, Nancy.

BLANCHET, J.-C., 1984, *Les premiers métallurgistes en Picardie et dans le nord de la France, Chalcolithique, Age du Bronze et début du premier Age du Fer*, Mémoire de la société préhistorique française, tome 17, pp. 441-442.

BOURHIS, J., GIOT, P.-R., BRIARD, J., 1975, *Analyses spectrographiques d'objets préhistoriques et antiques*, Troisième série, travaux du laboratoire «Anthropologie-Préhistoire-Protohistoire-Quaternaire armoricains», université de Rennes, Equipe de recherche du CNRS n°27.

BRIARD, J., 1965, Les dépôts bretons et l'âge du Bronze atlantique, Faculté des sciences de l'Université de Rennes, Becdelièvre, Rennes.

BRIARD, J., 1991, Le groupe de l'épée en langue de carpe en Armorique : une révision, *Actes du 1er colloque du Parc Archéologique de Beynac, du 10 au 14 sept. 1990 : l'âge du bronze atlantique, ses faciès, de l'Écosse à l'Andalousie et leurs relations avec le Bronze continental et la Méditerranée*, dir. C. Chevillot et A. Coffyn, Beynac-et-Cazenac, pp.125-144.

BRUN, P.,1988, L'entité «Rhin-Suisse-France orientale» : nature et évolution, *Actes du Colloque de Nemours 1986, Mémoires du Musée de Préhistoire d'Ile-de-France, 1*, Nemours.

BRUN, P., 1991, Le bronze atlantique et ses subdivisions culturelles : essai de définition, Actes du 1er colloque du Parc Archéologique de Beynac, du 10 au 14 sept. 1990 : *l'âge du bronze atlantique, ses faciès, de l'Écosse à l'Andalousie et leurs relations avec le Bronze continental et la Méditerranée, dir. C. Chevillot et A. Coffyn*, Beynac-et-Cazenac, pp. 11-24.

COFFYN, A., GOMEZ, J., MOHEN, J.P., 1981, *L'apogée de bronze atlantique, le dépôt de Vénat, L'âge du bronze en France 1*, Picard, Paris 1981, pp. 26-27.

HAMON, B., BLOUET, V., 1993, Falck, La petite Saule, *Bilan Scientifique Régional, Lorraine*, p. 60.

JACQUOT, E., 1857, Notice géologique et historique sur les mines de plomb et de cuivre des environs de Saint-Avold, de Hargarten et de Sarrelouis, *Mémoire de l'Académie de Metz*, 1857-58, pp.531-556.

KÖLLING, A., 1968, *Späte Bronzezeit an Sarr und Mosel*, Saarbrücker Beitrage zur Altertumskunde, Band 6, Bonn.

LINKENHELD, E., 1930, Les dépôts de l'âge du Bronze en Lorraine (Moselle) *Revue Anthropologique* 1-3, p. 77-86

MILLE, B., BOURGARIT, D., 2000, L'analyse des alliages anciens à base de cuivre : état des connaissances et développement d'un protocole d'analyse par ICP-AES, *Revue d'Archéométrie n° 24*, pp. 13-26.

MILLOTTE, J.P., 1965, Carte archéologique de la Lorraine (Âge du Bronze et du Fer), *Annales Littéraires de l'Université de Besançon, vol. 73.*

MOHEN, J.-P., 1977, *L'âge du bronze dans la région de paris, catalogue synthétique des collections conservées au Musée des Antiquités Nationales*, Paris, Réunion des musées nationaux.

REBOUL, R., MILLOTTE, J.P., 1975, Dépôts de l'âge du Bronze final en Lorraine et en Sarre, *Inventaria Archaeologia, France, fasc; 4*, F29-F49.

RYCHNER, V., KLÄNTSCHI, N., 1995, Arsenic, Nickel et Antimoine, *Cahiers d'Archéologie Romande n°63.*

VEBER, C., 1999, Introduction à l'étude du dépôt de Farébersviller et production métallique du BF IIIb en Lorraine. In *L'Atelier du Bronzier en Europe du XXe au VIIIe siècle avant notre ère*, C. Mordant, M. Pernot et V. Rychner éd., vol. 3, *Production, circulation et consommation du bronze*, CTHS, Paris, pp. 41-54.

VEBER, C., 2002, Le dépôt de Farébersviller (Moselle), ensemble du BF IIIb, *Archaeologia Mosellana*, Tome V.

VEBER, C., PERNOT, M., 2003, Technical study of some Vaudrevange type bracelets found in different areas (IX/VIIIth c. BC) metallographical examinations experimental wax model making, In *Proceedings of the XVth International Conference on Ancient Bronzes, Production and Technology*, 22-26 Maggio 2001, Grado and Aquileia.

VERNEY, A., BOCQUET, A., 1999, La composition élémentaire d'objets du Bronze final des Alpes françaises du nord : premiers résultats In *L'Atelier du Bronzier en Europe du XXe au VIIIe siècle avant notre ère*, C. Mordant, M. Pernot et V. Rychner éd., vol. 1, *Production, circulation et consommation du bronze*, CTHS, Paris, pp. 101-118.

L'ÉTAIN DANS LES EXAMENS MÉTALLURGIQUES DE QUATRE ENSEMBLES VILLANOVIENS

Anne LEHÖERFF

Résumé : Les tombes villanoviennes d'Étrurie, de Campanie ou de la plaine padane renferment des milliers d'objets métalliques de type et d'importance variés : parure, pièces de service à boire, armement. L'analyse du pourcentage d'étain de 30 objets de Tarquinia Monterozzi, Veio Quattro Fontanili et de collections britanniques met en évidence des différences selon les localités et non, comme on pourrait s'y attendre dans une logique technique, selon les objets (usage et fabrication). De plus, les observations des mobiliers et de la microstructure du matériau montrent un savoir-faire technique de très haut niveau ainsi qu'une corrélation entre le choix de l'alliage et le procédé de fabrication, là encore avec des distinctions selon les lieux, c'est-à-dire peut être des ateliers.

Abstract: Villanovians graves are rich in metallic objects: ornaments, vessels, weapons. Analysis of tin on 30 objects from the Tarquinia Monterozzi, Veio Quattro Fontanili, collections from the British Museum and the Ashmolean Museum shows differences between places. Metallographies show a high technological know-how level as well. The choice of alloy is in relation to production techniques and this technological choice seems to be connected to provenances, maybe to workshops.

INTRODUCTION

La richesse et la qualité de la métallurgie dans l'Italie villanovienne ne sont plus à démontrer. Depuis plus d'un siècle et demi des milliers d'objets ont été mis au jour dans les grandes nécropoles des IXe-VIIIe siècles avant notre ère en Etrurie, dans la plaine padane ou en Campanie (Bartoloni 1989, Biestti Sestieri 1996, Guidi et Piperno 1992, Italia Omnium...1988 pour une liste plus détaillée des travaux).

Dans sa dernière demeure, le défunt est presque systématiquement accompagné de parures de différents types parmi lesquelles les fibules masculines ou féminines s'illustrent par leur variété et leur abondance.

Au VIIIe siècle certaines tombes se distinguent par le nombre élevé de pièces associées au service à boire ou appartenant à une panoplie personnelle : des situles métalliques servent de réceptacle aux cendres de l'incinéré(e), de petits vases en nombre variable ou une gourde les accompagnent tandis que des pièces à caractère presque ornemental, tels que les trépieds, complètent beaucoup plus rarement cet ensemble. Un casque métallique surmonte dans certains cas l'urne funéraire, une épée, parfois même avec son fourreau métallique, atteste la présence d'un homme d'un certain statut social –un guerrier?— dans la tombe tandis qu'une ceinture ventrale décorée conduira à penser que les cendres appartenaient plutôt à une femme, elle aussi d'un rang élevé.

Au regard de la société villanovienne, et même seulement au regard de celle des morts, les tombes les plus riches comme la tombe du guerrier de Tarquinia-Monterozzi (Helbig 1874) ou la tombe AA1 de Veio (Veio....1970) restent très exceptionnelles. Pour les artisans métallurgistes ces mobiliers représentent un marché pour écouler leurs productions, soit aux vivants qui sont en mesure de les acquérir et les emportent ensuite avec eux au cours de leur dernier voyage, soit uniquement pour les défunts les plus fortunés.

Dans ces dépositions métalliques des objets techniquement variés coexistent : de petites pièces coulées comme des perles, des épées elles aussi obtenues pour l'essentiel en fonderie aussi bien que des vases réalisés a priori par martelage. Bien que ce mobilier ait attiré l'attention de nombreux chercheurs, il n'a pas fait l'objet d'études techniques dans son ensemble. Pourtant, les enjeux pour le monde artisanal des IXe et VIIIe siècles avant notre ère sont d'importance. Quels procédés de fabrication sont utilisés ? Avec quels matériaux ? Par quels artisans ? Et, pour répondre plus directement au thème ici traité, quelles quantités d'étain sont utilisées ? L'emploi de ce matériau rare et précieux n'est pas anodin (tableau 1). Indispensable dans l'alliage binaire du bronze, il apporte des qualités de fusibilité, contribue à éclaircir le rouge orangé du cuivre et à apporter des teintes plus jaunes au métal. Utilisé en quantité importante, il représente une contrainte pour un travail de mise en forme par déformation plastique de l'alliage. Sa présence, son importance selon les objets de ces riches dépositions funéraires villanoviennes sont donc autant d'indices pour évaluer et comprendre un artisanat métallurgique en pleine expansion.

LES MÉTHODES

Prendre la mesure de l'étain dans ces objets villanoviens mis au jour en contexte funéraire suppose la mise en œuvre d'études en laboratoire, d'analyses de composition mais aussi d'examens de la microstructure du métal. En effet, la seule mesure des pourcentages d'étain est insuffisante pour étudier et évaluer un artisanat métallurgique. Les analyses doivent être mises en relation avec d'autres résultats obtenus grâce à

Tableau 1 : tableau schématique des répercussions sur la fabrication des pièces en fonction d'un pourcentage élevé ou faible en étain.

Composition en étain	Mise en forme	Qualité de l'alliage
Pourcentage elevé (d'étain> 13%)	=> coulée facilité (point de fusion abaissé) => martelage difficile	Qualité "médiocre" possible Nécessité d'un matériau "propre"
Pourcentage faible (étain <8%)	=> coulée à température plus élevée => martelage plus aisé	Matériau de qualité "correcte"
	Mise en relation de ces 3 paramètres ===>	*évaluation de qualités techniques*

Commentaire : dans le premier cas de figure, le pourcentage égal ou supérieur à 13% facile la coulée et rend possible l'emploi d'un matériau comportant des impuretés et des conclusions ; ce même pourcentage d'étain rend en revanche plus délicat le martelage pour un matériau a-priori de bonne qualité. A l'inverse, un plus faible pourcentage d'étain augmente le point de fusion de l'alliage mais autorise un martelage plus facile à conduire, dans un matériau de qualité correcte sans être exceptionnel. L'analyse de ces paramètres donne une idée des qualités techniques artisanales.

Tableau 2

Objet	Prov.	Numéro	Analyses			Etat métallurgique
			Sn	Cu	Pb	
Gourde	T	P1-1967	14,5			Déformation plastique
Repose vase	T	P17-1092A	> 13		1	Déformé sous le décor
Petit vase	T	P4-179A	12,5			Déformé dans une recristallisation partielle ?
Petit vase	T	P3-179	12			Recristallisé
Vase à pied	T	P5-20A	7			Déformé ?
Fourreau	V	1	13			Recristallisé
Fourreau	V	2	>12			Déformé dans recristallisé
Casque	V	6 a	> 12		0,4	Recristallisé
Urne	V	3 bis	11,5		0,3	Déformé
Urne	V	7 et 7 bis	11,5			Recristallisé
Urne	V	8 a	11		0,5	Recristallisé
Urne	V	8 b	10,5		0,5	Recristallisé mais déformation « skin pass » ?
Urne	V	4 bis	10		0,3	Déformé
Casque	V	6 b	8		0,5	Recristallisé
Urne	AM	P11-1962-977	9		0,8	Déformé dans matrice recristallisée
Urne	AM	P9-1962-977	8		< 1	Déformé sous la bossette dans matrice recristallisée
Urne	AM	P8-1962-977	> 7		< 1	Recristallisé
Urne	AM	P12-1962-977	> 7		0,7	Recristallisé
Fourreau	AM	P1-275	7			Recristallisé à grain fin (recuit)

Légende : Prov. : Provenances. AM : Ashmolean Museum. BM : British Museum. T : Tarquinia. V : Veio

des observations sur les objets eux-mêmes, de préférence associées à des examens de la microstructure en laboratoire.

Plusieurs méthodes sont envisageables. Dans le cas présent, une attention particulière a été accordée aux examens métallurgiques afin de déterminer l'histoire thermomécanique des objets et de reconstituer la chaîne opératoire de fabrication[1]. Pour des études de ce type, l'obtention de résultats fiables passe par un travail de prélèvements de quelques mm^2 sur les objets sélectionnés, à des endroits stratégiques pour comprendre les techniques de fabrication. Un ou plusieurs plans d'observation par échantillon sont alors soigneusement choisis. Les examens sont effectués à l'aide d'un microscope métallographique. Les analyses de composition sont limitées ici essentiellement aux majeurs sans tenir compte de l'ensemble des éléments trace (tableaux 2 à 5 et figures 1 et 2) dans la mesure où l'objectif premier est de mettre en corrélation les techniques de fabrication avec l'alliage choisi pour des pièces typologiquement très variées (Lehoërff 1999 a). Les analyses sont obtenues grâce au système d'analyse par rayon X (programme sans étalon avec correction «ZAF») du Microscope électronique à balayage employé par ailleurs pour des observations complémentaires de la microstructure. Pour chaque échantillon, plusieurs mesures ont été effectuées et les résultats fournis correspondent à la moyenne établie individuellement.

[1] Ce travail a été réalisé dans le cadre d'un doctorat de l'université de Paris I, à paraître. Une première publication du travail en laboratoire est disponible dans Lehöerff 1999a.

Tableau 3

Objet	Prov.	Numéro	Analyses			Etat métallurgique
			Sn	Cu	Pb	
Fibule	T	P11-83	15			Brut de coulée et polissage
Fibule	T	P9-527A	11,5		<0,4	Recristallisé
Fibule	T	P10-529	11,4			Recristallisé
Fibule	T	P16-651A	**ca 11**			Recristallisé
Fibule	T	P7-659A	10		<0,3	Recristallisation à gros grains
Perle	T	P12-97	<1	45	>38	Brut de coulée
Fibule	AM	P3-1872-1165	7		<0,4	Recristallisation ?
Fibule	AM	P5-1872-1165	7		<0,4	Recuit de recristallisation
Fibule	AM	P4-1872-1165	5			Recuit de recristallisation à grain fin

Tableau 4

Objet	Prov.	Numéro	Analyses			Etat métallurgique
			Sn	Cu	Pb	
Ceinture	T	P18-250	8,5			Déformé
Ceinture	V	9b	10			Recristallisé
Ceinture	V	12	ca 10		0,6	Recristallisé
Ceinture	V	5	8,5		1	Recristallisation complexe
Rasoir	BM	P6-808-233	10			Déformé dans recristallisé à grain fin
Rasoir	BM	P7-808-233	<10		1	Recristallisé
Ceinture	AM	P6-1890-619	11		ca 1	Déformé
Ceinture	AM	P7-1890-619	11		<0,3	Déformé

Tableau 5

Objet	Prov.	Numéro	Analyses			Etat métallurgique
			Sn	Cu	Pb	
Hache	T	P6-1012A	1	+90	5	Brut de coulée
Epée	V	10	6,5		0,6	Recristallisé à gros grain dans brut de coulée
Epée	V	11	6,5		Inégal	Recristallisé à gros grain dans brut de coulée
Hache	BM	P11-1086	>10		ca 0,5	Recristallisé
Hache	BM	P3-1068	10		1	Macles mécaniques dans recristallisé
Hache	BM	P1-1068	9		0,5	Recristallisé sans déformation notable
Hache	BM	P2-1068	9		0,5	Recristallisé avec « skin pass »
Hache	BM	P4-1075	8		<0,5	Recristallisé avec peu ou pas de déformation
Hache	BM	P5-1075	8		<0,5	Déformé
Hache	BM	P9-767-611	<7		2	Brut de coulée
Hache	BM	P8-767-611	6		0,5	Recristallisé
Hache	BM	P10-767-611	<6		<0,5	Recristallisé
Hache	BM	P12-808-244	5,5		Traces	Brut de coulée

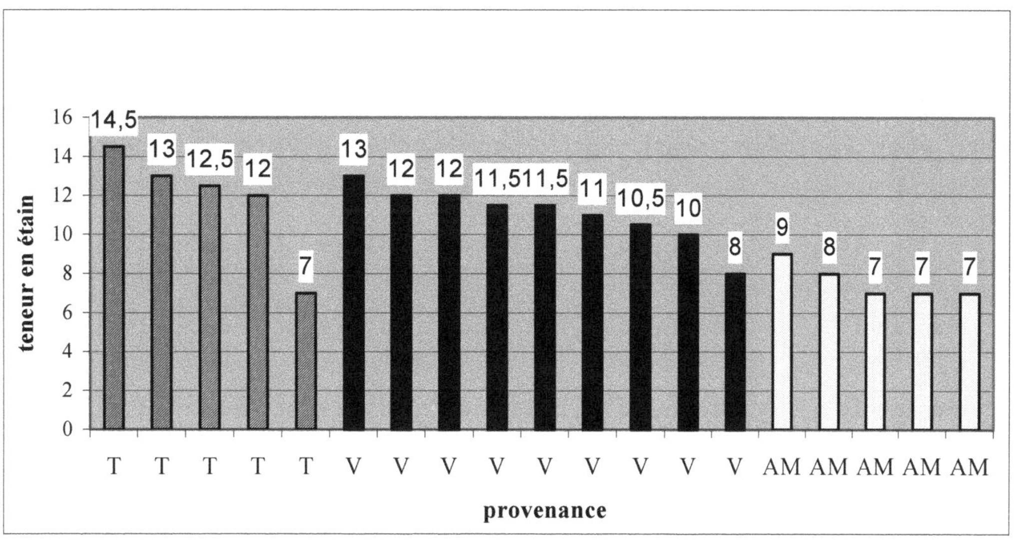

Figure 1.

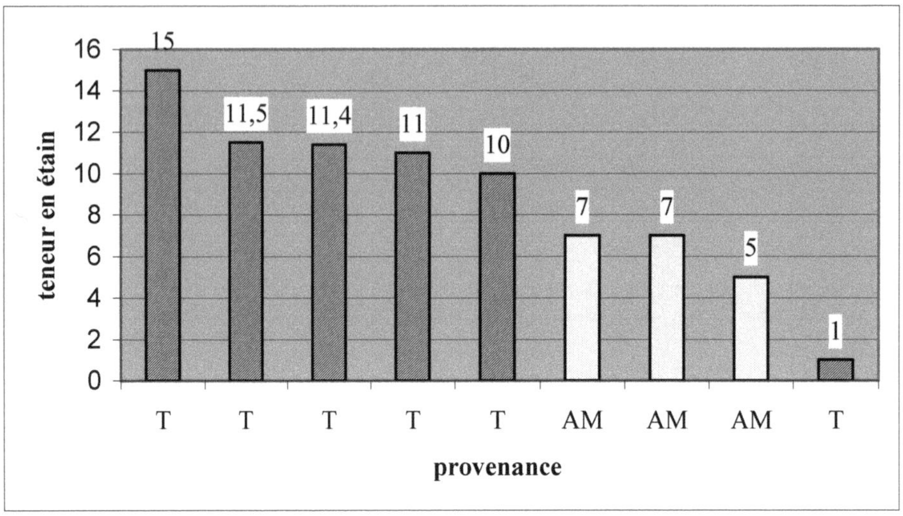

Figure 2.

LE CORPUS

Le corpus qui a été observé en laboratoire se compose de mobiliers de nature variée (fig.3), datés entre la fin du IXe siècle avant notre ère et le Villanovien IIB de Hencken (Hencken 1968). Ces pièces sont au nombre de 30, provenant de plusieurs nécropoles villanoviennes et conservées dans quatre musées : Tarquinia pour la nécropole de Monterozzi, Cività Castellana pour celle de Veio, le British Museum à Londres et l'Ashmolean Museum à Oxford. Pour ces deux derniers musées, les provenances précises sont souvent plus problématiques en raison de l'histoire des collections. Il semble cependant que le mobilier de l'Ashmolean Museum, de typologie villanovienne, puisse provenir non pas d'Étrurie mais de Campanie d'après les archives du musée.

A Tarquinia, 12 objets de typologie variée ont été sélectionnés (Lehoërff 1999a, p. 799) : une hache à ailerons, une ceinture ventrale féminine, une perle tubulaire, 5 fibules, 2 vases de petite taille, un trépied, une gourde décorée (Fig. 6). Dans 4 cas, plusieurs prélèvements ont pu être effectués (gourde n° 1967, vase n° 179, fibule hérisson n° 527, ceinture n° 250) et les 16 prélèvements ont permis d'obtenir après redécoupage un total de 22 échantillons.

A Veio, une certaine variété typologique a pu être préservée, pour des objets moins nombreux (8 au total) mais dont les contextes sont bien connus : l'urne de la tombe AA1 (Veio...1970) et celle de de la tombe M9B (Veio...1963), la ceinture de la tombe OP 4-5 (Veio..1972) et AA12A (Veio...1965), l'épée de la tombe Z15A (Veio...1965), fourreau de la tombe EE10B (Veio...1967) et casque de la tombe M9B (Veio...1963). De plus, plusieurs prélèvements ont été réalisés pour la majorité de ce mobilier. Ainsi, 16 prélèvements ont permis d'obtenir les 19 échantillons observés.

Au British Museum, à l'exception d'un rasoir (n° 76-7-6-11), les prélèvements ont été réalisés uniquement dans 5 haches. Les 6 objets ont permis un total de 12 échantillons. Enfin, les prélèvements effectués à l'Ashmolean ont conduit à des comparaisons avec des mobiliers d'autres provenance mais de typologie identique : un fourreau d'épée décoré, une fibule hérisson, une ceinture féminine, une urne décorée de motifs de canards. Le nombre des prélèvements, comme celui des échantillons, s'élève à 12. Au total, sur les 30 objets qui ont pu être retenus pour cette étude, 56 prélèvements ont permis l'observation et l'analyse de 65 échantillons (Lehöerff 1999a pour les illustrations photographiques de l'ensemble des objets).

LES ANALYSES DE COMPOSITION

Sur l'ensemble du corpus, l'alliage binaire cuivre-étain est très largement majoritaire. On peut seulement relever à Tarquinia une hache en cuivre (tableau 5) et une perle cuivre-plomb (fig.3). Pour cet objet de parure l'emploi du plomb présente des avantages techniques et économiques : il abaisse le point de fusion avec un métal blanc qui permet d'obtenir un alliage de couleur jaune pâle tout en économisant de l'étain, matériau rare, sans porter préjudice à la fabrication de l'objet puisque celui-ci n'est pas martelé mais coulé, probablement en série si l'on juge par la qualité et le nombre de ces perles dans les tombes.

Dans le détail, les résultats d'analyse les plus significatifs ont été obtenus dans les tôles métalliques, vaisselles et casques (tableau 2, fig. 1). Les teneurs en étain dans l'alliage de ces objets varient du simple au double entre 7 et 14%. Plus que des différences selon les objets, les analyses mettent en évidence des variations en fonction des provenances. Ainsi, pour des urnes de même typologie, à Veio l'étain varie entre 10 et 11,5% selon les pièces analysées alors que l'exemplaire conservé à l'Ashmolean Museum ne contient qu'entre 7 et 9% quelle que soit la partie du vase concernée. A l'inverse, les teneurs en plomb sur ce vase avoisine en moyenne les 1% tandis les échantillons de Veio n'affichent que des valeurs inférieures ou égales à 0,5%. À Tarquinia, les tôles qui ont pu être analysées présentent les teneurs les plus élevées, jusqu'à 14,5% pour la gourde. Le plomb n'est de surcroît présent que sous forme de traces. Dans ce dernier cas, l'alliage

Principaux types d'objets étudiés
(l'échelle de représentation varie d'un dessin à l'autre)

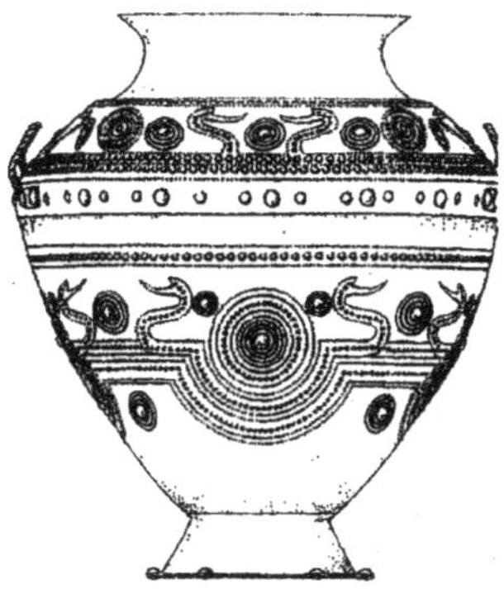

Urne décorée de Veio Quattro Fontanili. Tombe AA1 (d'après Veio...1970)

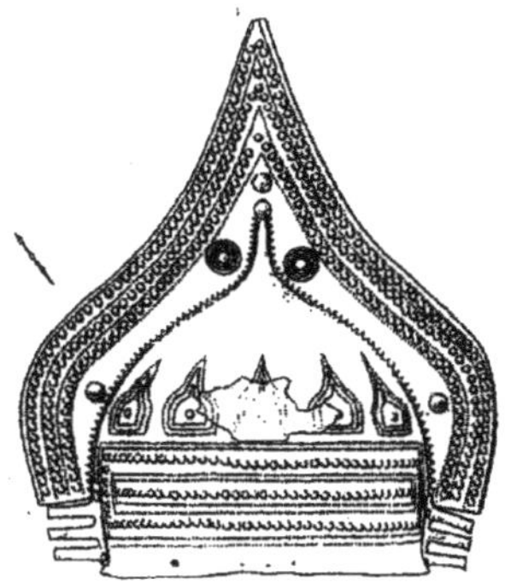

Casque décoré de Veio Quattro Fontanili. Tombe AA1 (d'après Veio...1970)

Gourde de Tarquinia Monterozzi. Tombe du Guerrier M1 (d'après Helbig 1874)

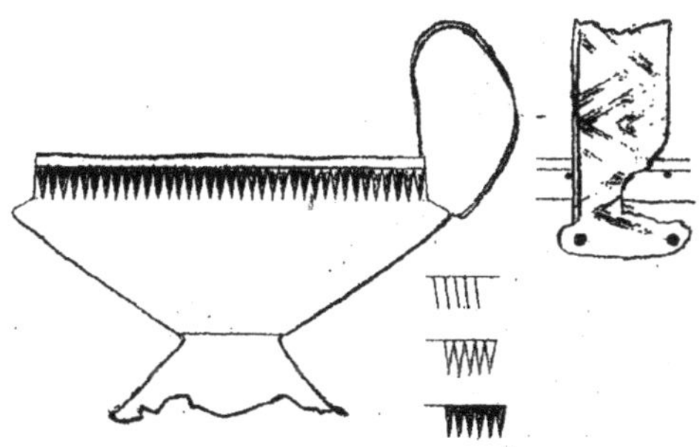

Coupe à une anse de Tarquinia Monterozzi. Probablement de la tombe dite M6

Epée d'antennes de Tarquinia Monterozzi. Probablement de la tombe dite M6

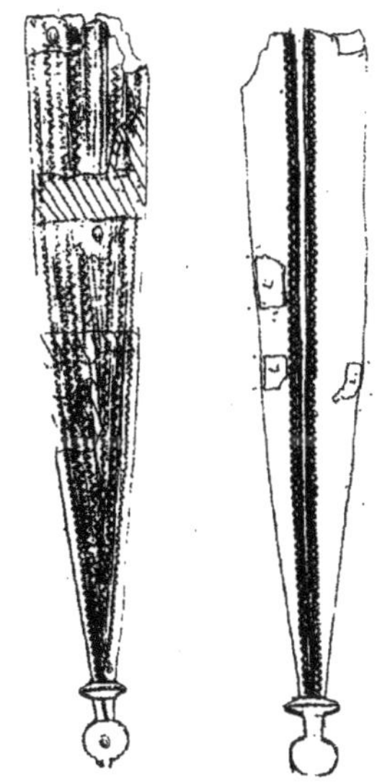

Fourreau d'épée conservé à l'Ashmolean Museum

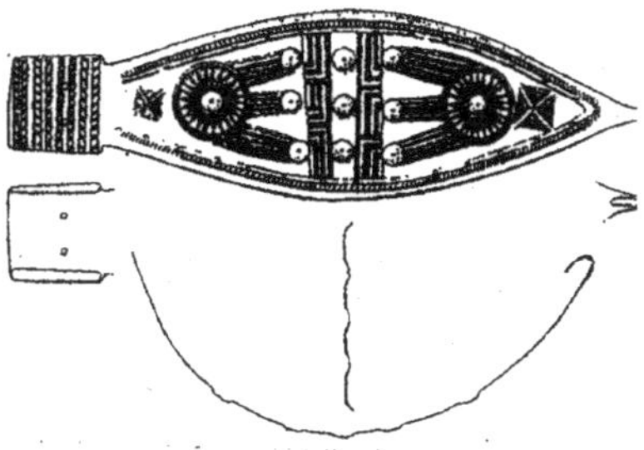

Ceinture décorée de Veio Quattro Fontanili. Tombe AA12A (d'après Veio...1963)

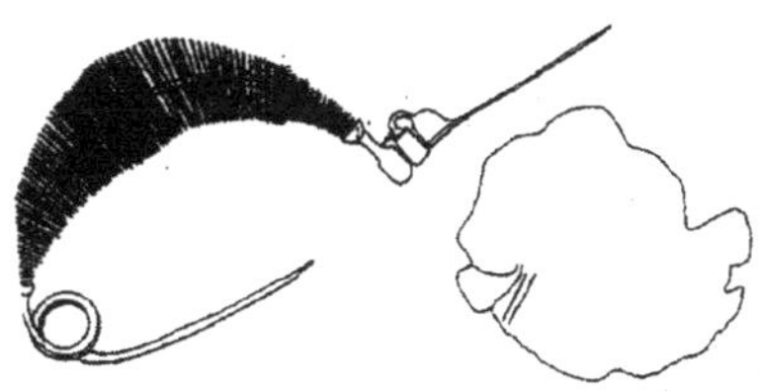

Fibule hérisson conservée à l'Ashmolean Museum

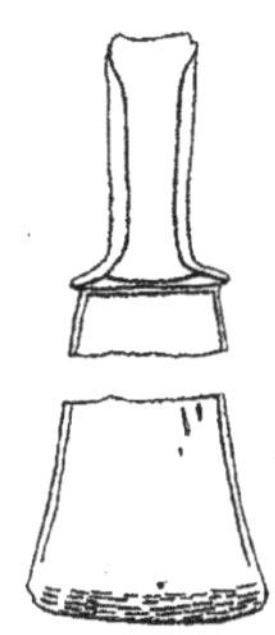

Hache villanovienne conservée au British Museum

Figure 3 : Principaux types d'objets étudiés (échelles différentes).

est donc particulièrement propre mais avec des teneurs d'étain telles que le travail de martelage ne peut être envisageable que pour un artisan très qualifié et maîtrisant parfaitement l'ensemble des opérations de fabrication de ces tôles, quelles que soient les pièces réalisées. L'urne de l'Ashmolean Museum atteste l'emploi d'un alliage de moins bonne qualité qui contient plus d'impuretés mais également de l'étain en moindre quantité. Le matériau ainsi obtenu permet un travail de déformation plastique moins délicat que celui des objets déposés dans les tombes de Tarquinia.

Ces différences dans les teneurs en étain entre un lieu et un autre se retrouvent également pour d'autres types et géométries de pièces (tableau 3 et fig. 2). Les petites pièces à géométrie complexe comme les fibules ont des teneurs à Tarquinia entre 10 et 15% d'étain alors que celle de l'Ashmolean Museum atteint au maximum les 7% pour le corps de la pièce et 5% pour l'un des disques de cette fibule hérisson. Simple coïncidence? Certes, l'échantillonnage est numériquement restreint. Il n'en reste pas moins que l'emploi d'un alliage plus riche en étain à Tarquinia semble se vérifier pour un mobilier dont la fabrication, le nombre mis au jour ou la valeur symbolique et marchande (y compris en poids de métal) sont différentes. Enfin, les mêmes remarques peuvent être formulées pour les deux fourreaux d'épée : là encore le fourreau conservé à l'Ashmolean affiche 7% d'étain tandis que son homologue de Veio est aux alentours de 12-13% selon la pièce analysée.

Ces remarques ne peuvent cependant pas être formulées pour la totalité des pièces analysées. Ainsi, les ceintures décorées mises au jour dans des tombes féminines représentent un lot quelque peu atypique (tableau 4). Les analyses ont, comme pour les tôles, été effectuées dans des mobiliers de Tarquinia, Veio et conservés à l'Ashmolean Museum. L'étain y est compris dans une fourchette restreinte entre 8,5 et 11% dans un matériau qui comporte parfois jusqu'à 1% de plomb sans que de basses teneurs ne se vérifient ici spécifiquement à l'Ashmolean Museum.

Enfin, le reste du mobilier (tableau 5) ne livre pas de résultat concluant concernant les teneurs en étain. Le seul exemplaire de rasoir qui a été analysé affiche un 10% d'étain standard qui n'autorise guère d'ample conclusion. Les pièces massives, haches et épées sont elles aussi de provenance trop inégalement réparties pour que les résultats soient significatifs : la seule hache de Tarquinia est en cuivre et n'a jamais servi, celles du British Museum sont de typologie variable avec des alliages binaires dans lesquels l'étain est assez peu élevé, de 5,5 à 10%, dans des matériaux qui peuvent contenir jusqu'à 2% de plomb. Enfin, seule une épée de Veio a pu être analysée. Elle contient seulement 6,5% d'étain dans la poignée comme dans la lame.

LES PROCÉDÉS DE FABRICATION ET LES ÉTATS MÉTALLURGIQUES

Dans le cadre d'une étude des procédés artisanaux, les analyses de composition ne peuvent être dissociées d'un travail d'examen des objets et du matériau proprement dit. En effet, sans s'étendre sur des questions méthodologiques, la microstructure garde en mémoire de nombreuses informations sur le type de travail réalisé dans le cadre de la fabrication de l'objet. Pour cette raison, il est donc indispensable de faire des prélèvements dans chaque pièce d'un même objet et de localiser de manière précise le prélèvement.

Aussi, ce type examen de la microstructure a été systématiquement effectué sur l'ensemble des échantillons préparés. Les résultats ont été confrontés à ceux qui ont été obtenus lors des analyses de composition. Les teneurs en étain sont alors loin d'être anodines. De manière schématique, plus le pourcentage d'étain est élevé et plus il est difficile de marteler le métal. En revanche, si l'on dispose d'assez d'étain, la coulée est facilitée, même dans le cas d'un matériau de «mauvaise» qualité. Cette aide à la fusion peut également être apportée par le plomb pour un matériau qui ne peut plus guère être martelé et pour une pièce d'usage très restreint.

Dans une logique technique, un matériau est donc en rapport direct non seulement avec un usage spécifique de l'objet mais aussi avec un travail de fabrication d'une pièce, tout au moins dans le cadre dans d'un artisanat de qualité. Aujourd'hui en France un bronzier ne s'aventurerait guère à employer un alliage à 14,5% d'étain s'il l'utilise pour une tôle martelée alors que ce même matériau –voire avec des teneurs supérieures en étain comme dans les bronzes à cloches– lui semblera adéquat pour une pièce coulée.

Qu'apportent ces connaissances relatives aux contraintes techniques inhérentes au matériau et quelles logiques peut-on dégager de cet artisanat villanovien ? Distingue-t-on des pratiques spécifiques ? Existe-t-il une adéquation technique entre le choix du matériau et celui de la fabrication? Ce sont autant de questions que l'on se pose au cours d'une telle étude.

Les tôles ont été travaillées principalement par martelage. En se basant sur l'épaisseur finale de la tôle et la morphologie des inclusions de sulfures on peut même dresser un ordre de grandeur de la réduction de l'épaisseur de l'ébauche au cours de cette déformation plastique (Piccardo et Pernot 1997 p. 46-47 pour la méthode, Lehoërff 1999a, p. 827-828 pour le détail des résultats de cette étude). Les tôles de vaisselle et d'armement (casque, fourreau) ont toutes subi des réductions d'épaisseur supérieures à 80% de l'ébauche initiale, avec une moyenne qui se situe plutôt vers 90% et des maxima à 100% (fourreau de l'Ashmolean Museum, fig. 4). Ces grandes déformations ont été réalisées quel que soit le pourcentage d'étain dans l'alliage. La gourde de Tarquinia qui contient le plus fort pourcentage d'étain (14,5%) a ainsi connu une réduction d'épaisseur de l'ordre de 90%. L'urne de Veio (échantillon 4 bis), avec 10% d'étain affiche une réduction d'épaisseur de 95% tandis que celle de l'Ashmolean Museum, dont les tôles des 3 échantillons s'échelonnent entre 7 et 9% d'étain, ont subi une réduction d'épaisseur de 85%. Ces estimations ne correspondent donc pas exactement à une logique technique telle qu'un artisan d'Europe occidentale l'envisagerait actuellement. Elles confirment par ailleurs

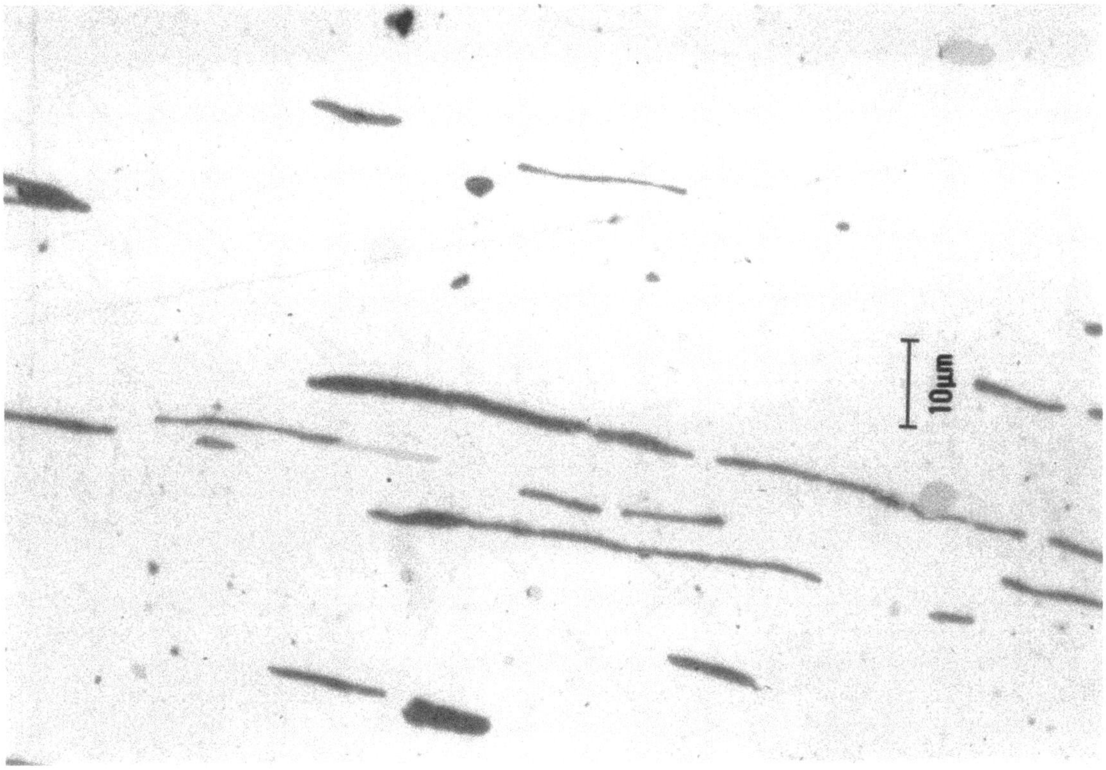

Figure 4 : Métallographie du fourreau de l'Ashmolean Museum (échantillons P1-275 et P2-275). Cette photographie avant attaque chimique met en évidence des sulfures très allongés qui suivent les courbures de la déformation plastique de la tôle à cet endroit. (photographie A. Lehöerff)

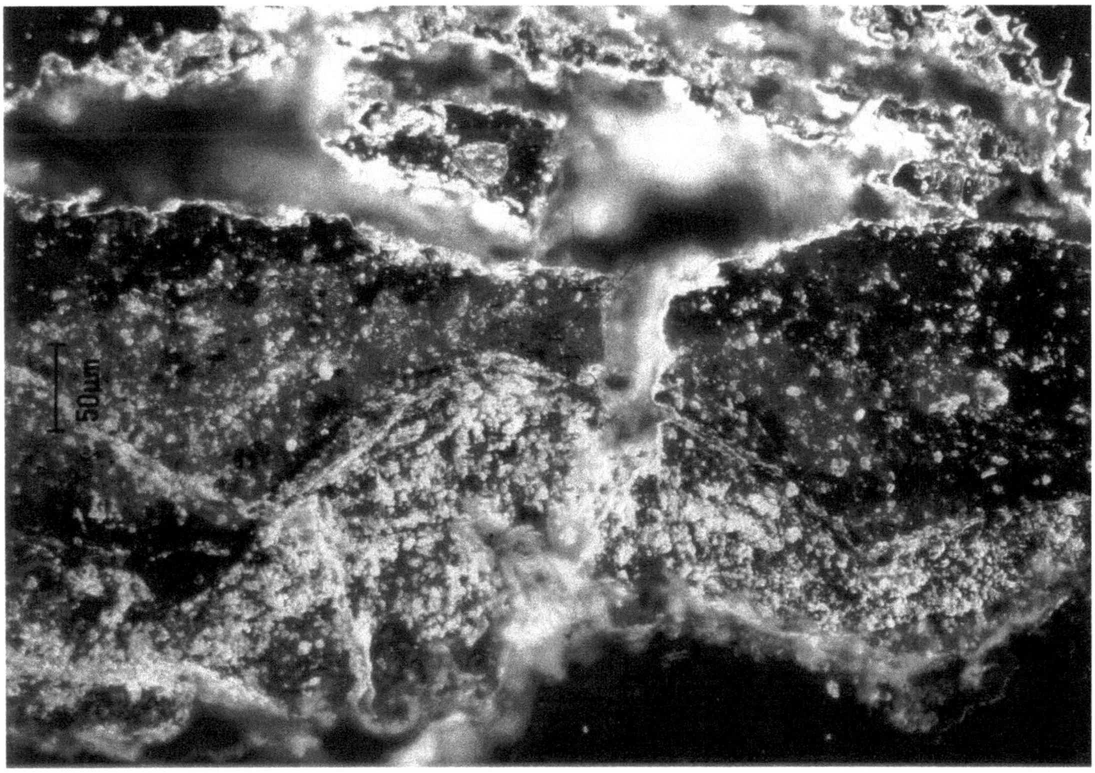

Figure 5 : Métallographie d'une ceinture ventrale de Tarquinia (échantillon P19-250). La prise de vue a été effectuée dans une coupe transversale d'une rainure du décor. Malgré l'abondante corrosion à cet endroit qui a interdit toute analyse de composition dans cet échantillon, on distingue le changement de direction des sulfures de forme peu allongée. (photographie A. Lehöerff)

l'excellence de cet artisanat villanovien en matière de tôlerie, en particulier à Tarquinia où de telles déformations plastiques sont inattendues dans des matériaux contenant plus de 11-12% d'étain.

Dans cet ensemble de pièces martelées, un groupe se distingue à nouveau, celui des larges ceintures féminines décorées. Avec des teneurs en étain inégales, et qui correspondent à une logique différente de celle qui a été relevée pour les autres mobiliers, cette catégorie de parure affiche des taux de déformation plastique inférieurs. Les pourcentages de réduction d'épaisseur de l'ébauche sont en effet de l'ordre de 60%, quelle que soit la provenance de l'objet.

Les états métallurgiques observés dans les microstructures sont très variables. Ils dépendent non seulement du type de fabrication des pièces mais également du traitement subi par l'objet après sa mise en service, en particulier au cours de la cérémonie funéraire. Lorsque la pièce comporte un décor (Lehoërff 1999 b) et que l'échantillon a été examiné à cet endroit précis (fig. 5), on constate qu'il s'agit d'un décor réalisé après la coulée pour les pièces retravaillées par martelage (ceintures, fourreaux, vaisselles), par ciselure et non par gravure, mais que quelques décors de fonderie existent également (haches). Aucun lien ne peut être établi entre les pourcentages d'étain et la présence ou le type de décor. En revanche, lorsque celui-ci est effectué par déformation plastique, le matériau est plutôt laissé dans un état déformé et non recristallisé après un recuit d'homogénéisation. Lorsque l'on observe un état recristallisé dans une plage sans décor (ceintures de Veio, certains échantillons de vaisselle à Veio), on peut supposer qu'il s'agit de l'avant-dernière phase du travail, avec une homogénéisation du métal avant la mise en place du décor. Certaines recristallisations du matériau appellent cependant d'autres explications liées à l'histoire des objets et en particulier au rituel funéraire lui-même. Certains mobiliers comme les fibules présentent en effet des recristallisations qui ne peuvent être liées ni à la fabrication de ces attaches ni à leur utilisation (ressorts en particulier trop mous et non élastiques). En revanche, dans ces incinérations villanoviennes, les fibules sont déposées avec le défunt sur le bûcher et se trouvent ainsi modifiées alors que la vaisselle ne subit visiblement pas le même sort au cours de la cérémonie funéraire. Enfin, il faut noter que certains bruts de coulée présentent des recristallisations partielles ou complexes présentées comme a-priori impossibles dans les traités de métallurgie contemporaine dans la mesure où il n'y a pas eu écrouissage et que le matériau est donc théoriquement une cristal parfait dans lequel aucun glissement ne s'est opéré [2].

CONCLUSIONS

Quelques conclusions s'imposent ici. On retiendra tout d'abord que la mesure en étain contenue dans un alliage n'a aucune valeur en elle-même mais n'acquiert une signification que lorsque ce chiffre est rapporté au type de pièce concerné et au travail de mise en forme retenu par l'artisan pour la fabrication de l'objet. Ensuite, la bonne corrélation entre les trois données –type d'objet/mise en forme/usage– confirme, sans surprise sur un tel mobilier, un véritable contrôle artisanal du pourcentage d'étain, rare, précieux mais utile à la fois pour la fabrication et/ou pour l'utilisation de l'objet.

[2] Les remarques faites ici sont le résultat d'observations réalisées sous le microscope métallographique. Les implications qu'elles entraînent dépassent cette présentation sur l'étain. Il s'agit d'un sujet de sciences des matériaux à part entière qui réclamerait des études et des expériences. Des études de Michel Pernot sont en cours.

Par ailleurs, dans le détail, les teneurs en étain ne varient pas comme on aurait pu s'y attendre selon des impératifs techniques liés aux procédés de fabrication pour tel ou tel type de pièce mais plutôt selon les lieux de fabrication. Même si le corpus reste numériquement trop limité pour des conclusions d'ampleur, ces choix de matériau (composition et propreté de l'alliage) en fonction des lieux ne sont-ils pas synonymes de choix individuels ou d'atelier ? A travers ce type d'examens et d'analyses, des perspectives sur l'artisanat métallurgique s'ouvrent donc, qui valent la peine d'être développées sur l'Italie centrale mais aussi pour d'autres aires culturelles. Car s'il s'agit de chiffres et de mesures, s'ébauche surtout ici la figure d'hommes souvent bien discrets derrière leurs réalisations, les artisans.

Addresse de l'auteur

A. LEHOËRFF
Maître de conférence
Université de Lille 3

Centre HALMA - UMR 8142
Domaine du Pont de bois
BP 149, F- 59653 Villeneuve d'Ascq Cedex FRANCE
Email : Lehoerff@univ-lille3.fr

Bibliographie

BARTOLONI, G., 1989, La cultura villanoviana. All'inizio della storia etrusca, Rome, 1989.

BIETTI SESTIERI, A. M.1996, Protostoria. Teoria e pratica, Rome, 1996.

GUIDI, A., PIPERNO, M. (dir.), 1992, Italia preistorica, Rome-Bari, 1992.

HELBIG, W., 1874, Scavi di Corneto dans Bulletino del Istituto di corrispondenza archeologica, 1874, p. 54-63 et 236-242.

HENCKEN, H., 1968, Tarquinia, Villanovans and early Etruscans, Cambridge, 1968.

ITALIA OMNIUM..., 1988, Italia omnium terrarum alumna. La civiltà dei Veneti, Reti, Celti, Piceni, Umbri, Campani e Iapigi, Milan, 1988.

LEHOËRFF, A., 1999 a, Le travail en laboratoire au service de l'histoire de l'artisanat métallurgique du début du premier millénaire avant notre ère en Italie. Quelques résultats sur des mobiliers de Tarquinia, Veio et des collections villanoviennes britanniques, dans Mélanges de l'École française de Rome, 111, 1999-2, p. 787-846.

LEHOËRFF, A., 1999 b, A. Lehoërff, La fabrication de fibules en Italie centrale entre le XIIe et le VIIIe siècle avant notre ère. Questions méthodologiques et premières études d'un corpus, dans M. Pernot, C. Rolley (dir.), Techniques antiques du bronze 2, Dijon, 1999 (centre de Recherches sur les Techniques gréco-romaines 15), p. 45-78.

PICCARDO, P., PERNOT, M., 1997, Studio analitoco strutturale di alcuni vasi celtici in bronzo, dans La metallurgia italiana 11, 1997, p. 43-52.

VEIO...1963, Veio (Isola Farnese). Scavi in una necropoli villanoviana in località Quattro Fontanili, dans Notizie degli Scavi 1963, p. 77-297.

VEIO...1965, Veio (Isola Farnese). Scavi in una necropoli villanoviana in località Quattro Fontanili, dans Notizie degli Scavi

1965, p. 49-263

VEIO...1967, Veio (Isola Farnese). Scavi in una necropoli villanoviana in località Quattro Fontanili, dans Notizie degli Scavi 1967, p. 87-280.

VEIO...1970, Veio (Isola Farnese). Scavi in una necropoli villanoviana in località Quattro Fontanili, dans Notizie degli Scavi 1970, p. 178-329.

VEIO...1972, Veio (Isola Farnese). Scavi in una necropoli villanoviana in località Quattro Fontanili, dans Notizie degli Scavi 1972, p. 195-384.

THE USE OF TIN AND LEAD IN BRONZE AGE METALLURGY

Margarita PRIMAS

Abstract: The technology of alloying is reviewed, based on evidence from central Europe, which is compared with the situation in the Aegean on one hand and western Europe on the other. The results do not support neither the hypothesis of independent invention nor the primary importance of availability. In a long-range perspective, more insight can be gained when the full spectrum of available metals is considered. This is exemplified with tin bronze and ternary lead-tin-copper alloys, which are key issues of innovation separated by a Millennium approximately.

Résumé: La technique de l'alliage est discutée en considérant à titre de comparaison des dates rélévantes d'Europe centrale avec celles des régions égéennes et atlantiques. Les résultats ne supportent ni l'hypothèse d'une invention indépendante ni celle de l'importance primordiale des ressources régionales disponibles. Dans une perspective de longue durée l'éventail complet des métaux devrait être considéré. On constate que l'alliage binaire cuivre-étain et l'alliage ternaire cuivre-étain-plomb manifestent deux phases innovateurs séparées d'un millénaire approximativement.

INTRODUCTION

Archaeologists face considerable problems when they pass from the level of empirical observations to the analysis of social processes. The adoption of innovations in metallurgy offers relevant examples, and among these, the decisive success of classical 5–10% tin bronze all over Europe is a much debated case.[1] From a technological point of view, the motivation was not immediately evident, for the preceding technology of arsenical copper fulfilled extant needs perfectly. Furthermore, it is not very likely that a shortage in suitable copper ores forced a change in technology. Tin bearing granite formations do not abound in Europe, which means that the shift from arsenical copper to tin bronze was connected with new strategies in exchange relations. In other words, tin bronze was not an option for secluded communities that detested mobility. The emergence of tin as an alloy has been explained in three different ways:

1. There are advocates of the availability hypothesis. They propose an independent invention in regions where tin and copper ore are associated. This is a testable hypothesis, and its evaluation could profit, theoretically, from the fact that the number of occurrences is limited.
2. The spread of a learning process has been proposed. In this option, the technology of alloying is considered as a non trivial invention that did not occur spontaneously.
3. Artefacts made of tin bronze could have been transmitted across a certain distance through one of the mechanisms of trade, gift giving, or kinship ties.

I propose to review the early use of tin and lead together, for these two metals share certain properties. Their melting point is low, and they form alloys with several non ferrous metals. In contrast to most copper ores, neither galena nor cassiterite or stannite show an attractive colour. Therefore, the presence of lead in the early Neolithic of the Near East[2] is more astonishing than the relatively late introduction of tin in the European metallurgy.

EARLY METAL ALLOYS IN EUROPE

Few bronze artefacts in the European spectrum of metal finds can be dated reliably into the first half of the 3rd Millennium BC. The highest tin content established for this period amounts to 7.62%. The object in question is a double-edged knife (fig. 1) from the Montenegin tumulus burial "Velika Gruda", where it was associated with an axe and a razor, both of arsenical copper.[3] Radiocarbon dates fix the grave at 4335±80 bp, i.e. 3080–2880 cal. BC, and the accompanying pottery is well in accordance with this date. Furthermore, additional alloyed metal items were found in a second burial mound nearby, "Mala Gruda": A tri-metallic dagger of gold, silver and 5.1% copper is just as unusual in the European repertory of the period as an axe cast of a silver-copper-alloy. The combination of three different alloys, all of them without parallels in the Adriatic region and in continental Europe as well, is distinctly more relevant than a single case could be. The most consistent explanation would be the provenance from one of the Near Eastern areas, where the technology of alloying was becoming familiar at this time.

The Montenegrin material raises the question of how much related evidence escaped the archaeologist's eyes? If we assume that other items of good 7–8% tin bronze were transmitted to Europe and recycled after use, low tin bronze artefacts coud be expected – and there are some finds. Hence, the third of the above mentioned explanations for precocious alloys should be considered further. Lead isotope

[1] Cf. the discussion by Pernicka 1998 and Craddock 1999.

[2] E.g. Yalçin & Pernicka 1999.

[3] Primas 1996, p. 100 ff. The metal analyses have been performed by J. P. Northover (Oxford).

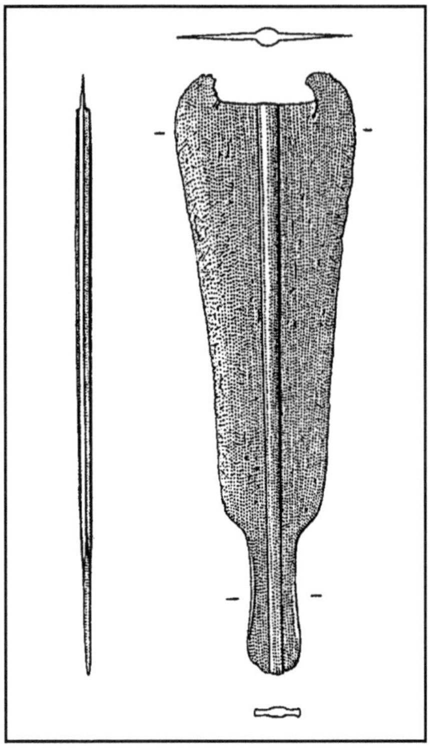

Figure 1. Velika Gruda, Montenegro: double-edged bronze knife, from Primas 1996. (1:2)

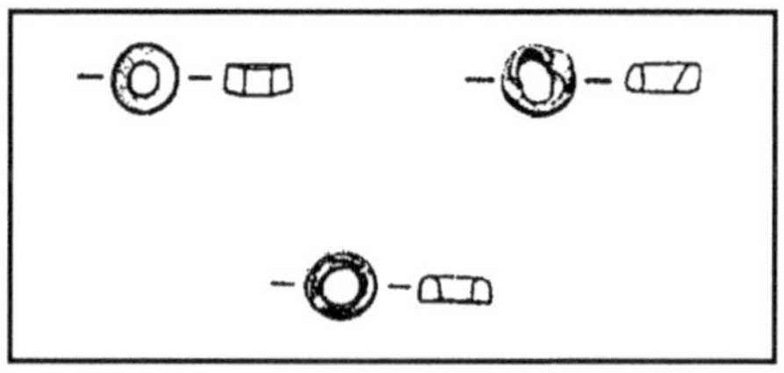

Figure 2. Wädenswil-Vorder Au, Switzerland: tin beads, from Eberschweiler 1999. (1:1)

determinations of the Velika Gruda dagger have been performed[4] and the results can be compared to the material of other regions, following the suggestions of Begemann et al. (1999). The situation changed during the last centuries of the 3rd Millennium BC. Innovation in metallurgy occurred now in distinct regional settings.

Table 1. The Progress of Tin Bronze.

BC cal.	Near East	Britain	Central Europe
3000–2500	Tin Bronze		
2200		Tin Bronze	selective use
1800			systematic use

According to a growing body of evidence, local copper and tin sources were found and exploited in western Europe since the last centuries of the third Millennium BC. In Britain a *systematic* alloying practice was established.[5] In central Europe, where large series of artefacts were analyzed, a time-lag became obvious. The *selective* use of tin bronze continued for another 500 years, but on a completely different level of production. Outlyers with a high tin content were regularly found in assemblages with otherwise low tin. This pattern can be observed in the rich hoards that characterize the so-called classical Unetice metallurgy in eastern central Europe. The Bresinchen hoard from eastern Germany is a goood example (Breddin 1969). Low flanged axes are the dominating type group, and heavy rings the second one, according to metal weight. The date of the assemblage is between 2000 and 1800 BC approximately. Bronze with more than 2% tin occurred very selective, as the following table clearly shows. Alltogether, only three objects in a total of 146 had more than 2% tin. Such was the situation in the neighbourhood of the tin bearing regions of Saxony and Bohemia.

Table 2. The Bresinchen Hoard.

Type	number	Tin: median value	maximum
Axes	103	0.1 %	0.45%
Daggers	8	1.1 %	2.0 %
Halberds	2	1.2 %	3.1 %
Torcs	10	0.5 %	1.2 %
Heavy Rings	22	0.23%	10 %

EARLY BRONZE AGE CONNECTIONS

In the Aegean area, metallurgy proceeded probably by the combination of two different incentives: One was the information transfer across the eastern Mediterranean, and the other the force of local conditions, as for instance the lead bearing ore bodies at Laurion. As a result, polymetallurgy flourished, with gold, silver, copper, and lead. In contrast to the availability of lead, tin had to be obtained from outside and appeared with a delay. As Muhly (1985) and others convincingly showed, a north-eastern source of the tin used in the Levant is likely, and for the Aegean, an eastern provenance is probable, too.[6] These assumptions have to be explored further with fieldwork and lead isotope analyses.

With regard to continental Europe, it was incidentally assumed that an incentive from the Aegean had triggered the course of innovation. If this were true, a similar spectrum of polymetallurgy could be expected in the Carpathian basin and the adjacent regions. This is clearly not the case, though plenty of lead ore in different associations would have been available. Though lead minerals did not remain completely unnoticed, their initial use remained restricted and brief. Along the Mediterranean coast of France, galena as well as metallic lead were found in certain chalcolithic contexts. The collective graves of the Fontbouisse group in the Languedoc are outstanding for their rich assemblages of beads in many different materials. According to Hélène Barge (1982), galena and lead as well as copper and gold were worked, in addition

4 Primas, 2002.

5 Needham, Leese, Hook & Hughes, 1989, termed the phase in question MA III. Needham 1996, fig. 2, proposed a revised system that included different cultural characteristics and an enhanced set of radiocarbon dates. The first metalworking tradition with consistently high tin levels started at 2300–2200 cal. BC.

6 Muhly 1985; Pernicka 1999.

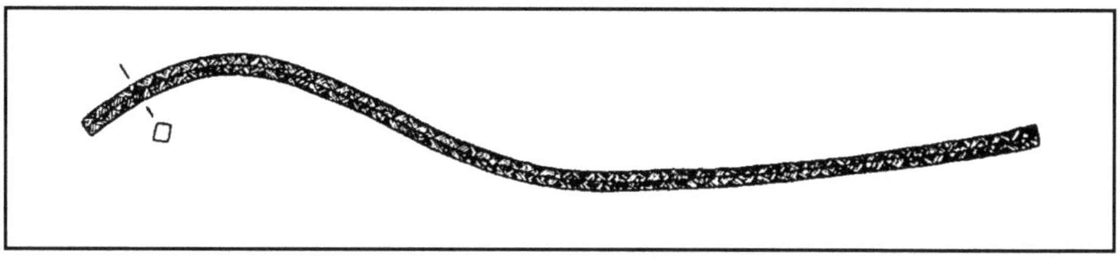

Figure 3. Auvernier, Switzerland: tin wire, from Rychner 1979. (1:2)

to various stone materials and shells. However, this tradition vanished in the course of the bell-beaker period, and no continuous production followed.

Lead was not used regularly in France or elsewhere in continental Europe until the Late Bronze Age (see below). On the contrary, it remained as rare as silver. The first of the before mentioned hypotheses, availability, is therefore not relevant for lead, and the question is, whether the case of tin was analogous. As already mentioned, the situation in the Aegean was completely different, which does not support the conception of a direct transfer of metal technology from this region to the continent.

I propose to have a closer look at the coeval networks of communication. Along the western facade of the continent as well as in large parts of central and south-western Europe, the bell-beaker phenomenon shaped the transition between the Chalcolithic and early Bronze Age. According to a tree-ring date from lake Zurich (Eberschweiler 1999), bell beaker pottery of the international or maritime style was present in the north-alpine area at 2450 BC, which is in good agreement with radiocarbon dates from the Netherlands. Though regional differentiation followed, there is evidence for an interregional levelling process in metal use. The spectrum of metal artifacts of this period included copper, gold and a few items with more than 1% tin in central Europe. In Britain, as mentioned before, the copper–bronze transition occurred at 2300/2200 BC (Needham 1996). Lead is absent and silver, which was the familiar associate to lead in the Aegean, remained exceptional until the Iron Age. I am inclined to conclude that a different pattern would have emerged, if the Aegean had been involved in interaction with continental Europe during this crucial period. Indeed, one of the characteristics of the so-called "Beaker Phenomenon" is its rather strict borderline in the Adriatic, in the eastern Alps and in the Danube valley around Budapest.[7]

TIN OR TIN ORE?

It has been questioned that alloying was performed with metallic tin. Though experiments[8] showed that cassiterite as well as stannite can be used for the production of good tin bronze, it is not likely that such procedures were very relevant in the practice of systematic and controlled alloying. Archaeologists became aware of the appearance of weathered tin during the last years, and the resulting learning process begins to amplify the spectrum of finds. Previously, only larger tin objects were recognized as such. They occurred mainly in Late Bronze Age contexts. In the north-alpine lake-side settlements of this period, metallic tin is not a rarity (cf. fig. 3). It occurred in different forms, though large ingots are missing, as almost everywhere, except on the well-known wrack of Ulu Burun. However, in settlement sites the same situation could be observed for copper, too. Hence, it is fairly obvious that fire based artisans rarely worked inside a village with wooden houses.

Recent finds showed that metallic tin was present in central and western Europe during the first centuries of the 2nd Mill. BC. The relevant finds are very small items, but the size of finds in the important source area of southern Britain is by no means more spectacular. Tiny ornaments prevailed, as for example the segmented beads from Buxheim in Bavaria (Möslein & Rieder 1997). I suspect them to be the tip of an ice-berg of unregistered evidence, for they were not detected during the excavation of the grave, but recognized later in the laboratory. They belonged to a neacklace of a probably female individual, according to the gender-related rites of burial that are typical for the Danube region. Neither the grave nor its furniture were otherwise exceptional.

The composition of a metal assemblage is sensitive to context. As many of the early finds of metallic tin come from graves and hoards, the dominance of ornaments is not astonishing. The well-known segmented tin beads from Exloermond near Odoorn in the Netherlands were associated with amber and fayence beads (Butler 1990). Hence, their chronological position cannot precede the date of fayence, which was widely used in central and western Europe from about 1800 BC onwards. Recently, several small, non segmented tin beads (cf. fig. 2) were uncovered in two settlements along lake Zurich. In one of the sites, Wädenswil-Vorder Au, the tin beads were found below the Early Bronze Age (EBA) layer with tree-ring dates around 1600 BC. The stratigraphic situation does not exclude that the beads came from the overlaying EBA layer.[9]

To sum up: The circulation of metallic tin in central and western Europe during the EBA is an established fact. The

[7] Cf. the detailed regional assessment in Strahm 1995.

[8] E.g. Herdits, Keen & Steinberger 1999.

[9] For Wädenswil-Vorder Au cf. Eberschweiler 1999 and Conscience 2000. In her thesis, Anne-Catherine Conscience was attempting a more consistent metalworking chronology for the Zurich area. Her untimely death stopped this promising work. She informed me of the second site with tin beads, Rapperswil-Technikum, where the context is Early Bronze Age, without a beaker settlement below. Furthermore, a gold wire was found there.

beaker tradition that preceded this period, had opened connections between Britain and the continent, and these connections could be influential for the tin bronze metallurgy of the western part of central Europe. The presence of tin and gold in one of the sites on Lake Zurich is suggestive, too. Gold had long before been appreciated as a rare commodity, but in the course of the EBA, its use multiplied. It may be significant that the earliest gold vessels of central and western Europe appeared at this moment. The well-known handled vessel from Rillaton in Cornwall and the beaker found at Eschenz on the upper Rhine[10] were produced in the same technology. As regards tin, I hope that excavators sharpen their eyes for the characteristics of tin and tin ore, for we need a more solid empirical basis.

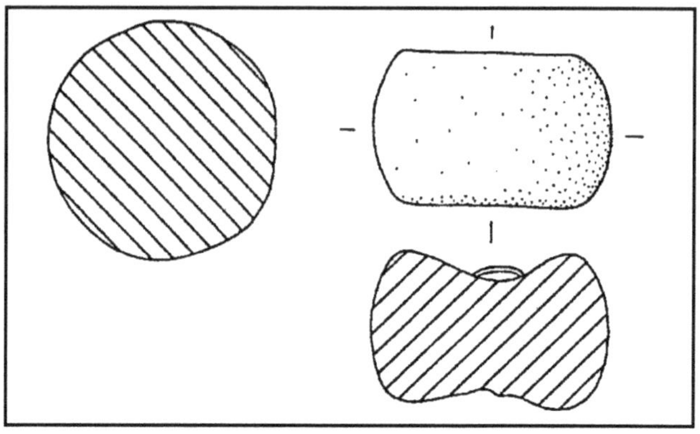

Figure 4. Auvernier, Switzerland: lead weight, from Rychner 1975. (1:2)

THE LATE ADVANCE OF LEAD

In a long-range perspective, more insight can be gained when the complete spectrum of available metals is considered. Hence, lead and iron are subjects of concern. They diversified the metal production of central Europe rather late, in the course of a major advance in metallurgy and metalworking techniques that occurred around 1300 cal. BC. Paul Craddock (1999) called special attention to the innovative aspects of the smelting technology that characterized the Late Bronze Age period in certain parts of the continent, for example in the alpine area. He proposed to discuss the new slag smelting process and the inception and spread of iron as connected phenomena. In addition to the data discussed in his paper, there is additional, supportive evidence, which shall be reviewed briefly.

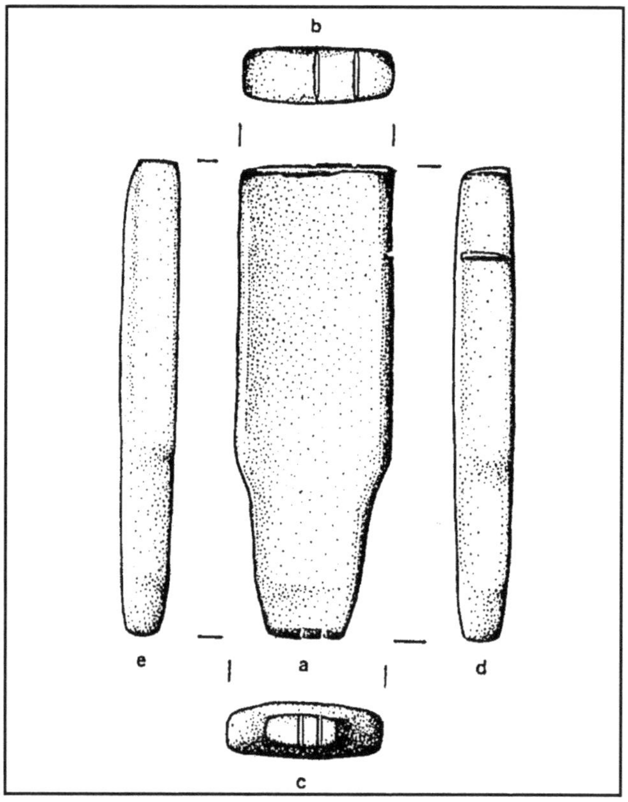

Figure 5. Lachen-Speyerdorf, Germany: anvil, high tin bronze, from Sperber 2000. (1:2)

The grave of a bronze smith uncovered at Lachen-Speyerdorf in western Germany and recently re-evaluated by Sperber (2000) shows paradigmatically the emergence of new standards in technology and in the organization of craft. By comparative evidence, the assemblage can be dated around 1100 BC. In addition to the common metal implements of the period, i.e. a razor and a knife, it contained an anvil cast of a ternary copper-tin-lead alloy (fig. 5) and part of a lead ingot of originally double-edged shape (fig. 6). Anvils and hammers with a high tin content appeared in connection with the advance of sheet bronze vessels and body armour, which are among the new products of the Late Bronze Age period, starting around 1300 BC on an interregional scale. The use of sheet bronze expanded rapidly, in tune with the requirements of ritual and elite display. The relevant techniques were based on a more thorough understanding of metal properties. Ternary alloys are not attested from the start, but followed about two centuries later. It seems that lead was not added with the intention of saving tin, but for sound technical reasons, and the same holds for the binary tin-lead-alloys mainly used for ornaments. In the lake-side settlements of the circum-alpine regions these new alloys are well represented. A balance weight of lead found at Auvernier

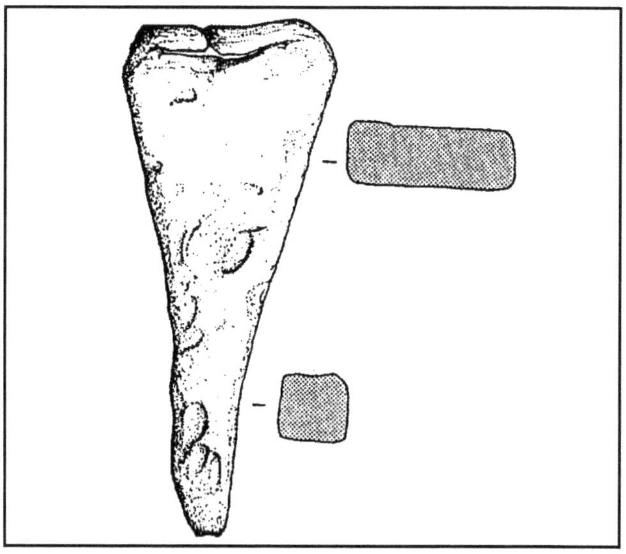

Figure 6. Lachen-Speyerdorf, Germany: lead ingot, fragment, from Sperber 2000. (1:2)

[10] Cf. Furger & Müller 1991. The beaker was found during works of construction nor far from the river. No docmentation of its stratigraphic position or context exists.

(cf. fig. 4) indicates that lead must have been available in large quantities, for it weighs 731.7 g. At present, the provenance of the lead is unknown and will be a nice test case for the availability hypothesis.

It is an interesting coincidence, to be followed further, that very high leaded bronze axes appeared coeval with the advance of iron working. In Slovenia, where an impressive bulk of material was analyzed,[11] these implements were found in contexts of the 9th century BC. At the same time, iron objects, mainly spear-heads and knives of probably local manufacture, have their debut in the north-alpine region.[12] Where the iron came from, is still an open question, for iron working does not imply iron smelting.

CONCLUDING REMARKS

The chronological basis strongly indicates a rapid transmission of alloying techniques to different regions outside the Near East and Anatolia. There is no unequivocal evidence for a second, independent invention. We should have a closer look at the potential sea routes of the bell-beaker period and certainly not forget overland traffic on horse-back either. The technology of systematically alloying copper and tin was probably transmitted to continental Europe not in a one-way road, but along existing networks of communication in the east and west. Polymetallurgy of Aegean type did probably not influence the process.

The situation changed significantly at the peak of the Late Bronze Age, ca. 1100 BC. Lead appeared on the scene, and ternary alloys of copper, tin and lead occurred in certain regions quite frequently. Leaded bronze is found in selected types of artefacts, and the control of alloying processes increased distinctly. Again, there is good evidence for an interregional network of communication, which was probably not based mainly on economic interests.

Author's address

Prof. Dr. Margarita PRIMAS
Universität Zürich, Abteilung Ur- und Frühgeschichte
Karl-Schmid-Strasse 4
CH-8006 Zürich SWITZERLAND
e-mail: primas@ggaweb.ch

Bibliography

BARGE, H., 1982, *Les parures du Neolithique ancien au début de l'âge des métaux en Languedoc*. Paris: CNRS.

BEGEMANN, F., KALLAS, K., SCHMITT-STRECKER, S. & PERNICKA, E., 1999, Tracing ancient tin via isotope analysis. In *The Beginnings of Metallurgy*, edited by A. Hauptmann, E. Pernicka, Th. Rehren, & Ü. Yalçin. Der Anschnitt, Beiheft 9. Bochum: Bergbau-Museum, p. 277–284.

[11] Cf. Trampuz-Orel & Heath 1999.

[12] Cf. the discussion and metallography in Tarot 2000, 16 ff.

BREDDIN, R., 1969, Der Aunjetitzer Hortfund von Bresinchen, Kr. Guben. *Veröffentlichungen des Museums für Ur- und Frühgeschichte Potsdam* 5, p. 15–56.

BUTLER, J. J., 1990, Bronze age metal and amber in the Netherlands I. *Palaeohistoria* 32, p. 47–110.

CRADDOCK, P. T., 1999, Paradigms of metallurgical innovation in prehistoric Europe. In *The Beginnings of Metallurgy*, edited by A. Hauptmann, E. Pernicka, Th. Rehren, & Ü. Yalçin. Der Anschnitt, Beiheft 9. Bochum: Bergbau-Museum, p. 175–192.

EBERSCHWEILER, B., 1999, Die jüngsten endneolithischen Ufersiedlungen am Zürichsee. *Jahrbuch der schweizerischen Gesellschaft für Ur- und Frühgeschichte* 82, p. 39–64.

FURGER, A. & MÜLLER, F., 1991, *Gold der Helvetier, Ausstellungskatalog*. Zürich: Schweizerisches Landesmuseum.

HERDITS, H., KEEN, J. & STEINBERGER, M., 1995, Wie kommt das Zinn in die Bronze? Ein Beitrag zur Experimentellen Archäologie. *Archäologie Österreichs* 6/1, p. 78–85.

MÖSLEIN, S. & RIEDER, K. H., 1997, Zinnperlen aus einem frühbronzezeitlichen Grab von Buxheim. *Das Archäologische Jahr in Bayern*, p. 68–70.

MUHLY, J. D., 1985, Sources of Tin and the Beginning of Bronze Metallurgy. *American Journal of Archaeology* 89, p. 275–291.

NEEDHAM, S., 1996, Chronology and periodisation in the British Bronze Age. In *Absolute Chronology. Archaeological Europe 2500–500 BC*, edited by C. Randsborg. *Acta Archaeologica* 67, Suppl. Vol. I. Copenhagen: Munksgaard, p. 121–140.

NEEDHAM, S. P., LEESE, M. N., HOOK, D. R. & HUGHES, M. J., 1989, Developments in the Early Bronze Age metallurgy of southern Britain. *World Archaeology* 20, 3, p. 383–402.

PERNICKA, E., 1998, Die Ausbreitung der Zinnbronze im 3. Jahrtausend. In *Mensch und Umwelt in der frühen Bronzezeit Europas*, edited by B. Hänsel. Kiel: Oetker/Voges, p. 135–147.

PRIMAS, M., 1996, Velika Gruda I. Tumulus burials of the early 3rd Mill. BC in the Adriatic: Velika Gruda, Mala Gruda and their context. *Universitätsforschungen zur Prähistorischen Archäologie* 32. Bonn: Habelt.

PRIMAS, M., 2002, Frühe Zinnbronzen in Mittel- und Südeuropa – Zum Stand der Integration naturwissenschaftlicher und archäologischer Daten. In *The Beginnings of Metallurgy in the Old World*, edited by M. Bartelheim, E. Pernicka, R. Krause. Rahden/Westf.: Leidorf, p. 303-314.

RYCHNER, V., 1979, L'âge du Bronze final à Auvernier (lac de Neuchâtel, Suisse): typologie et chronologie des anciennes collections conservées en Suisse. *Cahiers d'Archéologie Romande* 15–16. Lausanne: Bibliothèque Cantonale.

SPERBER, L., 2000, Zum Grab eines spätbronzezeitlichen Metallhandwerkers von Lachen-Speyerdorf, Stadt Neustadt a. d. Weinstrasse. *Archäologisches Korrsepondenzblatt* 30, p. 383–402.

STRAHM, Ch., 1995, *Das Glockenbecher-Phänomen*. Ein Seminar. Freiburg i. Br.: Institut für Ur- und Frühgeschichte.

TAROT, J. (2000) Die bronzezeitlichen Lanzenspitzen der Schweiz, unter Einbeziehung von Liechtenstein und Vorarlberg. *Universitätsforschungen zur Prähistorischen Archäologie* 66. Bonn: Habelt.

TRAMPUZ-OREL, N. & HEATH, D. J., 1998, Analysis of Heavily Leaded Shaft-Hole Axes. In *Mensch und Umwelt in der frühen Bronzezeit Europas*, edited by B. Hänsel. Kiel: Oetker/Voges, p. 237–248.

YALÇIN, Ü. & PERNICKA, E. (1999) Frühneolithische Metallurgie von Ahikli Höyük. In *The Beginnings of Metallurgy*, edited by A. Hauptmann, E. Pernicka, Th. Rehren, & Ü. Yalçin. Der Anschnitt, Beiheft 9. Bochum: Bergbau-Museum, p. 45–54.

IRON AGE TIN IN THE ORIENTAL ALPS

Alessandra GIUMLIA-MAIR

Résumé : Aucun dépôt d'étain n'est connu dans la zone alpine ; néanmoins, les données archéologiques suggèrent que les artisans de l'Âge de Bronze et de Fer n'avaient pas des problèmes d'approvisionnement d'étain. Jusqu'au présent, dans la région autour des Alpes orientales, d'étain métallique n'a pas été découvert, mais des analyses de groupes d'artéfacts des sites différents montrent clairement que les forgerons ont été capable de différencier leur production et de choisir l'alliage plus convenable aux objets et aux outils de fonction diverse, c'est-à-dire qu'ils avaient d'étain disponible à ajouter aux alliages bronzes. C'est particulièrement évident dans le cas d'objets caractéristiques. Ce texte évalue les résultats des analyses et suggère des éventuelles routes d'où venaient les différents métaux.

Abstract: No tin deposits are known in the Alpine area, however the archaeological record suggests that Bronze Age and Iron Age artisans did not have particular problems with the supply of tin. Up to now, in the region around the Oriental Alps no metallic tin was found, but several analyses of groups of artefacts from different sites clearly demonstrate that the smiths were able to differentiate their production and to choose the most suitable alloy for objects and tools of various function, i.e. that they actually possessed tin to add to their different bronze alloys. In the case of characteristic items this is particularly evident. The paper evaluates analysis results and suggests possible routes from which the different metals came.

INTRODUCTION

The long history of tin in ancient metallurgy is at the moment largely unknown and has still to be written. Despite many archaeometallurgical researches, carried out by several scholars in the last decades, we still have an extremely sketchy idea of the ways and the places, from which ancient civilisations got their tin and on the trade routes of prehistory and early history.

Even in the case of relatively late periods, such as the Hellenistic or the Roman times, on which we can collect much information from written sources, several questions on the tin supply and trade are still open.

Most of the studies and researches on tin discuss the earliest period of the use of this metal in metallurgy and in particular the question of the provenance of early tin, which, at least from the middle Bronze Age, has been used in huge amounts all over Europe, the Near East and beyond.

As the earliest bronzes we know come from the East, common sense tells us that the source of tin should not be looked for in the traditional tin-supplying sites of historical times in the West, i.e. Cornwall or Portugal. It is well known that the only large deposits of tin on the East of the Mediterranean lay in the regions around Northern Afghanistan and Tajikistan. It is common knowledge too, that also other goods, coming from those areas, were traded at very early times and are widely spread all over Mediterranean countries. The most striking example and the one which is always mentioned in connexion with the discussion on the origins of early tin, is of course the great quantity of lapislazuli, which is to be found as ornamental stone on luxury objects, jewellery and trinkets of different civilisations since the 5th mill.BC[1]. Of course the Afghanistan region of Badakshan (including Gorno-Badakshan) and the Chaghai district in Beluchistan are the only possible land of origin for early lapislazuli[2].

As recently pointed out elsewhere[3], other indicative items, which up to now were not much taken into consideration by scholars working on this topic, are the distinctive tapered vessels made of carved chlorite and often decorated with turquoise, which were traded from the very same areas and from Iran, through Syria to the Mediterranean, to the Indo valley and further, already in the second half of the 3rd Millennium BC, at the time in which bronze begins to become common. M.L.Carter[4] discusses the routes, through which these magnificent chlorite vessels were brought to other civilisations.

The same routes could also, without any difficulties, have been used also for the tin trade.

Up to now, the origin of the word *kassiteros* – tin in Greek – has not been considered in connection with the origins of tin. As Makkay[5] pointed out, the names of metals, as well as the myths and the rituals connected with the elitary world of metallurgy, are generally extremely ancient and are mostly almost religiously kept intact in their archaic form through many Millennia. The word *kassiteros* is not Greek, but comes from ancient Elamite and is also found in the Babylonian word *kassi-tira* and in the Sanscrit *kastiram*[6]. The origin of the word seems therefore to suggest a Transcaucasian provenance of the metal. Elam would of course only be one

1 CASANOVA 1995, 17.

2 TOSI 1974; CASANOVA 1993; 1995, 15-20; BOUQUILLON & POIROT 1995, 33-34.

3 GIUMLIA-MAIR 2002.

4 CARTER 1997, p.325-334.

5 MAKKAY 1992.

6 LIDDEL & SCOTT 1996, s.v.

of the trading points on the long tin route, beginning perhaps at the feet of the Pamir[7].

The provenance of early tin from the Afghanistan area is by no means confirmed, however recent researches in the area have shown traces of mining, The most impressive results are for sure those of the excavations of Weissgerber and Cierny[8], which have brought to light the galleries and shafts of enormous mines. However, at least up to now, evidence that the tin deposits in this area were indeed the mines from which the bronze workers of the 3rd millennium received their tin are still missing.

In the last decade the possibility of the occurrence of tin in the Anatolian area have also been discussed[9] and Ottoman archives[10] also record tin mining in Anatolia.

The provenance of the earliest tin remains an elusive problem and there is still a lot of work to do on this subject.

This paper deals with a much less impressive topic than the quest for the oldest tin in the Ancient World and is limited to a circumscribed area and to a much later and better defined period. In the last decades several studies on archaeological copper-based finds from various sites around the Oriental Alps were carried out, but up to now, no special attempt has been done to put together the data regarding the usage of tin and what can be inferred from them about tin trade.

The present study discusses the use and the trade of tin in the Oriental Alps in the time between the Final Bronze Age and the beginning of the Iron Age, and it is based on the data which can be drawn from metallurgical analyses carried out on archaeological material from this region in the last years. The aim is rather modest, nevertheless the task is not an easy one. As we will see, the picture reconstructed through archaeological data and chemical analyses, seems to be rather complex.

MATERIAL AND SITES

1) Variano

The earliest group of objects, recently analysed in this area, comes from a hillfort at Variano, in the Friuli plain, not far from Udine, on a site, which had been inhabited from the Middle Bronze Age to the Early Iron Age. The excavations, carried out by P. Càssola Guida and S. Corazza since 1997, and still ongoing studies, have shown that the earlier fortified village structure, dated to the 16th – 15th c. BC, presumably a terraced site surrounded by a wooden palisade, has been completely restructured in the Final Bronze Age. For this, huge amounts of earth and gravel were brought to the site and the fortified settlement was built at a higher level with a new, massive and very impressive earthwork[11].

The discovery of the hillfort is of particular importance, as the well preserved remains of the building allow, for the first time in Friuli, the reconstruction of the life inside of what seems to have been a well organized community. The outline of the houses can easily be recognized by means of the many post-holes and seem to be of large size. For example one of the best preserved buildings has a length of 11 m and a width of 5 m and shows interesting and rather complex architectural features. The function of the different rooms can be recognized and attributed to different usages. The finds of pottery and food remains give us a good idea of what was going on in the structure when it was inhabited[11]. In several stratigraphic unities, but especially in 44, several fragments of copper based objects belonging to the later period, i.e. to the final Bronze Age, were brought to light. The group is rather small and, because of the importance of the site for the understanding of the role played by the settlements in the Upper Friulian Plain in this period, it has been possible to carry out chemical analyses on all items in good conditions. XRF, a non-destructive method, was employed on all pieces as first check. Each object was superficially cleaned and analysed on different points (at least 5) to get a good average and avoid serious errors due to the presence of corrosion or segregation. The results were then cross checked with an analysis by SEM/EDX on the same cleaned and numbered spots. Finally, where possible, as a last control, several samples were analysed by ICP.

The finds from Variano are the only group of objects, discussed in this paper, which were not all analysed by destructive methods, i.e. by dissolving small samples in acids for analysis by ICP or AAS. Therefore only the data on major and minor elements are reliable. The trace elements are to be considered only indicative and are omitted from the table.

The data resulting from the analyses have shown the use of copper and of a variety of copper-based alloys (Histogram 1, Variano). The object with the highest tin content (13% Sn) is a fragmentary wheel-shaped pendant (inv.Nr.432889; Fig.1) of the type found in Northern and Central Italy, France, Switzerland and Slovenia from the Middle Bronze Age to the Early Iron Age[12]. From other analyses, but also from the silvery grey colour of many of these objects, we know that their tin content is in most cases very high, up to 20%. The example from Variano follows this broad trend.

The next highest tin content is that of a fragmentary knife with around 8% of tin, however, from other analyses of similar Alpine material we know that these classes of objects have in this period quite erratic compositions[13]. The second object containing around 8% of tin is a fairly long, well worked and polished fragmentary pin with a shiny and compact patina. The copper has been accurately refined.

7 GIUMLIA-MAIR 2002, 111.

8 see this volume.

9 YENER 2000.

10 KAPTAN 1983; 1984; 1986.

11 CASSOLA GUIDA & CORAZZA 2000.

12 cfr HEATH et al.2000.

13 TRAMPUŽ OREL 1996, p.186, fig.2.

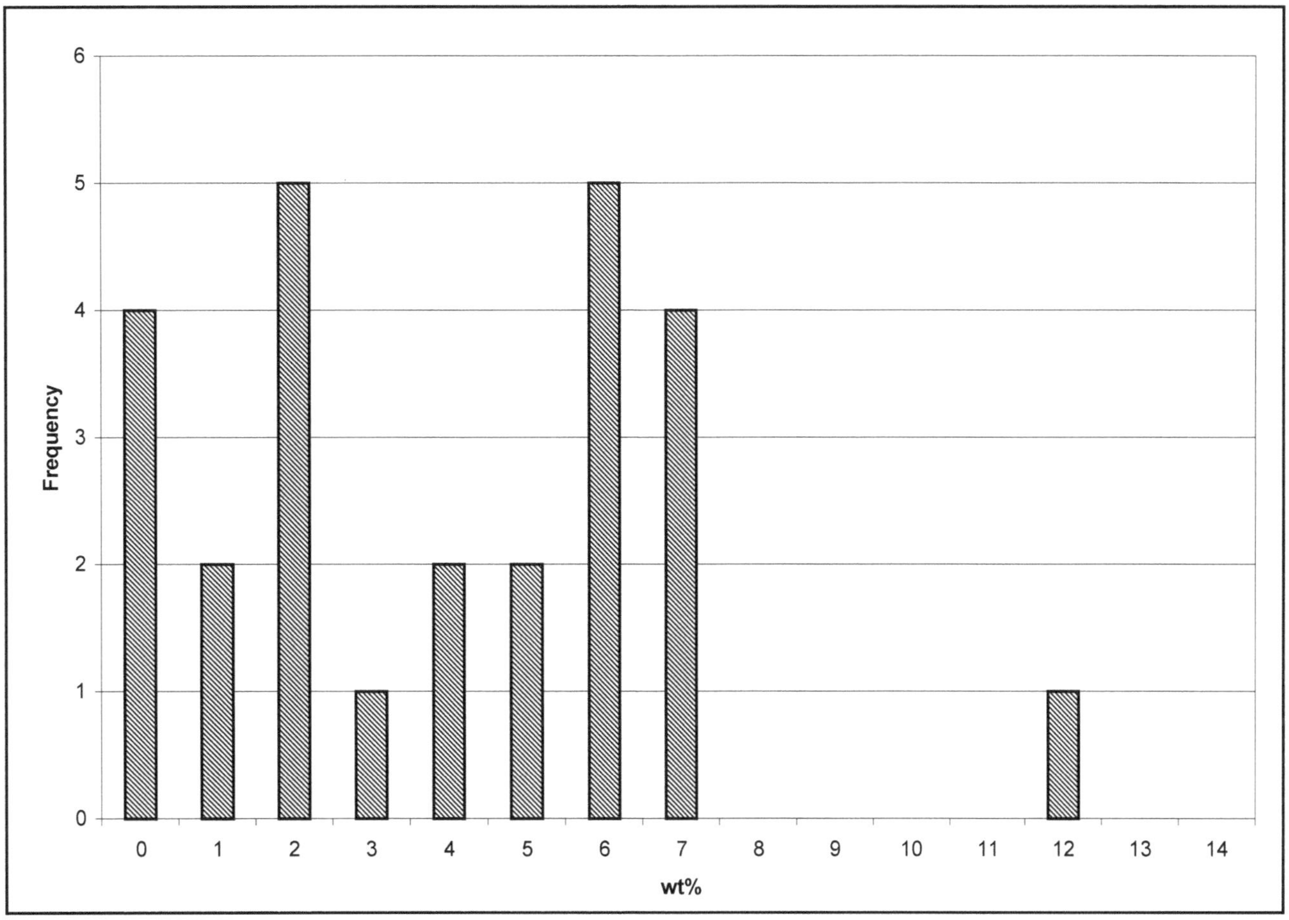

Histogram 1. Variano - Sn content

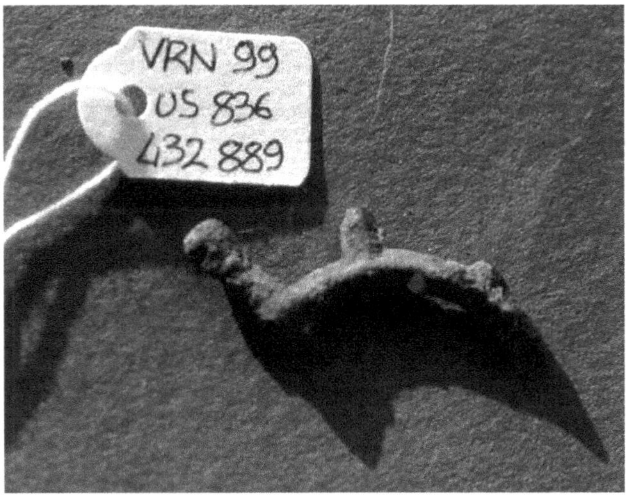

Figure 1. Wheel-shaped pendant fragment from the Variano hillfort. These distinctive ornamental objects are particularly fragile, because they contain very high tin percentages, from around 13% to around 20%. (Foto Giumlia-Mair)

Several drippings and some small and fragmentary rods show an Sn content around 7%, the same composition of 2 small "ingots". These finds suggest that some simple metallurgical activity was carried out on the hillfort.

Several small objects, which represent the most frequent finds, such as a fragmentary sickle, awls, points - small tools, presumably used for working organic materials such as wood, bark, leather and skins - show irregular, but very low tin contents, while particularly thin objects, such as needles or wire rings, are made of unalloyed copper. Some drippings of unalloyed copper were also identified. This means of course that tin was, in some form, available to the artisans of the site and that they added it sparingly to their copper.

The presence of two small ingots of bronze, containing 7% of tin, similar to those found during the excavation of the Iron Age workshop remains of the nearby site of Pozzuolo del Friuli[14], is rather interesting. The use of "standard bronze" with around 7% of tin, produced by recycling bronze scraps of different composition, has often been hypothesised for later times. It is of course rather difficult to prove that there really was an organized trade of recycled bronze: the ingots might also have been produced by the local smith for his own production, but the identification at Variano of two ingots with exactly this composition and the fact that only non-descript metal drippings and simple rods show a similar composition, but no "real" object has been produced with a comparable alloy, suggest that the employment of standard bronze as raw material might have been common also in the final Bronze Age.

As it is the case with all groups of final Bronze Age objects analysed up to now, none of the finds shows a lead content, which might suggest a deliberate and efficient use of this metal. From the analysis data obtained from local finds, before

[14] GIUMLIA-MAIR 1998b; 1998c p.53.

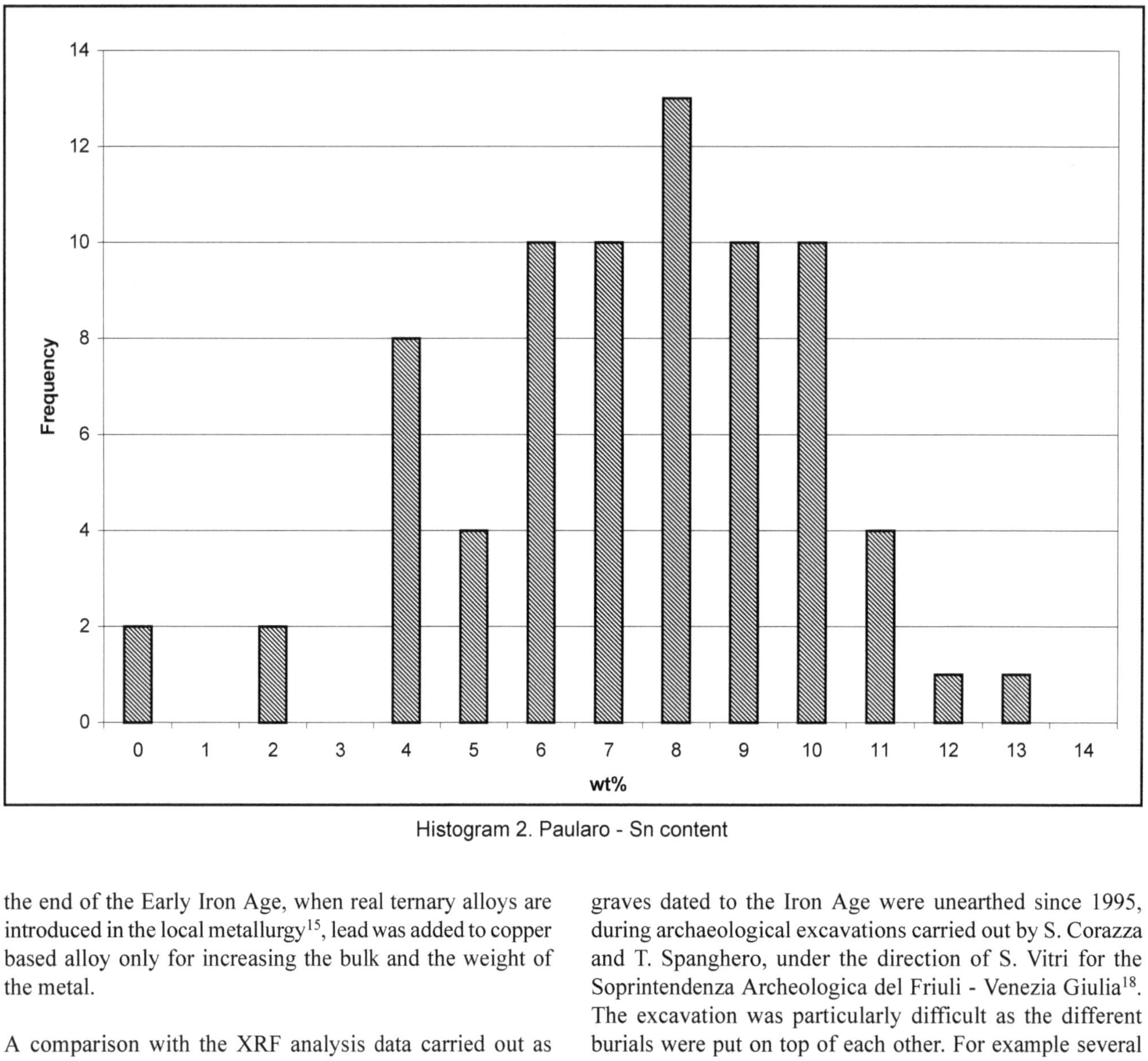

Histogram 2. Paularo - Sn content

the end of the Early Iron Age, when real ternary alloys are introduced in the local metallurgy[15], lead was added to copper based alloy only for increasing the bulk and the weight of the metal.

A comparison with the XRF analysis data carried out as preliminary check on the drillings taken from the bronze finds from the broadly contemporary site of Moscosi di Cingoli in the region Marche[16], i.e. in the area later inhabited by the Picenian, seems to indicate a much wider employment of bronze (and no unalloyed copper) in analogous small objects such as decorative pins, awls and similar small tools. Also the single objects from Moscosi seem to show higher and more regular tin contents, however proper bulk analyses have not been carried out yet, and the researches an the studies on the material are still ongoing and might perhaps reveal more different frequencies of composition in the copper-based finds from the region Marche.

2) Misincinis

Another recently analysed group of copper-based finds from this area are the finds from the necropolis of Misincinis at Paularo in Carnia[17] (Histogram 2, Paularo). Around 120 graves dated to the Iron Age were unearthed since 1995, during archaeological excavations carried out by S. Corazza and T. Spanghero, under the direction of S. Vitri for the Soprintendenza Archeologica del Friuli - Venezia Giulia[18]. The excavation was particularly difficult as the different burials were put on top of each other. For example several clusters of 107 graves were excavated in a section of 27 square meters only. Research and studies are still ongoing.

The site was discovered by chance in the valley of the river But, a narrow valley, perpendicular to the Alps, which goes up to today's Austrian border and is connected with the Gail - Valley in Carinthia (Austria) by a comfortable path through Ramaz, Mt. Lodin and the Val di Puartes Pass (where Venetic inscriptions have been found), and through the Sella Pizzul and the Lanza Pass with the valley of the river Fella. In the Iron Age this valley was apparently an important North-South route. No settlement belonging to the necropolis was discovered yet in the area and it might be located under the modern village. The finds can be compared with objects found in other regions in both directions, for example Certosa type fibulae point to the South, but the so-called Ostalpine Tierkopffibulae (Eastalpine animal head fibulae) show clearly their Northern influence. Some of the objects can be found in the large area, which goes from the Veneto region (occupied by the civilization of Este), to Eastern Slovenia, and from Trentino, through the Austrian regions of Tyrol, Carinthia and Styria and on both sides of the river Inn up to the Danube

[15] GIUMLIA-MAIR 1998b.

[16] SILVESTRINI 1991; 1992; 1994; SILVESTRINI & PIGNOCCHI in print.

[17] VITRI et al. 2001.

[18] CORAZZA & VITRI 2001.

in Lower Austria. Southern influences in the style of the objects are more recognizable in the ancient phase, between the 8th and the 6th c. BC, while in the second phase, from the 5th to the 4th c. BC the characteristics of the material seem to indicate the prevalence of Northern Alpine influences. These are still rather noticeable in the last centuries BC, in the romanization period, when the finds show the typical La Tène shapes.

In several graves were present containers, made of clay or of wood, for the cremation remains and a wooden structure, which held the grave walls. Only some of the burnt bone remains were collected from the cremation fire and put into the grave, together with the grave gifts. It is quite interesting to note that the wood for the cremation pyre was chosen depending on the sex of the dead: men were burnt on beech wood and women on apple- or pear wood. Among the cremation remains the ornaments worn by the deceased during the funerary ceremony can be clearly distinguished from other grave gifts, because of their generally bad condition. The gifts consist mainly of personal ornaments of copper based alloys, such as fibulae of different shapes, belt fittings, bracelets, ornaments for the hair, collars, but also glass beads and iron implements, such as knives, fibulae, horse trappings, belt-hooks etc.[18]. The objects made of iron are now also being studied.

Among the copper-based small decorative objects for personal use, found in the graves as grave gifts, it is relatively easy to distinguish imported material from local material, because of typological and stylistic differences, but the analyses have shown that the locally produced objects have also a different composition[19]. The items with the highest tin content (up to around 13 % Sn) are the so-called fibulae serpeggianti – serpentine – or snake fibulae, used by men– which are produced in one single piece and, being very hard and not elastic, were most probably used to fix the heavy folds of wool garments. The fold-stopper is a movable bead, cast with the addition of lead. Fibulae of this type are known all around the Alps and all examples in the region, analysed up to now, show a very similar composition. In this particular case they were most probably imported to Paularo from the East, i.e. from the *facies* of S.Lucia/Most na Soči.

Certosa type fibulae show a rather regular composition, with around 10 % of tin and are presumably imported from the Veneto area. One of the examples has been repaired by using an alloy of tin and lead in almost equal parts and only very little copper, characterized by a very low melting point, without doubt to avoid damage to the object during the application of the solder.

Fibulae of the type Castellin di Fisterre, were also most probably imported.

Some of the objects are without any possible doubt locally produced. A good example are ring pendants, decorated with 7 globules. This kind of decorative object is known from

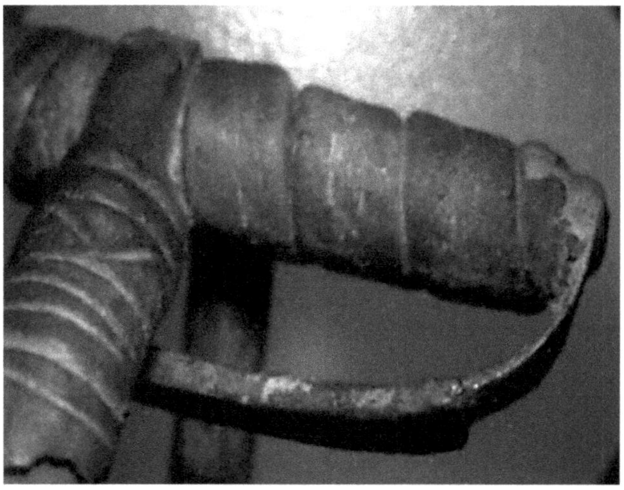

Figure 2. Detail of the spiral of a Paularo Type fibula at the microscope. The spiral is formed with a flat stripe of metal. The copper used by the local artisans contain relatively high trace elements. (Foto Giumlia-Mair)

Figure 3. Detail of the catchplate of a Paularo Type fibula at the microscope. The finial represents a stylized human face. (Foto Giumlia-Mair)

other sites, but elsewhere the number of globules is different. The variety with 7 globules is present only at Paularo and it can be safely concluded that they were a local product. The composition of the ring pendants resulted to be very similar to that of the Paularo type fibulae, i.e. of a further typical local item (Fig. 2 and 3). The copper used for the production of the alloys is not particularly well purified and the composition is rather irregular and not especially well prepared. The tin content is somewhat lower than in other objects of this kind.

Some of the items found in the necropolis of Paularo were also analysed by PGAA at the Nuclear Centre of Budapest and by this method low traces of mercury were identified in the objects. It is important to note that in the Slovenian Oriental Alps, in Idrjia, there is the second largest mercury deposit in the world and the traces of this element in the copper-based artefacts from Paularo might indicate a

[18] CORAZZA & VITRI 2001.

[19] VITRI et al 2001, 415.

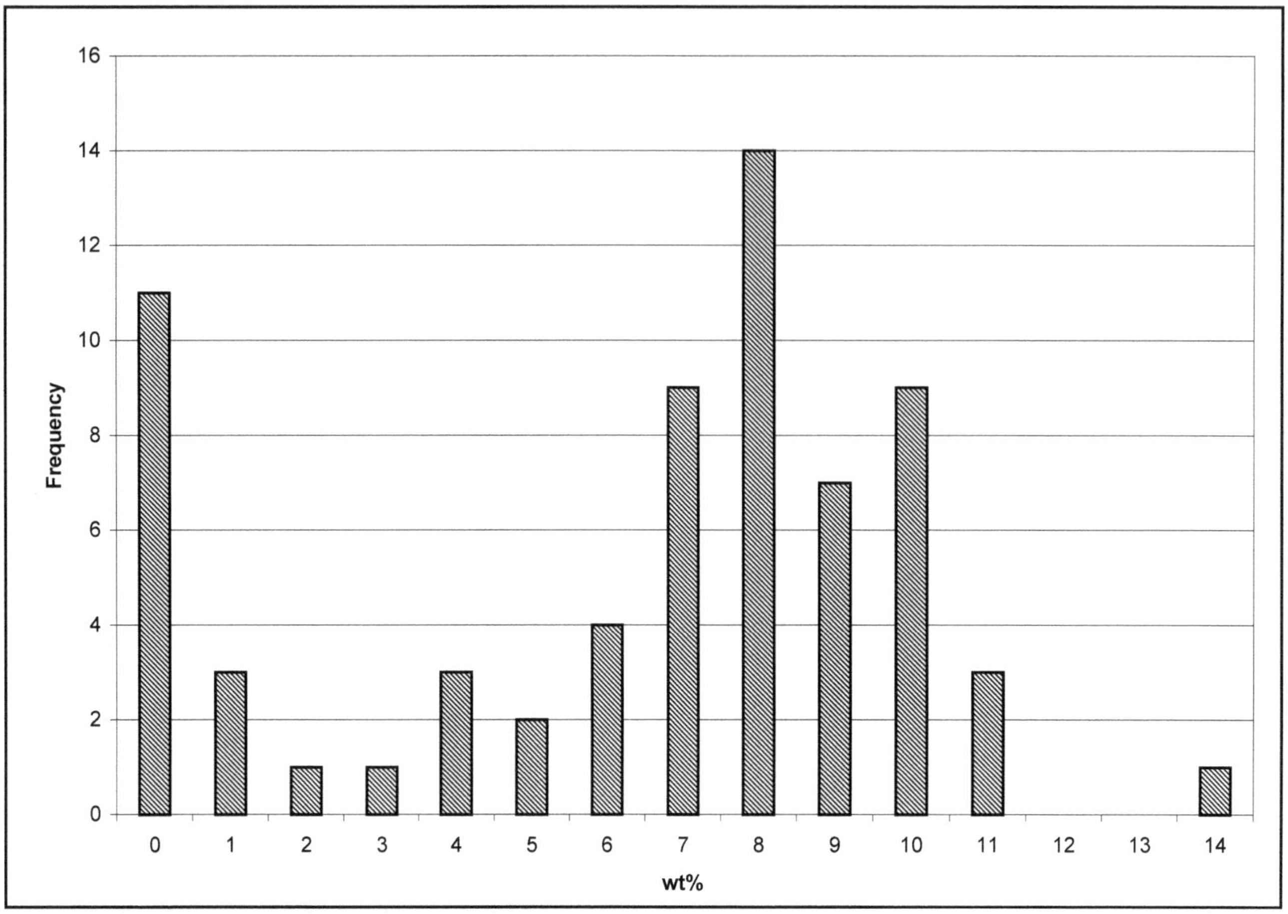

Histogram 3. Prealpine finds - Sn content

provenance from this area. The presence of this element might be a useful fingerprint for tracing down the possible provenance of the metal. Only very few of the objects analysed by this method contain mercury and some are fragments of fibula, rather difficult to assign to a particular type, but one is surely a Paularo type fibula and some of the fragments might also belong to the same type of fibula and therefore be of local production. This might of course signify that the Paularo metalworkers received their copper from the Eastern Alpine area, controlled by the S. Lucia / Most na Soči civilization, however the skills of the artisans were rather different.

The variable lead content is relatively high, if we consider the small size of the objects, and in some cases it gives the impression of having been employed as substitute for tin. The local level of technology seems in general to be lower than that of larger contemporary sites or of sites in which traces of metallurgical processes have been identified (see above). It appears that the local smiths had more problems with the tin supply, and that the tin they had at their disposal was scarcer than in other broadly contemporary sites of this region.

3) Prealpine finds

In one of the recently analysed groups of finds from the Friuli region are collected objects, which were found on different sites around the hills and lower mountains of the Friulian Prealpine region, mostly from the Carnia area[20]. They can be dated to the time between the Final Bronze Age and the Iron Age and comprise different items: small decorative objects for personal use, such as fibulae, pins, necklaces, bangles and pendants, tools, such as axes and knives, vessels and one belt hook with suspension rings for a sword. In this case it is difficult to say if and how the sites were related to each other, because the researches are now only at the first stages.

Up to now the analytical results suggest that the use of tin is not much different from what we know from the analyses of Slovenian material, however it is too early to be able to say if the same broader trends can be observed on this side of the Alps (Histogram 3, Prealpine finds). In this group of objects there are some, which are present also in the other groups of analysed objects and their composition is rather similar.

As a good example we can mention a fragment of a wheel pendant, which contains 14,7% of tin and little more than 1% of lead and the blades, which contain about 6-8 % and around 10-11 % of tin and compare very well with the data of analysis of objects belonging to this class from other sites in the region. Also in this case some of the finds seem to contain very low traces of mercury and suggest that also the Carnia settlements received their copper from the same mining areas, from which the metal artisans of Paularo received their copper supply. To have a better idea of the

[20] MUSCIO & GIUMLIA-MAIR 2001.

kind of copper used on these sites also several bun ingots and fragments of bun ingots were analysed. Their trace elements seem to be rather similar to those determined in the same classes of objects analysed in Slovenia in recent years (i.e. with a different trace elements pattern than the copper employed in Etruscan and Italic areas)[21].

The Bronze Age pieces show a composition which can remind of that of Slovenian material belonging to the same classes. No noticeable differences can be observed in the use of tin.

The pyxis from Raveo CV 37.453 is perhaps, among all finds, the object which has been more carefully worked, together with the now very corroded and fragmented patera. These two items show a slightly different trace elements pattern, without detectable nickel, manganese and bismuth, which are present in low amounts in most of the pieces belonging to this group of finds. Only 3 rivets of the patera seem to contain very small amounts of bismuth, however these rivets might also be a later repair and might not originally belong to the object. None of the many fragments and parts analysed, revealed any trace of mercury. The general picture suggests that the pyxis and the patera were perhaps imported from some other site. However it is also important to keep in mind that the lack of trace elements might also be due to the use of a better refined copper in the production of these objects, which were undoubtedly of better quality. The slight differences determined by the analysis are not sufficient to establish a sure foreign provenance for the two pieces.

Small decorative objects of personal use, such as fibulae, pins, collars and bracelets can be also compared with similar objects from other groups. Dragon fibulae (fibule a drago) and serpentine type (serpeggianti) fibulae show the characteristic high tin content. In the case of the fibula a drago the decorative antennae (or horns) attached to the bow by means of a rivet, are made of a copper-based alloy containing a significantly higher lead content, as previously determined in other similar fibulae from other sites. Navicella- armlets and fibulae, as well as Certosa type fibulae are also similar in composition to the examples from S. Lucia / Most na Soči (see below and histograms).

One of the Certosa type fibulae is decorated with a large bead made of copper without any tin, but with an addition of lead (around 9% Pb) to facilitate the casting and the reproduction of the ribbed surface decoration. The bead made of unalloyed copper might perhaps be a later addition. The more recent finds, for example fibulae La Tène seem to contain less tin, but only few La Tène objects have been analysed up to now and we do not possess enough observations yet to be sure of this. One of the La Tène type fibulae is made of a rather debased silver alloy with around 50 % of copper and some tin and lead. It should also be noted that one of the items included in the analysed group, most probably a small tool for knitting thin nets, does not belong to this period and should not be considered in the evaluation of the data.

[21] cfr. TRAMPUŽ OREL 1996; CRADDOCK 1977; 1998a; b.

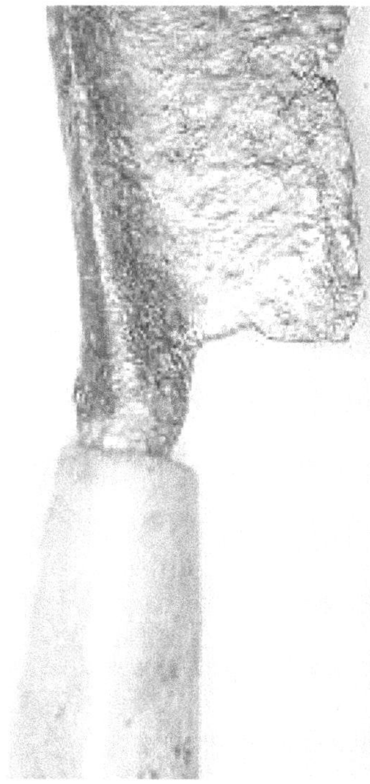

Figure 4. Detail of a bronze knife from Pozzuolo del Friuli, still with its bone handle. The blade contains around 6% of tin. (Foto Giumlia-Mair)

The results of the newly analysed groups of copper-based finds from different sites in Friuli-Venezia Giulia can be compared to the analysis data of previously analysed groups of finds from S. Lucia / Most na Soči and Pozzuolo del Friuli[22], which are chronologically broadly parallel to the material discussed above and comprise objects belonging to the same classes. The metallurgy of both Pozzuolo del Friuli and of S. Lucia / Most na Soči have been discussed in detail elsewhere, therefore in this paper only the data regarding the use of tin are going to be considered.

4) Pozzuolo del Friuli

The Iron Age Site of Pozzuolo del Friuli[23] is located on a low hill in the middle of the Friulian plain in an area which has been inhabited from the Neolithic to the Roman period. The excavations carried out by P.Càssola Guida and her team, revealed a castelliere structure, which overlooked a wide section of the plain and represented an important centre on one of the North-South axes of the communication net of the region. In the Southern part of the hillfort noticeable remains of metallurgical activities, dated to the 9th – mid 8th c. BC have been identified in a ditch with secondary concentration in distinct layers. Among the finds there are casting moulds, crucible fragments, a polishing wheel, purification slag, metal scrap, tools and finished and unfinished metal objects (Fig.4). The workshop had been apparently destroyed by fire and the debris swept into the ditch. The analysis of the finds gives a fairly good idea of the range of alloys in use on the site and information on the

[22] GIUMLIA-MAIR 1998a; b; c.

[23] PETTARIN 1998.

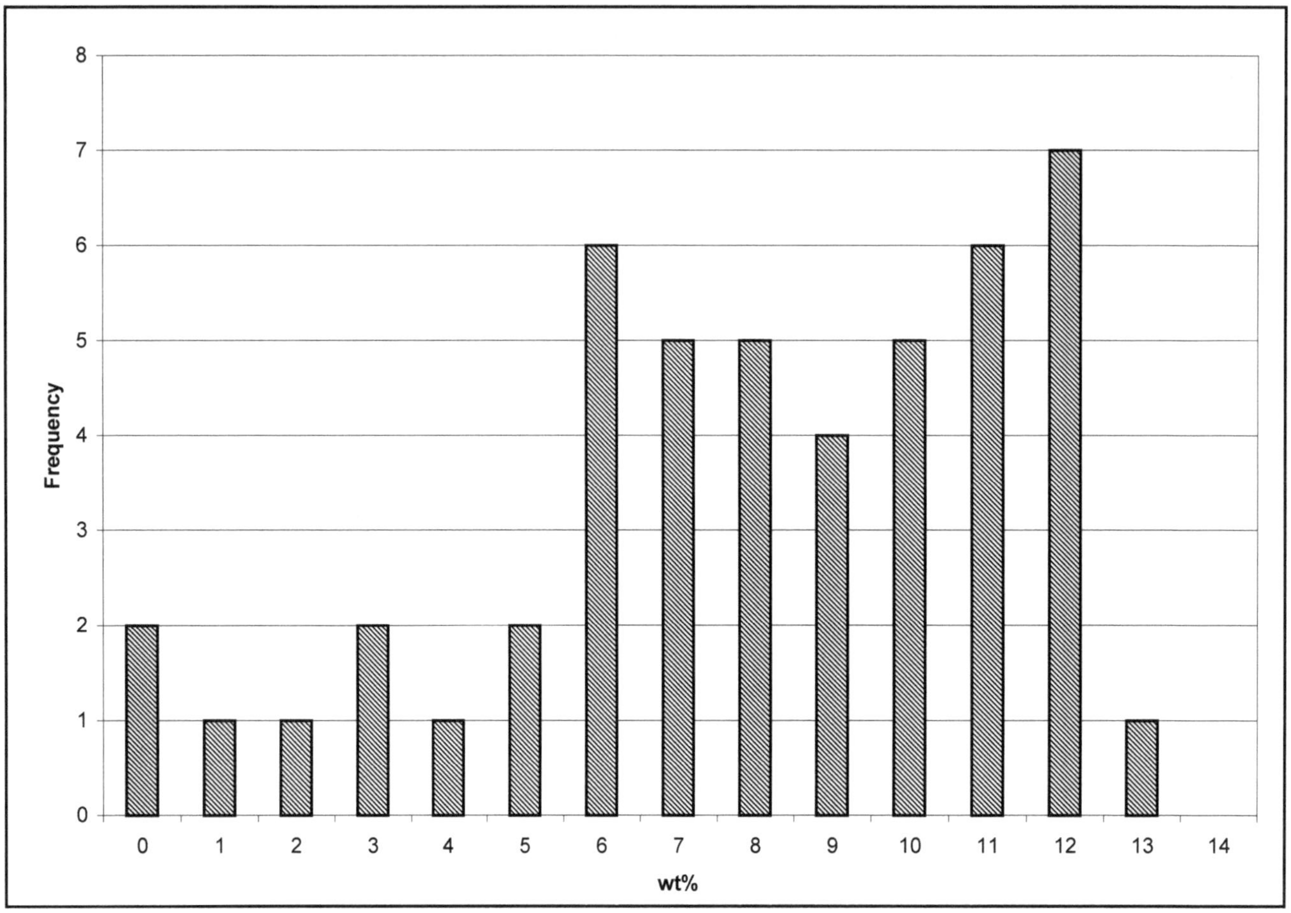

Histogram 4. Pozzuolo - Sn content

working organization of the smiths[24] (Histogram 4, Pozzuolo).

Apparently at the beginning of the Iron Age the technology of the local smiths was still very much at the Bronze Age level, without any correlation between the production technique and the use of lead. The objects contain rather high tin concentrations, up to over 12%. It does not seem that in this area there was any shortage of tin. The copper employed for the various alloys is mostly well purified and only low lead percentages are present in the metal. An important result of the analyses is the observation that the artisan seems to have collected metal scrap for recycling and that the fragments were all of bonze of good quality. It is quite clear that the smiths were easily able to distinguish good quality alloys from bad ones and kept the better quality scrap in the shop for the production of small decorative objects, for local use, which, being small, were easier to produce and for which there might have more local request.

Some purification slag and two pieces of well purified copper ingots were identified on the site and this suggests that the local workshop received raw copper from the mining districts and that the smiths had to carry out the purification in the workshop.

The identification of a fragment of a small bronze bar containing around 7% of tin (similar to the one identified on the Bronze Age site of the Variano hillfort) suggests that recycled bronze with this composition, resulting from remelting scrap with tin contents between 3 and 13%, was traded in the area. This had been hypothesised in the past, as in several sites, mainly located on rivers, larger hoards of scrap bronze have been found in different periods. This is for example the case of Gradisca sul Cosa (near the river Tagliamento)[25] and S. Pietro (near the river Isonzo)[26].

Now the identification of bronze bars with this composition on two different sites, where no bronze objects with similar tin content were found, seems to confirm this hypothesis.

A further interesting fact, revealed by the analysis of the metal, is the presence and the level of the trace elements in the objects from Pozzuolo del Friuli. The elements silver, arsenic, nickel and antimony are present in relatively high concentrations. The bismuth content is rather irregular, but, if present, shows noticeably high concentrations. Also cobalt has a rather distinctive pattern. As already discussed elsewhere, the mean content of the trace elements is similar to the distinctive trace elements pattern determined in the objects from S.Lucia / Most na Soči[27]. As the two sites were important centres on the routes, most probably beaten by metal traders already early in the Bronze Age, this is not very surprising.

[24] GIUMLIA-MAIR 1998b; c.

[25] CASSOLA GUIDA P.& BALISTA G., 1992; 1996.

[26] FURLANI 1996.

[27] GIUMLIA-MAIR 1998b, p.192-195.

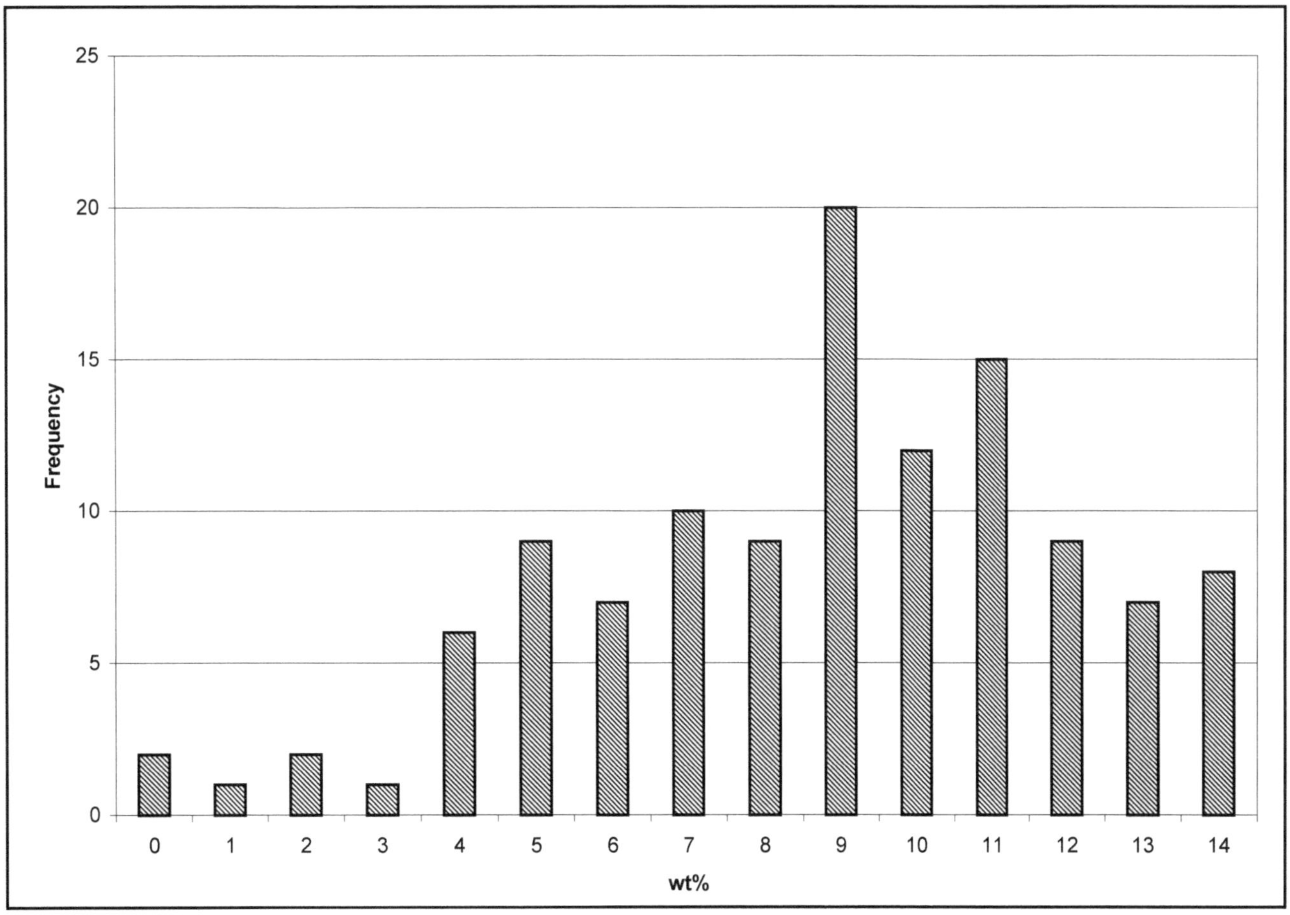

Histogram 5. S. Lucia - Sn content

An evident trace of an intensive and organized metal trade already in earlier times is the wide distribution of the pick-shaped ingots, which are found from France, Bavaria, Switzerland and the Western Alps, down to Central Italy and in the West in the Pannonian plain, but are noticeably concentrated in Austria and Slovenia[28]. The two settlements have similar, but not exactly the same traditions, had apparently commercial contacts and received their copper from the same sources.

5) S. Lucia / Most na Soči

The site of S. Lucia / Most na Soči already mentioned above, is surely the best known and one of the better studied Iron Age sites in the Eastern Alps. Its large necropolis was unearthed in the 19th c.[29] and the excavations brought to light over 6400 graves[30]. The Iron Age settlement was only partially excavated, also in recent times[31], because the larger part of it is under the modern village, but it is quite clear that this was the main centre of the archaeological *facies*, which controlled the passages which went through the Eastern Alps in all directions. The town was located on the convergence of several valleys and at the confluence of two navigable rivers and of a mountain torrent. The roads went up to the Drava valley to the Hallstatt settlements in Austria, to the Ljubljana plain in the East and down to the Po basin in the West. More settlements belonging to the same civilisation have been discovered and were partly excavated[32]. These controlled strategic crossings in the mountain passes and important roads. The territory reached its widest extension in the 6th – 5th c.BC.

The graves of the large necropolis contained mostly several grave gifts. These were mainly personal ornaments and vessels. No tools or weapons were found in the graves.

The metallurgy of the Iron Age site of S. Lucia / Most na Soči has been exhaustively discussed elsewhere[33] (Histogram 5. S. Lucia). The highest tin content determined in the fibulae analysed for the project reaches 13 %. This is due to the high frequency of serpentine fibulae (fibule serpeggianti) and dragon fibulae (fibule a drago), which require more strength and less flexibility and are the objects with the highest tin content.

The vessels (Fig. 5) show in their hammered part a very regular tin content, between 10 and 11%, with a clear peak around 10% of tin. None of the hammered vessel parts contained less than 9% of tin. The cast parts show a less regular use of tin and lower levels, however only one of the

28 BORGNA1992, p.75-82.

29 MARCHESETTI 1993.

30 GABROVEC & SVOLJŠAK 1983; TERŽAN et al. 1984/1985.

31 SVOLJŠAK 1974; 1977; 1979; 1980; 1983 a; MLINAR 2001.

32 SVOLJŠAK 1983b.

33 GIUMLIA-MAIR 1998a.

Figure 5. Example of situla from the necropolisof S. Lucia di Tolmino/Most na Soči. The vessels show in their hammered part a very regular tin content, between 10 and 11%. (Foto Giumlia-Mair)

samples contained less than 5%. The tin content of rivets and small functional parts of the vessels showed a very random composition.

Interestingly, both classes of finds contain in average higher tin concentrations than the same classes of objects belonging to the Greek (7,8% Sn), Etruscan (6% Sn), Sardinian (8% Sn) and Italic (6% Sn) civilisations analysed by Craddock[34].

The analyses of the broadly contemporary finds from Albacete (Spain), analysed by Rovira Llorens contain an average of 9,8% of tin[35]. The finds from Arbedo (Switzerland), an important site dated to the 6th-5th c. BC (Golasecca 3 A1, around 450 BC) analysed by Northover, have also a very high average (11,7 %) of tin[36]. However the higher tin results might be also due to the use of surface analyses (i.e. X-rays fluorescence and microprobe) and corrosion or segregation of the finds.

The finds from S. Lucia contain high tin levels. This fact can have different explanations. The Greek, Etruscan, Sardinian and Italic artisans were perhaps better skilled in the choice of their alloys and were able to reduce the amount of tin to the minimum required by the function of the objects and by

Figure 6. Ribbed cista (cista a cordoni) with concentric circles and water birds decorations from S. Zeno (Trento). The tin content is very high in all parts of the vessel, particularly in the handles and in the decorative ring pendants (Foto Giumlia-Mair)

Figure 7. Vessel with cross-shaped attachments (Kreuzattachenkessel) from S. Zeno. Also in the case of this class of vessels the tin content is particularly high. (Foto Giumlia-Mair)

their production technique, while the artisans at S. Lucia might have still applied the Bronze Age techniques and alloys, which required higher tin amounts. The second possibility might simply be that in this period tin was more readily available in the Oriental Alps, than in Central Italy.

6) San Zeno

A group of vessels (Fig. 6 and 7), found on the important site of San Zeno, in the province Trento, has also been recently analysed and represents an excellent comparison for the vessels of S. Lucia. Both are Alpine sites and the vessel types are very similar. Some of the vessels from S. Zeno come from excavations in the Non Valley, a few were accidental

[34] CRADDOCK 1977; 1986a; b.

[35] ROVIRA LLORENS 1992.

[36] NORTHOVER 1998.

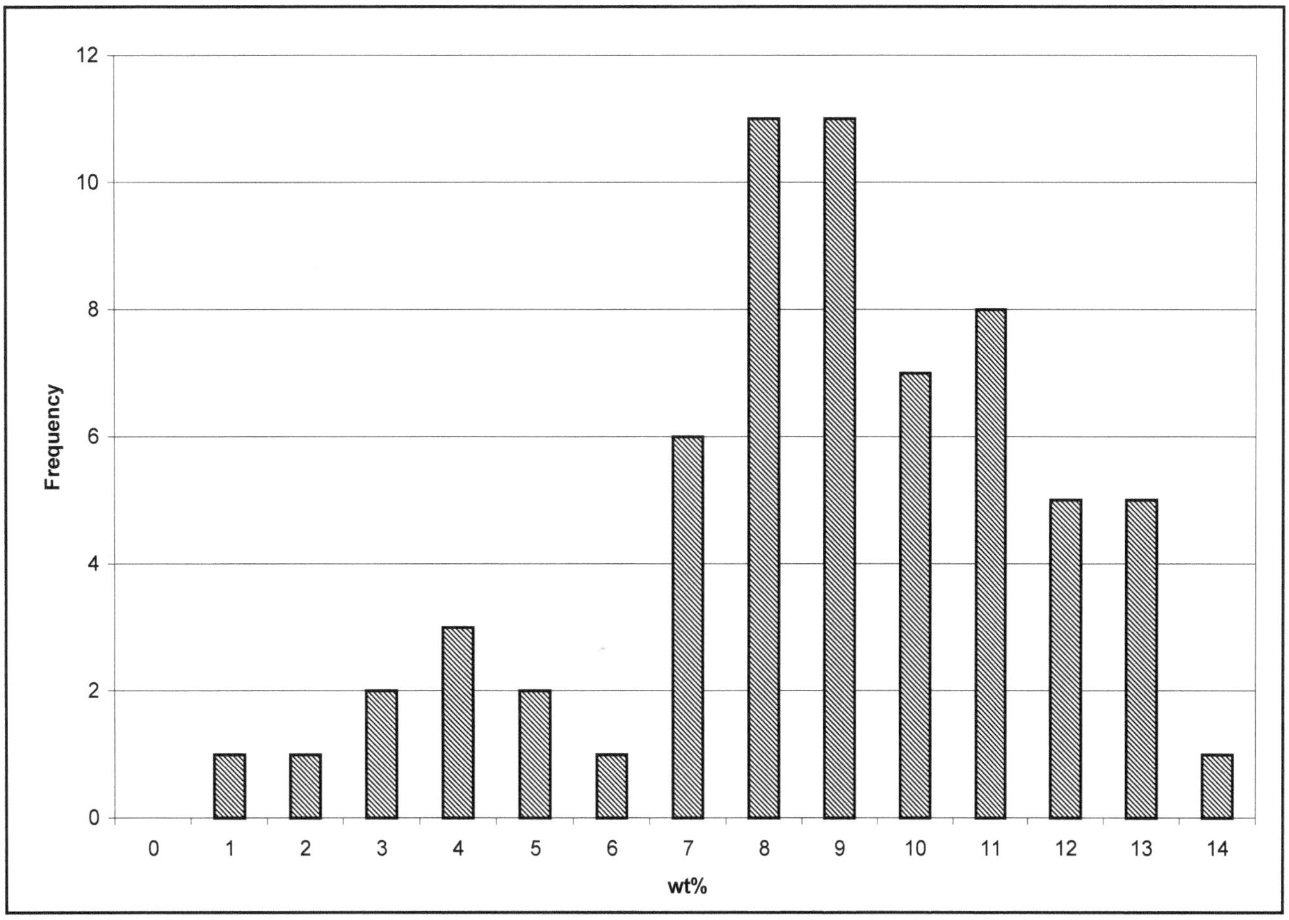

Histogram 6. Sanzeno - Sn content

finds. The pieces were examined by optical microscopy and SEM and analysed by ICP and AAS for the determination of their composition. The patina was also analysed by XRD before conservation.

The importance of the rich Iron Age archaeological site of S.Zeno with its sanctuary is well known and of course it is not necessary to mention its significance for Iron Age metallurgy[37].

For this project only vessels were analysed (Histogram 6, S. Zeno), with only two exceptions, one helm and one fibula. Among the vessels there are situlae with a protruding shoulder (situle a spalla carenata), similar to the ones analysed for the S. Lucia / Most na Soči project, which are dated to the 7th-6th c. BC, ribbed cistae (ciste a cordoni) (Fig. 6), dated here to the 6th c.BC, but which are found also in Hallstatt settlements, in Veneto and at S. Lucia between the 7th and the 6th c. BC. Their typical decoration, with water-birds and concentric discs, is also found in the same Transalpine areas and down the Po plain[38].

Cauldrons with cross-shaped attachments (bacili con attacchi a croce,) from S.Zeno were also analysed (Fig. 7). Up to now there is only one not completely certain analytical comparison in the Triveneto area, but the shape is found in the Caput Adriae and in the surrounding territories in Trentino, Austria, Slovenia, Istria and Hungary[39].

The histogram, which shows the frequencies of tin percentages in the analysed pieces from S.Zeno, should not be considered a statistical data evaluation, but only a graphic way of illustrating the results, as the number of observations is not sufficient for statistics. Nevertheless the number of analyses gives us a fairly good idea of the alloys employed for the vessels and there are peculiarities which need some discussion.

The histogram evidences a peak corresponding to rather high tin percentages in the area between 8-11 %, a second peak in the area around 12-13% and a small peak around 3-5% of tin.

By a closer look at the values and at the table of results it is quite clear that the values of the small peak correspond to the tin percentages which have been determined in small functional parts of the vessels, i.e. in the rivets, which fix together the different sheets of the vessel bodies and hold the attachments of the handles. It should also be noted that the components with low tin are mainly hidden rivets, while the decorative ones, with conical protruding ends show much higher tin percentages.

[37] cfr. for example DE MARINIS 1988; CIURLETTI & MARZATICO 1999.

[38] cfr. MARZATICO 1998.

[39] cfr . VON MERHART 1952; EGG 1996.

The highest tin percentages are those of the handles and of the decorative ring which hang from the handle attachments, i.e. the most visible parts contain very high tin percentages.

The tin content is in general very high if we compare it with that of the groups of objects from other sites. The more so, if we consider that most of the observations that produce the low tin peak, around the 4% area, refer to a vessel, which is with fairly good certainty to be dated to a later period and should therefore not be considered in this discussion. The difference with the other histograms results even more noticeable, if we consider that 96% of the samples from the S. Zeno group are taken from vessels. As mentioned before, vessels of the same kind are present in the group of finds from S.Lucia, however their tin content is in all their parts much lower than that in the vessels from S. Zeno. In the S. Zeno group some of the hammered sheet metal vessels contain more than 13 % of tin, while the tin content of the handles and the decorative rings reaches almost 15%.

This is never the case with the vessels from S. Lucia. As already mentioned in the paragraphs on the use of tin at S.Lucia, the only objects with tin contents higher than 12% are serpentine and dragon fibulae. As a comparison, similar vessels from France, a ribbed cista (cista a cordoni) from Estissac and a cauldron with cross-shaped attachments (bacile con attacchi a croce) can be mentioned. They were recently analysed by Piccardo and Pernot[40] and resulted to be similar to the S. Lucia examples. Their tin content goes from around 5 to around 12 % in the different parts. The same percentages resulted from the analyses of similar bronze vessels in Germany[41].

The lead content of the vessels from S.Zeno is on the contrary very low, also in the cast parts, which in the group of vessels from S. Lucia contain lead and sometimes show relatively high concentrations of this metal. The presence of lead facilitates the casting by lowering the temperature and by increasing the fluidity of the metal, however lead also darkens the colour of bronze and this can be the only reason for avoiding it in cast pieces. The intention of the artisans at S. Zeno was of course to produce objects that had a very golden colour, i.e. the reason for using such high tin percentages was for sure aesthetic, as the high tin content renders rather difficult the production of sheet. This must also mean that tin was rather readily available to the metalworkers of S. Zeno. Tin was, apparently, somewhat less common and less abundant for the artisans at S. Lucia at the outmost corner of the Oriental Alps.

LATE BRONZE AGE TIN IN THE ORIENTAL ALPS

By looking at the very numerous data of analysis carried out in the last 30 years on Bronze Age material[42] we see that bronze became common in the Middle Bronze Age. There were some regional peculiarities, for example in Denmark, where the tin content looks rather irregular and erratic, however in general in Europe the tin content goes from 8 to 11%, while in objects of lesser importance and quality tin is very low or even unalloyed copper is employed. In the working process the tin content does not have to be particularly precise, as the final hardness is achieved by hammering the objects.

The most regular tin content is that of swords, with around 10% of tin[43], however some particular objects can contain more tin. A good example are the Tüllenhämmer analysed by Riederer[44] with respectively over 14, 15 and 16 % of tin. In the Carpathians, in Reinecke A2 there is a decrease from 10 to around 6% in the tin content, which might signify the use of standardized recycled metal in this period.

The abundance of tin in the South-West of Britain can explain the high tin level in the bronzes of British production[45], with a constant 10% in the Early Bronze Age and 11% in the Middle Bronze Age. In Switzerland[46] the average tin content is 8-9%, but abruptly collapses in the period Hallstatt B2.

The analyses of Slovenian material seem to indicate a reduced employment of tin in the local objects, at least in the Late Bronze Age, a period for which analysis data are available[47]. The few data we posses from the analyses of the finds from Variano seem to reflect a similar situation. Apparently there was scarcity of tin, while down the Adriatic coast metal artisans could count on regular supplies. A credible explanation for the abundance of tin in the Marche is perhaps that this metal came possibly from the Etruscan mining area in the Colline Metallifere of Tuscany, but this hypothesis is by no mean proved yet.

There is also still the possibility that the tin came from the other side of the Adriatic or, more generally, by sea route.

In the Late Bronze Age there is also an increased presence of Baltic amber and finds of high magnesium glass (HMG) of Eastern Mediterranean provenance in the Veneto and Trentino area, but also in Transalpine sites in France, Switzerland, Germany and Bohemia, which seem to indicate an intensive long distance trade of Baltic amber from the North, bartered with glass from the Aegaean and confirmed by the presence of Mycenaean pottery around the Adriatic area[48]. The connection between the amber trade and bronze in East Germany, Poland, Lithuania and further areas on the

40 PICCARDO & PERNOT 1997, 47.

41 JACOB 1995, 142-164.

42 for example JUNGHANS et al 1960; 1968; CRADDOCK 1976.

43 cfr. CRADDOCK 1976; RIEDERER 1994/95.

44 RIEDERER 1992, 295.

45 CRADDOCK 1978, 376; cfr. also LIVERSAGE 1994.

46 RYCHNER 1990, 201-202; RYCHNER & KLÄNTSCHI 1995 1995, 61.

47 TRAMPUŽ -OREL et al. 1991, p. 267-277 ; TRAMPUŽ -OREL N., 1996, p. 165-242. TRAMPUŽ -OREL & HEATH, 1998, I, p. 237-248. TRAMPUŽ -OREL et al. 1998, p.223-237. HEATH & TRAMPUŽ -OREL 2000.

48 ANGELINI et al. in print.

Baltic, and down the Danube has been pointed out long ago by Penhallurick[49].

More recently Bellintani[50] argues convincingly that in the Final Bronze Age there must have been a complex trade and barter system in which amber, metals, glass and also antler played an important role, particularly on the routes through the Oriental Alps and through the Adige Valley. Another useful indication comes from the finds of cowry shells from Africa, India and the Red Sea in prehistoric Baltic sites and in Central Europe, for which B. Johansson[51] suggests long distance trade routes or a network of tradesmen exchanging precious items.

On the same routes also the tin from the European Hercynian Tin Belt running from the rich Erzgebirge in Central Europe to The Scilly Island off Lands End in England might have been traded.

Since palaeolithic times fire was used in mines for the extraction of minerals, therefore also the Erzgebirge granites deposits might have been used, but we also know from Agricola[52], that "Zinngraupen", cassiterite pebbles, were washed in Dippoldiswalde and Altenberg im 1512[53]. Placer cassiterite might also have been traded in earlier times on preferential routes.

EARLY IRON AGE TIN IN THE ORIENTAL ALPS

In several contexts, for example in the Greek, Etruscan and Italic metallurgy the tin content diminishes with the introduction of leaded alloys in the Early Iron Age, but this tendency seems to be much less noticeable in the most important sites of the Oriental Alps. As already pointed out, the tin content in the Iron Age finds from Eastern Alpine sites seem to be higher than in the same classes of objects belonging to the Etruscan, Greek and Italic civilisations.

We have to note that the objects analysed in the Eastern Alpine sites are the typical products of what we might call a koinè of the larger region around the Caput Adriae, which covers a wide area with common traditions, is characterised by intensive trade of goods of all sorts and goes from Trentino, down to the Palaeovenetian settlements of the Este culture in Veneto, passing also through the Pusteria and Valsugana Valleys, to Friuli-Venezia Giulia, Slovenia and Istria with the facies of S. Lucia / Most na Soči and up North to the Transalpine Hallstatt settlements all over the Oriental Alps and to the Danube. However many of the same finds and decorative elements are found also in Raetia and in France, in the East in the Pannonian plain, and even in the Bologna area and in Picenum in the South.

The abundance of tin in the vessels from S. Zeno might simply be a sign of richness of the settlement, but might also point to the role played by the site on the control of the trade route, on which the tin came from the North, crossing over the Alps at the Brenner Pass to descend the Adige Valley and be bartered with other goods around the Po plain, but particularly to serve the huge Eastern Mediterranean market from the Northern Adriatic shores, together with precious amber, after Diodorus Siculus, from the fabulous island Basileia (Diod., 5, 23), in exchange with glass, shells, ivory and ostrich eggs[54].

In the Iron Age, amber, glass and other exotic goods were traded also up and down other fluvial routes- the motorways of antiquity - between the Baltic and the Adriatic. The most important rivers belonging to the trade net were the Oder, which, from a spring in Moravia, flows into the Baltic near Szczecin in Poland, after crossing the wide Polish plain, and the Elbe, which from the Sudety mountains and through the important mining district of Kutná Hora, flows into Helgoland Bay, after crossing the Erzgebirge and a large section of Germany. From the Oder and the Elbe and from their tributaries it is rather easy to reach the Danube Valley and some of the largest tributaries on the Southern riverbank of the Danube. Inn, Drava and Sava are only the largest affluent streams and would be sufficient to reach all corners of the "koinè" territory around the Oriental Alps: the Maloja Pass in the Raetian Alps, the Resia Pass, the sources of the Adige, the smaller passes in Friuli and the Predil Pass, near the Isonzo spring. The river Isonzo, on which S.Lucia is situated, was the last segment of the long amber, glass and metal route. Along its stream also copper and iron, the metals from the North-East of the Alps, were apparently brought to the plain and perhaps also to the Adriatic. The characteristic trace elements pattern of the copper from the North, with silver, bismuth, antimony, nickel and other elements is very different from that of the copper used by Etruscan and Italic populations.

Some remembrance of the early tin trade on the Adriatic can still be found in the ancient literature. As recently mentioned elsewhere[55], there is a Greek text, called the *Periegesis ad Nicomedem regem*, known also as the Περιήγησις of the Pseudo-Skymnos or *Orbis Descriptio*, which refers that tin was present on two islands in the territory of the Istri, a population which lived in the area apparently comprised between the Danube and Istria (*orbis descr.* 391-393), i.e. the territory of our koinè. The text is dated to around 110 BC and is based on Ephoros' Ιστοριαι of the 4th c. BC.

Of course there is also the possibility that the tin mentioned by the ancient sources came to the markets on the islands in the territory of the Caput Adriae from the tin deposits on the Serbian mountains of Bukulie and Cer, along the streams of the Danube, the Drava or, even more probably, along the Sava[56].

[49] PENHALLURICK 1986, 64; cfr also HAWKES 1940, 330.

[50] BELLINTANI, 2000; 2002.

[51] JOHANSSON 1995, 348 and 350-351.

[52] AGRICOLA 8, 270.

[53] cfr. also PENHALLURICK 1986, 75-77.

[54] cfr. also CASSOLA GUIDA 1999, 492; BORGNA 2000, 46-47.

[55] GIUMLIA-MAIR 2000, 79-80.

[56] cfr. PENHALLURICK 1986, p.61, map 12; DURMAN 1997, p. 7-14.

The distribution centre of tin (and of course of other goods), might have been located on two of the Dalmatian islands or perhaps on the delta of the river Po on the other side of the Adriatic.

Indeed a passage in the Pseudo-Aristotle work "On marvellous things heard"[57] mentions copper tin and amber on Islands in the Adriatic gulf.

This paragraph has been often interpreted as an indication of the existence of an emporion for these goods[58].

"In the Electrides Islands, which lie in the gulf of the Adriatic, they say that two statues have been dedicated, one of tin and one of copper, wrought in the old-fashioned style <...>They say that the river Eridanus (the Po) silted up these islands. There is a lake apparently near the river, containing hot water<...>. The local inhabitants say that Phaethon fell into this lake when he was struck by a thunderbolt. There are many poplars in it, from which oozes the so-called electron. They say that this is like gum, and hardens like a stone; it is collected by the inhabitants and brought to the Greeks."

Author's address

Alessandra GIUMLIA-MAIR
AGM Archeoanalisi
Via E. Toti 8
39012 MERANO (BZ)
ITALY
e-mail: giumlia@tiu.it

Bibliography

AGRICOLA G., 1556, Georgii Agricolae de re metallica libri XII, Froben, Basileae, DTV (1977), München.

ANGELINI I., ARTIOLI G., BELLINTANI P., DIELLA V., POLLA A., RECCHIA G., RESIDORI G., in print, Materiali vetrosi da Grotta Manaccora e Coppa Nevigata: inquadramento archeologico e archeometrico nell'ambito dell'età del Bronzo italiana, Atti del convegno Il vetro in Italia meridionale, II incontro di studi, AIHV e Università di Napoli (4/6.12.2001).

BELLINTANI P., 1995, Breve guida al Museo Civico di Fratta Polesine, Centro Polesano di Studi Storici, Archeologici ed Etnografici di Rovigo, Rovigo.

BELLINTANI P., 2000, I bottoni conici ed altri materiali vetrosi delle fasi non avanzate della media età del Bronzo dell'Italia Settentrionale e Centrale, Padusa XXXVI Nuova Serie, 95-107.

BELLINTANI P., 2002, Bernsteinstraâen, Glasstraâen Uber die Alpen. Menschen, Wage, Waren, Almanach 7/8; Archaeologisches Landesmuseum Baden - Württemberg, Stuttgart, 45 -58.

BORGNA E., 1992, Il ripostiglio di Madriolo presso Cividale e i pani a piccone del Friuli-Venezia Giulia, ed.Cassola Guida P., Roma.

[57] PSEUDO-ARISTOTLE, de mirabilibus auscultationibus, 81: 836a, 24-27, 836b 2-5.

[58] BELLINTANI 1995, p.18.

BORGNA E., 2000, Osservazioni sulla presenza egeo-cipriota nella metallurgia delle regioni comprese tra l'Alto Adriatico e l'Europa Centrale, in : Ancient metallurgy between Oriental Alps and Pannonian Plain, Giumlia-Mair A. ed., Quaderni dell'Associazione Nazionale per Aquileia 8, 41-51.

BOUQUILLON A. & POIROT J.-P., Les minéraux et leur origine, in Les pierres précieuses de l'Orient ancient des Sumériens aux Sassanides, F. Tallon ed., Musée du Louvre, Exposition-dossier des Antiquités orientales 49, p.33-38.

CARTER M.L., 1997,The Ancient Near East, in: Miho Museum (J.N.Newland), Nissha, p.325-334.

CASANOVA M., 1993, Lapis lazuli beads in Susa and Central Asia: a preliminary study, South Asian Archaeology 93, p. 137-145.

CASANOVA M., 1995, Le lapis-lazuli dans l'Orient antique, in Les pierres précieuses de l'Orient ancient des Sumériens aux Sassanides, F. Tallon ed., Musée du Louvre, Exposition-dossier des Antiquités orientales 49, p.15-20.

CASSOLA GUIDA P.& BALISTA G., 1992, Il Castelliere di Gradisca di Spilimbergo (Pordenone) (Scavi 1987-´89), Tipologia di insediamento e distribuzione antropica nell´area veneto-istriana dalla protostoria all´alto medioevo, Atti del Seminario di Studio Asolo 1989, Edizioni della Laguna, Monfalcone, p.7-16.

CASSOLA GUIDA P.& BALISTA G., 1996, Gradisca di Spilimbergo (località Castellieri, Pordenone), Preistoria e protostoria, Guide archeologiche, N.7, Forlì, p.207-213.

CASSOLA GUIDA P., 1999, Indizi di presenze egeo-orientali nell'Alto Adriatico alla fine dell'età del bronzo, in: Epi ponton plazomenoi, Simposio italiano di studi egei in onore di L.Bernabò Brea e G.Pugliese Caratelli, Roma, febbraio 1998, Roma, p. 487-497.

CASSOLA GUIDA P., CORAZZA S., 2000, Il castelliere di Variano presso Brasiliano (Udine, Italia Nord-Orientale), Aquileia Nostra LXXI, p.30-51.

CIERNY J. & WEISSGERBER G., in print, Bronze Age Tin Mines in Central Asia, in: Colloquium The Problems of Early Tin, Giumlia-Mair A & Lo Schiavo F. eds., Proceedings of the XIV UISPP Conference in Liège, September 2001.

CIURLETTI G. & MARZATICO F., 1999, I Reti / Die Räter, Atti del simposio 23-25 settembre 1993. Castello di Scenico, Archeoalp 5, Trento.

CORAZZA S. e VITRI S., 2001, La necropoli di Misincinis dopo lo scavo, Primi risultati delle indagini 1995-1997, Tielle grafiche, Sequals.

CRADDOCK P.T., 1976, The composition of the copper alloys used by the Greek, Etruscan and Roman civilisations. 1. The Greeks before the Archaic period, Journal of Archaeological Science 3, p. 93-113.

CRADDOCK P.T., 1977, Copper alloys used by the Greek, Etruscan and Roman civilisations, 2. The Archaic, Classical and Hellenistic Greeks, Journal of Archaeological Science 4, p.103-123.

CRADDOCK P.T., 1978, Deliberate alloying in the Atlantic Bronze Age, in The Origins of Metallurgy in Atlantic Europe, Ryan M. ed., Proceedings of the 5th Atlantic colloquium, Dublin, 369-385.

CRADDOCK P.T., 1986a, The metallurgy and composition of Etruscan bronze, Studi Etruschi 52, p.211-271.

CRADDOCK P.T., 1986b, The Metallurgy of Italic and Sardinian Bronzes, Italian Iron Age artefacts, BMP, London, p. 143- 149.

DE MARINIS R., 1988, Le popolazioni alpine di stirpe retica, in Italia Omnium Terrarum Alumna, Milano p. 25-40.

DURMAN A., 1997, Tin in South-Eastern Europe?, Opuscula Archaeologica 21, pp.7-14.

EGG M., 1996, Das hallstattzeitliche Fürstengrab von Strettweg bei Judenburg in der Obersteiermark, (R.G.Z.M.), Mainz.

FURLANI U., 1996, Il ripostiglio di S.Pietro presso Gorizia, in Depojske in posamezne kovinske najdbe bakrene in bronaste dobe na Slovenskem/Hoards and Individual Metal Finds from the Eneolithic and Bronze Ages in Slovenia II, B.Ter an ed., Katalogi in Monografije 30, Ljubljana, p.73-88.

GABROVEC S. & SVOLJŠAK D., 1983, Most na Soči (S.Lucia) I, Katalogi in Monografije 22, Ljubljana.

GIUMLIA-MAIR A., 1998, The metallurgy of the copper based artefacts from the Iron Age necropolis of S.Lucia /Most na Soči Proceedings of the XIII International Congress of Prehistoric and Protohistoric Sciences, Forlì 1996, 4 Sections, p.665 - 672.

GIUMLIA-MAIR A., 1998, Studi tecnici sui bronzi dall'officina dei *Ciastjei*, in: Paola Cassola Guida, Silvia Pettarin, Gabriella Petrucci, Alessandra Giumlia-Mair, Pozzuolo del Friuli, II, 2, La prima età del ferro nel settore meridionale del castelliere. Le attività produttive ed i resti faunistici. p. 181-230.

GIUMLIA-MAIR A., 1998, Iron Age metal workshops in the Eastern Alpine area , Metallurgica Antiqua, Festschrift Maddin/ Bachmann, Der Anschnitt, Beiheft 8, Bochum, p. 45-55.

GIUMLIA-MAIR A., 2000, Archaeology and metals between Alps and Pannonia. Proceedings of the "Workshop on Ancient Metallurgy between Oriental Alps and Pannonian Plain", (A.Giumlia-Mair ed.), 29-30 Ottobre, Quaderni di Aquileia Nostra, 8, 17-23.

GIUMLIA-MAIR A., 2000, Bronze technology in the Eastern Alpine Regions between the Final Bronze Age and the Early Iron Age. Proceedings of the "Workshop on Ancient Metallurgy between Oriental Alps and Pannonian Plain", (Giumlia-Mair ed.), 29-30 Ottobre, Quaderni di Aquileia Nostra, 8, p.77-91.

GIUMLIA-MAIR A., 2000, Argento e leghe "argentee" nell'antichità, Atti del 7° Convegno Le Scienze della Terra e l'Archeometria, Bollettino dell'Accademia Gioenia di Scienze Naturali 33, N.357, 295-314.

GIUMLIA-MAIR A., 2002, Il mistero dello stagno, In: Catalogo della Mostra Le arti di Efesto- Capolavori in metallo dalla Magna Grecia, Scuderie del Castello di Miramare, Trieste, p.110-111.

HAWKES C.F.C., 1940, The Prehistoric Foundations of Europe, Methuen.

HEATH D.J., TRAMPUŽ OREL N., MILIČ Z., 2000, Wheel-shaped pendants: evidence of a Late Bronze Age metal workshop in the Caput Adriae, in Ancient Metallurgy between Oriental Alps and Pannonian Plain, edited by A.Giumlia-Mair, Trieste, Quaderni dell'Associazione Nazionale di Aquileia, 8, pp.53-70.

JACOB Ch., 1995, Metallgefäße der Bronze- und Hallstattzeit in Nordwest-, West- und Süddeutschland, Prähistorische Bronzefunde, Abt.II, Band 9, Steiner Verlag, Stuttgart.

JOHANSSON B., 1995, Sub-fossil Cowry Shells in Northern Europe – Traces of Eastern and Western contacts with the Orient, in Robertsson A.-M., Hicks S., ?kerlund A., Risberg J., Hackens T. eds., Landscapes and Life. Studies in honour of Urve Miller, PACT 50, p. 347-352.

JUNGHANS S., SANGMEISTER E., SCHRÖDER M., 1960 & 1968, Studien zu den Anfängen der Metallurgie, Mann Verlag, Berlin.

KAPTAN E., 1983, The significance of tin in Turkish Mining History and its Origin, Bull. Of the Mineral Research and Exploration Institute of Turkey 95/96, p.106-114.

KAPTAN E., 1984, New Discoveries in the Mining History of Turkey in the Neighbourhood of Gümüþköy, Kütahya, MTA Bulletin 97/98, p.140-147.

KAPTAN E., 1986, Ancient Mining in the Tokat Province, Anatolia: New Finds, Anatolica 13, p.19-36.

KAWAMI T.S., Vessel, in: Miho Museum (J.N. Newland), Nissha, 42, 15.

LAMBERG-KARLOVSKY C.C. and KOHL P.L., 1971, The Early Bronze Age of Iran as seen from Tepe Yahya, Expedition 13, 3-4, 14-21.

LIDDEL H.G. & SCOTT R., 1996, A Greek-English Lexicon, Oxford.

LIVERSAGE D., 1994, Interpreting composition patterns in ancient bronze: The Carpathian Basin, Acta Archaeologica 65, København, p. 57-134.

MAKKAY J. 1992, Ancient metal names and the first use of metal, Balcanica, XXIII, p.311-318.

MARCHESETTI C., 1993 (repr.), Scritti sulla necropolis di S.Lucia di Tolmino (Scavi 1884-1902), E. Montagnari Kokelj ed., Civici Musi di Storia ed Arte, Trieste.

MARZATICO F., 1998, Recipienti preromani in lega di rame (bronzo) dal territorio atesino, in Rame d'arte, dalla preistoria al XX secolo nelle Alpi centro orientali, U. Raffaelli ed., Trento, p.13-36.

MLINAR M., 2001, Nove Zanke Svetolucijske uganke, arheološke raziskave na Mostu na Soči: 2000 do 2001, Tolminski Muzej, Ljubljana.

MUSCIO G. & GIUMLIA-MAIR A., 2001, Presenze celtiche – Indagini territoriali, in I Celti in Friuli: archeologia, storia e territorio, I 2001., Aquileia Nostra LXXII, 377-384.

NORTHOVER J.P., 1998, Analysis of copper alloy metalwork from Arbedo TI, in Schindler M.P., Der Depotfund von Arbedo TI und die Bronzedepotfunde des Alpenraums vom 6. bis zum Beginn des 4.Jh. v. Chr., Antiqua 30, Basel, 289-315.

PENHALLURICK R., 1986, Tin in Antiquity, London.

PETTARIN S., 1998, Le attività produttive in: Paola Cassola Guida, Silvia Pettarin, Gabriella Petrucci, Alessandra Giumlia-Mair, Pozzuolo del Friuli, II, 2, La prima età del ferro nel settore meridionale del castelliere. Le attività produttive ed i resti faunistici. p.9-95.

PICCARDO P. & PERNOT M., 1997, Studio analitico e strutturale di alcuni vasi celtici in bronzo, La metallurgia Italiana 11/97, p.43-52.

RIEDERER J., 1992, Die Metallanalysen der Bronzeobjekte des Hortfundes von Lengyeltóti, Acta praehistorica et Archaeologica 24, p. 295-300.

RIEDERER J., 1994/95, Die Metallanalysen der Schwerter des Hortfundes von Berlin-Buch, Acta praehistorica et Archaeologica 26/27, p. 129-131.

RYCHNER V., 1990, L'analyse chimique du bronze préhistorique: pourquoi?, Zeitschrift schweizerischer Archäologie und Kunstgeschichte 47, p. 201-212.

RYCHNER V.& KLÄNTSCHI N., 1995, Arsenic, nickel et antimoine. Un approche de la metallurgie du Bronze moyen et final en Suisse par l'analyse spectrometrique I-II, Cahiers d'archeologie romande 63-64.

ROVIRA LLORENS S., 1992, Un estudio arqueometalúrgico in Las fíbulas de la provincia de Albacete, Serie I, N. 66, Albacete, p. 293-312.

SILVESTRINI M., 1991, Cingoli (MC), loc. Moscosi: insediamento dell'età del bronzo e del ferro, in Scavi e ricerche nelle Marche. Introduzione alla mostra, Urbino, p. 15-17.

SILVESTRINI M., 1992, Marche. Cingoli (Fraz.Moscosi, loc.Piano di Fonte Marcosa, Prov.di Macerata). Scoperte e scavi preistorici in Italia nelgli anni 1990 e 1991. Notiziario, Rivista di Scienze Preistoriche XLIV, p.259.

SILVESTRINI M., 1994, Marche. Cingoli (Fraz.Moscosi, loc.Piano di Fonte Marcosa, Prov.di Macerata). Scoperte e scavi preistorici in Italia nell'anno 1994. Notiziario, Rivista di Scienze Preistoriche XLVI, p.229.

SILVESTRINI M. & PIGNOCCHI G., in print, L'insediamento dell'età del Bronzo di Moscosi di Cingoli (MC) (campagna di scavo 1992), Rivista di Scienze Preistoriche.

SVOLJŠAK D., 1974, Raziskovanje prazgodovinske naselbine na Mostu na Soči, Goriški letnik 1, Nova Gorica, p. 5-32.

SVOLJŠAK D., 1977, Abitato preistorico a Most na Soči, in Le età del bronzo e del ferro nell'Isontino, Atti del convegno, Gorizia, pp. 23-37.

SVOLJŠAK D, 1979, The prehistoric settlement at Most na Soči, Archaeologia Iugoslavica 17, Beograd, pp. 13-20.

SVOLJŠAK D, 1980, Prazgodovinska naselbina na Mostu na Soči, Situla 20-21, Ljubljana, pp. 187-197.

SVOLJŠAK D., 1983a, Santa Lucia - Most na Soči in Preistoria del Caput Adriae, (ed. Comune Trieste), pp. 159-164.

SVOLJŠAK D,1983b, Most na Soči (S.Lucia) e i suoi sistemi di difesa in Preistoria del Caput Adriae, Atti del convegno, Trieste, pp. 115-118 .

TERŽAN B., LO SCHIAVO F., TRAMPUŽ -OREL N., 1984/1985, Most na Soči (S.Lucia) II, Katalogi in Monografije 23/1-2, Ljubljana.

TOSI M., 1974, The Lapis Lazuli trade across the Iranian Plateau in the 3rd Mill. BC., Gururajamanjarika, in onore di G. Tucci, I, p.3-22.

TRAMPUŽ -OREL N., MILIČ Z., HUDNIK V., OREL B., 1991, Inductively coupled Plasma-Atomic Emission Spectroscopy Analysis of metals from Late Bronze Age hoards in Slovenia, Archaeometry 33, pp. 267-277.

TRAMPUŽ -OREL N., 1996, Spektrometri ne raziskave depojskih najdb pozne bronaste dobe/Spectrometric research of the Late Bronze Age Hoard Finds, in Depojske in posamezne kovinske najdbe bakrene in bronaste dobe na Slovenskem/Hoards and Individual Metal Finds from the Eneolithic and Bronze Ages in Slovenia II, B.Ter an ed., Katalogi in Monografije 30, Ljubljana, p. 165-242.

TRAMPUŽ -OREL N., HEATH D.J., 1998, Analysis of heavily leaded shaft-hole axes, in Mensch und Umwelt in der Bronzezeit Europas/Man and Environment in European Bronze Age, B. Hänsel ed., Kiel, p. 237-248.

TRAMPUŽ -OREL N., HEATH D.J., HUDNIK V., 1998, Chemical analysis of Slovenian bronzes from the Late Bronze Age, in L'atelier du bronzier en Europe du XXe au VIIIe siècle avant notre ère, Vol. I, Proceedings of the International Conference Bronze '96, C.Mordant, M.Pernot, V.Rychner ed., Neuchatel et Dijon, I, Dijon, p.223-237.

VITRI S., CORAZZA S., GARNA G., GIUMLIA-MAIR A., 2001, Scavi sistematici, Paularo, Misincinis. Scavi 2001. in I Celti in Friuli: archeologia, storia e territorio,I 2001., Aquileia Nostra LXXII, 409-415.

VON MERHART G., 1952, Studien über einige Gattungen von Bronzegefäßen, Hallstatt und Italien, in: Gero von Merhart, Hallstatt und Italien, Gesammelte Aufsätze zur frühen Eisenzeit in Italien und Mitteleuropa, Festschrift des RGZM, II, Mainz.

YENER K.A., 2000, The domestication of metals: the rise of complex metal industries in Anatolia, Leyden.

IIème Section – Archéométallurgie dans les grandes iles de la Méditerranée Chypre, Sardaigne, Sicile

IIIrd Section – Archaeometallurgy on major Mediterranean Islands: Cyprus, Sardinia, Sicily

THE TRADE OF TIN AND THE ISLAND OF COPPER

Vasiliki KASSIANIDOU

Résumé: Le rôle de Chypre en tant que producteur et exporteur de cuivre est bien connu. Ce rôle est attesté à la fois par les témoignages archéologiques provenant de l´île et d´ailleurs, et par les témoignages textuels (si l´on accepte l´identification d´Alashiya avec Chypre). Le cuivre chypriote, sous la forme de lingots en forme de peau de bœuf, a été trouvé depuis la côte bulgare au nord jusqu´au Delta du Nile au sud, et depuis la côte de Haiffa à l´Est jusqu´en Sardaigne à l´ouest. Cependant, au lieu de cuivre, pour qu´il soit utilisable, il devait être mélangé avec un autre métal. Jusqu´à l´âge du Bronze Moyen le métal utilisé avec le cuivre pour produire du bronze était l´étain. Puisque l´étain n´ était pas disponible sur l´ île, on devait l´importer. Le but de cette communication a deux aspects: elle va considérer l´effet que l´introduction de l´étain bronze a eu sur la société de Chypre, et par conséquent le rôle possible que l´ île ait joué dans le commerce de l´étain à la Méditerranée orientale. Que Chypre a pu être invoquée dans le commerce de l´étain dans la Méditerranée orientale est sous-entendu par le fait qu´un certain nombre des lingots d´étain connus, par exemple ceux transportés par le bateau d´Uluburun et ceux trouvés à la côte de Kfar Samir, près de Haiffa, sont inscrits avec des symboles dont quelques-uns ont été identifiés comme Cypro-Minoens.

Abstract: The role of Cyprus as a copper producer and exporter is well known. This is supported both by archaeological evidence from the island and abroad and by textual evidence (if one accepts the identification of Alashiya with Cyprus). Cypriot copper in the form of oxhide ingots has been found from the Bulgarian coast in the north, to the Nile Delta in the south, and from the coast of Haifa in the east, to Sardinia in the West. In order for copper, however, to be useful it had to be alloyed with another metal. By the Middle Bronze Age the metal used with copper to produce bronze, was tin. Tin was not available on the island, and therefore, it had to be imported. The scope of this paper is twofold: it will consider the effect the introduction of tin-bronze had on the society of Cyprus and subsequently the possible role of the island in the trade of tin in the Eastern Mediterranean. That Cyprus may have been involved in the trade of tin in the Eastern Mediterranean is implied by the fact that a number of the known tin ingots, for example those carried by the Uluburun ship and those found at the coast of Kfar Samir, near Haifa, are inscribed with symbols some of which have been identified as Cypro-Minoan.

INTRODUCTION

The role of Cyprus as a copper producer and exporter is well known (Fig.1). This is supported both by archaeological evidence from the island and abroad, and by textual evidence, if one accepts the identification of Alashiya with Cyprus (Muhly, 1972: 217; Knapp, 1996b: 3, 11; *contra* Merrillees, 1972, 1987; Bass, 1997: 156; Bass, 1967: 77). Cypriot copper in the form of oxhide ingots has been found from the Bulgarian coast in the north (Dimitrov, 1979: 73), to the Nile Delta in the south (Pusch, 1995: 123), and from the coast of Haifa in the east (Galili et al., 1986: 32) to the island of Sardinia in the West (Lo Schiavo, 1989; 1998) (Fig. 2). In order for copper, however, to be turned into a harder and, therefore, more useful material, it had to be alloyed with another metal. The first alloys to be used were those of copper and arsenic but by the Middle Bronze Age the metal used with copper to produce bronze, was tin (Charles, 1980: 168, 172,). Tin was not available on the island, and therefore, it had to be imported. The scope of this paper is twofold: on the one hand it will consider the effect that the introduction of tin-bronze had on the Bronze Age society of the island and on the other it will contemplate the possible role of Cyprus in the trade of tin in the Eastern Mediterranean.

CYPRUS AND COPPER

Cyprus, the third largest Mediterranean island, is still today considered to be one of the richest countries in copper per surface area in the world (Constantinou, 1982: 15). The exploitation of these copper deposits must have started in the Early Bronze Age, although the earliest archaeological evidence we have relating to the mining and smelting of copper ores dates to the Middle Bronze Age or the so-called Middle Cypriot Period (1900-1600 BC) (see below). Copper, however, is but one of Cyprus's natural resources. Equally significant although often forgotten, are the now somewhat depleted forests of Troodos, which in Antiquity provided fuel for the smelting furnaces, the pottery kilns, and other installations, but also timber for the construction of buildings and most importantly of ships. According to Muhly (1986: 49): *"What the Amarna letters from the king of Alashiya, emphasize most of all is Cyprus as a source of wood and a centre of shipbuilding"*. The fertile Mesaoria plain provided arable land for the cultivation of grain, while the foothills of Troodos ideal conditions for the cultivation of the vine and the olive. This natural wealth is made even more important by the island's geographic position in the Eastern Mediterranean, in the crossroads between the East and the West. The natural wealth in metal and timber and the geographical position of the island could not but lead it, eventually and rather belatedly, to the centre of the intricate system of trade which started to develop in the Middle Bronze, peaked in the Late Bronze Age and finally collapsed during the 12th century BC. Central to this trade was the export of copper from Cyprus to the neighbouring countries both East and West and the import of metals absent from the island such as gold, silver, lead and of course tin.

TIN AND TIN-BRONZE IN CYPRUS

Tin-bronze makes its first appearance in Cyprus in the transition between the Early Bronze Age and the Middle

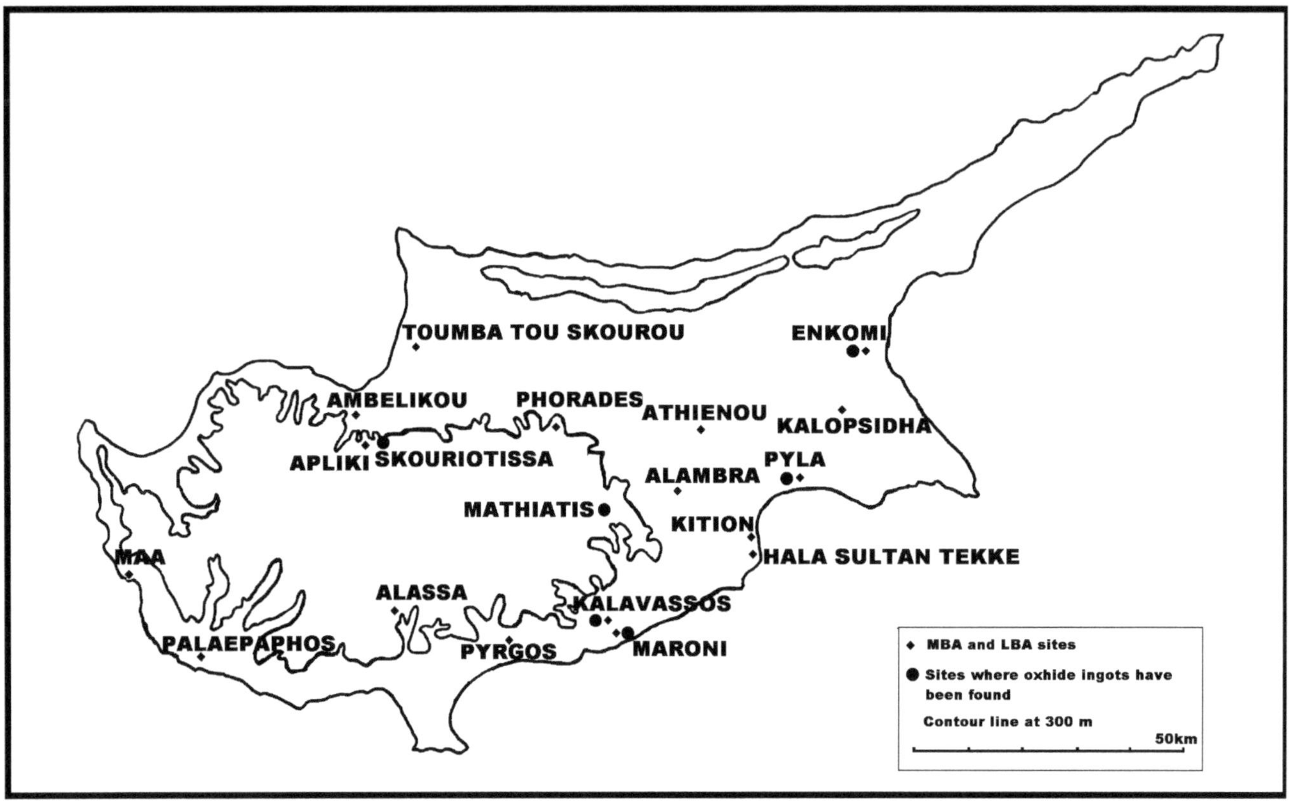

Figure 1. Map of Cyprus showing MBA and LBA sites mentioned in the text, and sites where copper oxhide ingots have been found.

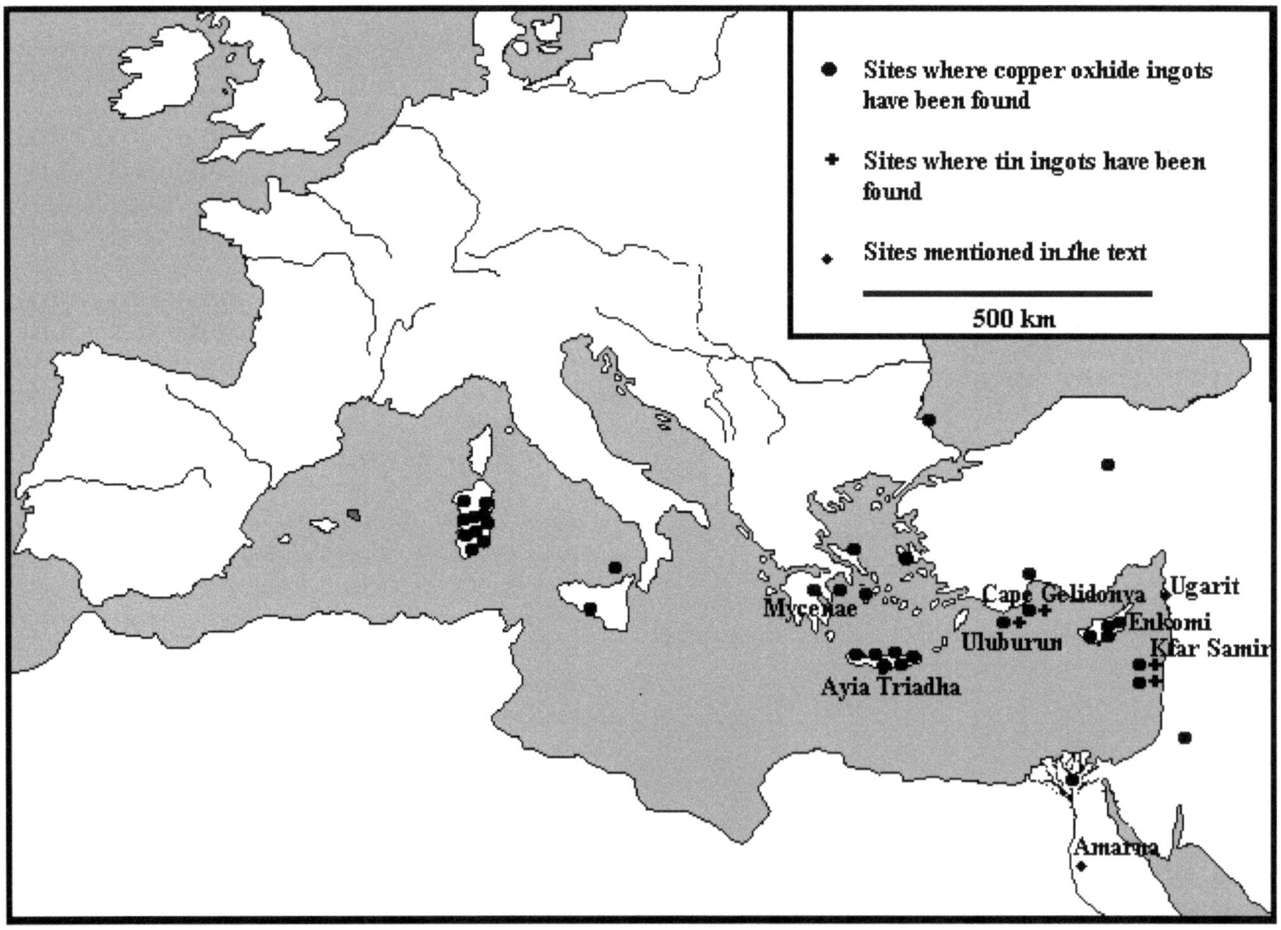

Figure 2. Map of the Mediterranean showing sites mentioned in the text, and sites where copper oxhide and tin ingots have been found.

Bronze Age (i.e. Early Cypriot III (2000-1900) to Middle Cypriot I (1900-1800 BC), but the few early objects which on analysis were found to contain tin, are thought to have been imported (Weinstein Balthazar, 1990: 161). Arsenical copper and unalloyed copper, however, retain their predominance during the Early Bronze Age and most of the Middle Bronze Age. Lead isotope analysis indicates that the copper, with only few exceptions, is local (Stos-Gale and Gale, 1994: 116). It was already known, through Dikaios excavations at Ambelikou, that the exploitation of the rich copper deposits of the Troodos had started in the Middle Bronze Age (Dikaios, 1946; Merrillees, 1984). Recent excavations by Belgiorno (1999; 2000) at the site of Pyrgos-Mavrorachi, where significant amounts of copper smelting slag have been uncovered in what looks like an industrial setting, strongly support the early findings. It is also significant that the analysis of the slags from this site undertaken by Giardino (2000: 23) have shown that the copper was extracted from sulphide ores.

As for the arsenic, although the Cypriot massive sulphide-copper ores have a relatively low arsenical content, there are polymetallic ores rich in arsenic in what is called the Limassol Forest area (Bear, 1963: 36). According to geological reports the arsenic content in the polymetallic ores of Pevkos mines ranges between 0,5 and 7,6 % (Gass et al., 1994: 185). It has, therefore, been suggested that these may have been exploited in antiquity, and that they were mixed with copper metal to produce arsenical bronze (Swiny, 1982: 71). This would make Cyprus self sufficient, as far as, base metals are concerned, which is quite significant since in the Early Bronze Age and the beginning of the Middle Bronze Age, Cyprus has only limited contacts with its neighbouring lands (Knapp and Cherry, 1994: 42).

It is in the early 2nd millennium, the Middle Cypriot I (1900-1800 BC), that the first objects of tin-bronze, which may be locally made, appear. However, according to Weinstein Balthazar (1990: 161), these may in fact have been made of recycled tin-bronze deriving from imported finished objects. That tin-bronze was recycled, seems to be indicated by the significant number of tin-arsenic-copper alloys, which are believed to have been the result of mixing local arsenical copper with imported tin-bronze (Weinstein Balthazar, 1990: 162). Tin-bronze takes quite some time to get established and by Middle Cypriot II (1800-1725 BC) or Middle Cypriot III (1725-1600 BC) it is still only about half the objects produced that were made of tin-bronze or tin-arsenic-copper alloys. Furthermore, the smiths seem to reserve this new alloy only for certain types of the limited repertoire of metal objects (Weinstein Balthazar, 1990: 431). At the beginning tin-bronze was used for riveted knives and toggle pins, while in the Middle Cypriot III period shaft hole axes are also made of this alloy. Hook tang weapons, scrapers, awls, tweezers and pins were still made of arsenical bronze until the end of the Middle Bronze Age.

Interestingly enough, based on the archaeological evidence both from Cyprus and abroad, the second half of the Middle Bronze Age marks the beginning of Cyprus' involvement in the complex, international, state controlled and centred on palace economies, trade system of the Eastern Mediterranean. Cypriot pottery has been recovered from various sites of this period in the Levant, Egypt and Anatolia (Muhly, 1985: 29; Knapp and Cherry, 1994: 42). Although pottery, understandably forms the majority of Cypriot exports (since ceramics, unlike copper, are much more visible archaeologically, due to the nature of the material which is durable and of course cannot be recycled), the written sources suggest that Cyprus (if we accept the identification of Cyprus with Alashiya) has already started to supply copper to the states of the Near East. Some of the earliest references to the copper and bronze from Alashiya comes from Mari and is dated to the first half of the eighteenth century BC (Muhly, 1972: 204).

It is in the Late Bronze Age (1600-1050 BC) that tin-bronze is finally established and, I believe that this is intimately associated with other major changes and developments that take place on the island. In the Late Bronze Age large urban centres are established on the coast, mainly in the south, south-east and south-west in sharp contrast with the previous period where the majority of sites is located further inland in arable lands and in close proximity to the Troodos (Knapp, 1994: 283). Monumental public and religious architecture, writing (in the form of the Cypro-Minoan script which until today remains undeciphered), foreign imports and burials of individuals who clearly belong to a different class mark the emergence of a complex society (Knapp, 1994: 282). The sharp increase in the amount and variety of foreign imports clearly show Cyprus' active involvement in the trade mechanisms of the Eastern Mediterranean (Knapp and Cherry, 1994: 43-6). Finally, at this time Alashiya appears in Egyptian, Hittite, and Near Eastern texts commonly as a producer and exporter of copper (Muhly, 1972: 209-212; Knapp, 1996b: 8). These texts indicate the presence of a king who corresponded with the Pharaoh of Egypt and the king of the Hittites and one who used state agents (tamkaru) to conduct foreign trade (Knapp, 1994: 287).

The prosperity of the island is believed to have derived from the ever-increasing trade in ceramics, raw materials, and organic products, with the Near East and the Aegean (Karageorghis, 1996: 66). The main product exported from Cyprus was, however, copper and, in the Late Bronze Age there is a marked increase in the number and variety of metallurgical finds from practically all excavated sites (Knapp, 1994: 287; Knapp 1986, 44; Muhly 1996, 47; Muhly 1989, 301-2). The fact, that copper workshops have been found in the precincts of sanctuaries, in Kition and Athienou, as well as, the discovery of two bronze cult figures shown standing on an copper ox-hide ingot [the Ingot God (Schaeffer 1965) and the Bomford Goddess (Catling 1971)] have led some to argue that the copper industry was intimately associated with religion (Karageorghis and Demas, 1985: 253-4; Karageorghis, 1976: 170). Knapp (1996a: 10) has opposed this idea and has suggested instead that:

> *To organize and legitimise control over an island culturally if not politically divided, developing elites established unequal access on essential resources or prestige goods, and adopted legitimising symbols (e.g. miniature 'Votive'*

> *ingots, bronze figurines and stands, clay cylinder seals) that enabled them to co-opt goods and labour, ostensibly on the community's behalf, for their own political and economic goal* (Knapp, 1996a: 17).

I believe that the introduction of tin-bronze, a material that although had comparable mechanical properties to arsenical bronze was nevertheless preferable because it enabled the smith to produce a more consistent and reliable alloy (Charles, 1978: 29), strengthened the status of the ruling class. By introducing an alloy, which requires the use of a metal that was not locally available but could only be acquired through international trade, the elites could control one of the essential commodities for the smooth function of the economy. The control of the import and therefore access to tin, the material with which bronze tools, weapons and other objects necessary for all aspects of the economy were made, was a much more critical factor in this effort to establish power over the rest of the population, than the control of other exotic prestige goods such as foreign pottery, ostrich eggs, faience etc.

With the introduction of tin-bronze, the craft of metalworking was greatly developed and apart from a greater variety of tools and weapons than the limited one which characterized the Middle Bronze Age, the workshops of Enkomi, Hala Sultan Tekke and other sites produced true masterpieces such as the tripod and four sided stands, which were widely exported, and the Horned God which remains one of the largest cast bronze statues of this period in the Mediterranean world (Papasavvas, 2001: 214-5). These objects gave further prestige to their owners, the elites who were based in the urban centres (both coastal and inland), and made them stand apart even more from the inhabitants of the small inland agricultural and industrial villages, who produced the agricultural products and the copper on which the economy was based (for discussions of the proposed model of a threefold settlement system in Late Bronze Age Cyprus, see: Catling, 1962; Keswani, 1993; and Knapp, 1997).

But this is only half of the story. I believe, as I will try to show, that, the same elite, which was controlling the import and distribution of tin within Cyprus, was also involved in the distribution and trade of tin to other major centres of this period.

CYPRUS AND THE TRADE OF TIN

In a paper published in 1977, Maddin, Wheeler and Muhly (1977: 45) announced the discovery of two tin ingots, on the surface of which were inscribed two marks (Fig. 3). The two ingots had been bought for the Haifa Municipal Art Museum and lacked secure archaeological context (Artzy, 1983: 52). However, the marks had been identified as belonging to the Cypro Minoan script and, therefore, the ingots were dated to the Late Bronze Age (Maddin et al., 1977: 46).

Although, analysis had shown that tin was among the materials carried by the ship that sank off the Cape Gelidonya (Bass, 1967: 83; Maddin et al., 1977: 44), it was not certain that it

Figure 3. The two inscribed tin ingots discovered in Haifa (Wheeler, 1977: 25).

was in the form of ingots and some did suggest that it may have been the tin ore, cassiterite (tin oxide), thus fuelling the debate once again on whether tin-bronze was produced by mixing copper with metallic tin or with cassiterite (Charles, 1978: 26).

The discovery of the tin ingots, therefore, was extremely significant because here was finally proof that tin was traded in its metallic form. Even more significant, for the purpose of this paper, was that the marks were identified as Cypro-Minoan, and this led the three scholars, R. Maddin, T. Wheeler and J.D., Muhly, to the following conclusion (Maddin et al., 1977: 46):

> *Secondly it is probable that, both metals necessary in the making of bronze, were distributed by an administrative complex centred on Cyprus. Although the source of the tin is unknown, it passed through Cyprus where it received the markings, which are also found on some copper ingots of LBA date.*

In other words, from the moment of the first discovery of tin ingots, the question of Cyprus' involvement in the trade of this metal was also raised, and it is this issue that I will develop in the remainder of this paper.

A few years later, Artzy (1983: 52) published two more ingots, which were found in a car workshop in Haifa, where they were being used for soldering broken radiators. The new ingots were not only identical in size and shape with the previous two, but they were also engraved with two marks. In fact, one of the ingots had identical marks to one of the earlier discovered ones (Artzy, 1983: 52). The four ingots, therefore, were rightly identified as belonging to the same group or cargo of a ship. There was one difference, however, and that was the presence of a moulded head on one of the new ingots, which clearly had been produced when

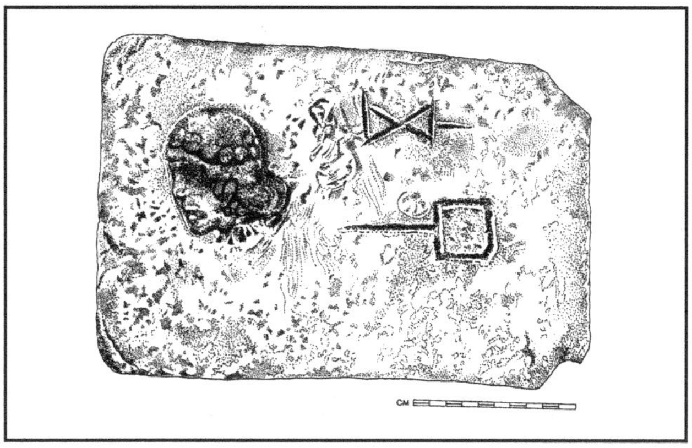

Figure 4 Inscribed tin ingot with a moulded head, from Haifa (Artzy, 1983: 53).

the ingot was cast (Fig. 4). On stylistic grounds Artzy (1983: 52) dated the figure to the fifth century BC and identified her as Arethusa, a fountain goddess who appears on Syracusan coins of this period. She then proceeded to search for the engraved symbols on syllabaries of the fifth century BC which led her to suggest that the symbols may in fact be Iberian (Artzy, 1983: 53). She, therefore, cautiously suggested that the ingots might have been produced in Iberia, sometime in the first quarter of the 5th century BC (Artzy, 1983: 54). I have not located any published reactions to this proposition, which refuted both the Late Bronze Age date of the tin ingots, and the allocation of the marks to the Cypro-Minoan script.

The sea of Haifa in two different locations, however, produced even more inscribed tin ingots, which were found together with other objects clearly dating to the Late Bronze Age. The first group of five tin ingots was discovered at the site of Kfar Samir (at the Hishule Carmel coast) in association with a copper oxhide ingot, as well as, perforated stone anchors of a type which is usually identified as coming from either Cyprus or the Syro-Palestinian coast (Galili et al, 1986: 25). The ingots weigh between 2,2 and 4,2 kg and have an irregular plano-convex shape (Galili et al, 1986: 25). The marks were inscribed on the flat side of three of the ingots when the metal had hardened (Fig. 5a-c); although not certain, the excavators tentatively identified them as belonging to the Cypro-Minoan script (Galili et al, 1986: 32).

In the same general area of the Kfar Samir coast, another group of tin and lead ingots was discovered together with perforated stone anchors and other objects, some of which are of Egyptian provenance (Raban and Galili, 1985: 326). The finds were dated to the 14th-13th century BC (Raban and Galili, 1985: 327). A total of ten tin ingots were recovered, eight of which were bar-shaped (Raban and Galili, 1985: 327). The ingots were badly corroded and only one of them clearly bore inscribed marks, which unfortunately could not be clearly seen. The other two ingots were much larger, weighing 27 and 37 kg each, and hemispherical in form. One of the two had been sawed in half in antiquity. As they are heavily concreted, it was very difficult to see whether they had also been inscribed. The five small lead ingots, of varied shape and weight ranging from to 0.68-6.94 kg, on the other hand, all bore inscribed marks, which were tentatively identified as Cypro-Minoan (Raban and Galili, 1985: 327; Misch-Brandl et al., 1985: 9).

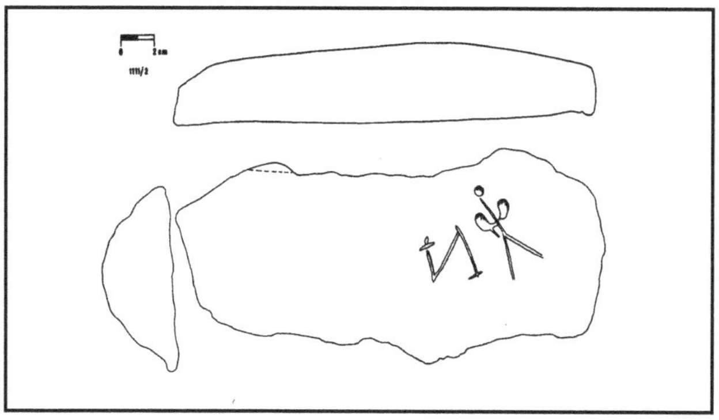

Figure 5a. Inscribed tin ingot, from Hishule Carmel (Galili, et al. 1986: 29).

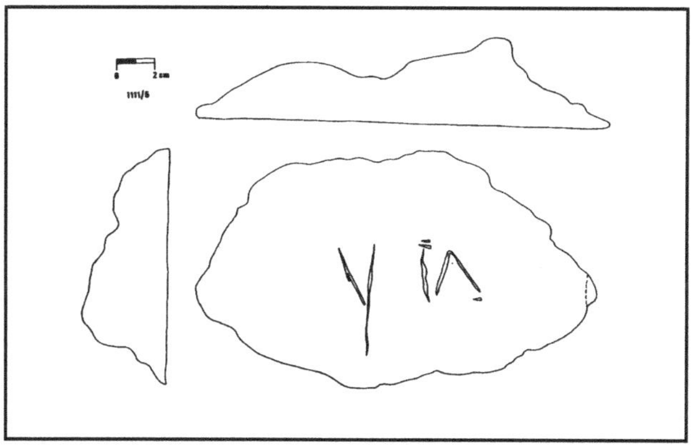

Figure 5b. Inscribed tin ingot, from Hishule Carmel (Galili, et al. 1986: 29).

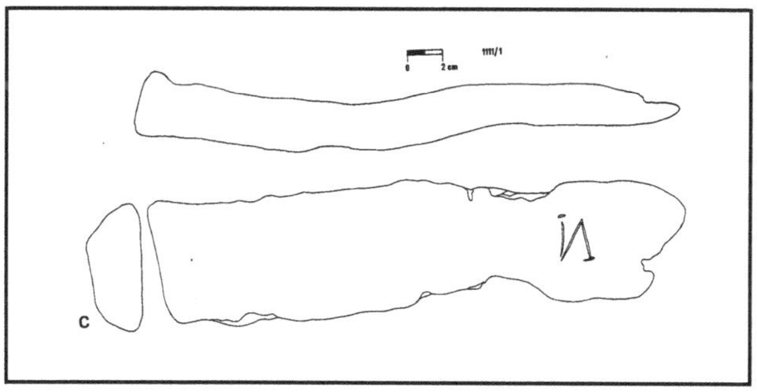

Figure 5c. Inscribed tin ingot, from Hishule Carmel (Galili, et al. 1986: 30).

Thus, apart from the initial four ingots whose Late Bronze Age date had been dismissed, a total of fifteen tin ingots, which dated to the Late Bronze Age and were inscribed with marks that could very possibly belong to the Cypro-Minoan script, were recovered from the Kfar Samir coast. Although the question of Cyprus′ involvement in the trade of tin during the Late Bronze Age was back on the table for discussion, it has not really been addressed again and the discoveries in Haifa received less attention than they deserve.

This is possibly because in the same period a much more impressive shipwreck was discovered, and subsequently excavated, off the south-western coast of Anatolia, at the site of Uluburun also known as Kas (Bass, 1986; Bass et al,

1989; Pulak, 1997; Pulak, 1998; Pulak, 2000a; Pulak, 2000b; Pulak, 2001; Sibella, 1996). The wealth and variety of the cargo led to the conclusion that this could only have been a royal ship, which loaded with finished objects and raw materials in a port in the Eastern Mediterranean was heading West (to the Aegean?) to an unfortunately unknown destination (Pulak, 2001: 48-9; Pulak, 1998: 218-9; Pulak, 1997: 255-6). As it is well known, the ship carried almost ten tons of copper in the form of 354 oxhide ingots, and 121 bun and oval ingots (Pulak, 1997: 235; Pulak, 1998: 193; Pulak, 2000a: 140-5). More significantly it carried almost a ton of metallic tin in the form of oxhide ingots, bun ingots, rectangular slabs, as well as, sections of large thick disks and a unique example similar in shape to a stone anchor with a large hole at the end (Pulak, 1997: 239; Pulak, 1998: 199; Pulak, 2000a: 150-1: Maddin, 1989: 101-4). Many of the tin oxhide ingots had been sectioned in antiquity to halves, and quarters (Pulak, 1997: 239; Pulak, 1998: 199; Pulak, 2000a: 150; Pulak, 2001: 22).

The following discussion is based on the latest available publications on the shipwreck and may have to be modified as new results on this material are published, especially since at the time of the publications conservation and study of the ingots was still underway. At least 160, of the 176 copper oxhide ingots that had been cleaned and conserved, bear marks on the rougher side (Pulak, 2000a: 141). The oxhide ingots were incised with one, and some with as many as three marks (Pulak, 1997: 235; Pulak, 1998: 194, Pulak, 2000a: 146) and they had been stored in such a way on the ship, that the marks would have been visible (Pulak, 2000a: 141). More than half of the bun ingots were also inscribed, but they bear only single marks incised on the smooth mould side (Pulak, 1998: 134; Pulak, 2000a: 146). Only one of the three complete tin oxhide ingots recovered from the site was conserved and it was inscribed, as were the sectioned tin ingots (Pulak, 1997: 240; Pulak, 2000a: 150, 153). In the latter the mark was centred on the fragment, indicating that the incision had been made after the ingots had been sectioned (Pulak, 1997: 240; Pulak, 2000a: 153).

Thirty-two marks have thus far been identified on the Uluburun ingots (for a published table of the marks see Sibella, 1996: 9; Pulak, 1998: 196; Pulak, 2000a: 146). These are often repeated on the different shaped copper oxhide ingots and the copper bun ingots, while one mark is found both on a tin oxhide ingot and a copper oxhide ingot (Sibella, 1996: 10; Pulak, 1998: 194, 196). It is remarkable that among the 32 marks, there are some that appear on copper oxhide ingots found in Ayia Triadha in Crete, in San Antioco di Bisarcio in Sardinia and in Enkomi in Cyprus (Sibella, 1996: 10). Unlike most of the oxhide ingots, from the Cape Gelidonya (Bass, 1967: 52), and those found in various other Late Bronze Age sites such as the ones from Enkomi (Catling, 1964: 267-8), Mycenae (Buchholz, 1959: 36) Sardinia, (Vagnetti and Lo Schiavo, 1989: 225) and elsewhere, which bore impressed marks, the ingots from Uluburun all bore inscribed marks.

As stamped marks have to be produced when the metal has not yet fully hardened, it has been suggested that these must have been produced just after the casting of the molten metal, in other words at the place of production (Bass, 1967: 72). There were, therefore, termed "primary" marks by Bass (1967: 74). Inscribed marks, on the other hand are considered to be "secondary", as they could have been produced either at the place of production, or at the point of export and even at the point of import. The question of the meaning of the primary and secondary marks had been extensively discussed by Bass (1967: 72-4), who suggested that the primary marks identified the source of copper, while the secondary marks, may have been incised by a person inspecting the material ready for export, by the merchant who received them or even by the merchant who had bought them. The excavators of Uluburun, where literally hundreds of inscribed ingots were recovered, also tried to address the same question (Sibella, 1996: 10; Pulak, 2000a: 146). According to Pulak (1998: 196):

> *That the Uluburun marks were all incised and not stamped during casting or cooling (as were many from Cape Gelidonya) leads to the belief that they probably were made at some point of receipt or export rather than at the primary production centre or centres. This supposition finds a basis in the form of a similar mark on a tin ingot from Uluburun. As tin and copper would have been mined in different geographical regions, it is highly unlikely that the same mark would have been incised on ingots of dissimilar metals and of diverse origins unless this was done at a centre that handled both ingot types during the distribution process.*

Thus it is significant to assign the marks to a script as this may give us some clues regarding the point of manufacture of the metal ingots, as far as the primary marks are concerned, and the centres involved in the trade of metals as far as the secondary marks are concerned.

According to Sibella (1996: 10), the member of the Uluburun team who has been studying the ingots and their inscribed marks, of thirty marks identified so far, at least five belong to the Cypro-Minoan script. Pulak´s (1998: 195; 2000a: 46) only comment on the marks, however, is the following:

> *"The marks vary in shape from a simple cross to more complicated forms. It is of interest to note that an appreciable number are associated with the sea and ships: the fishhook in several forms, a trident, a fish, possibly quarter rudders and even a sailing boat". This suggests that the people who incised the marks were closely involved with the sea. Therefore, it is more likely that the marks were placed on the ingots at locations near the sea rather than inland".*

This I believe is a very simplistic argument; especially if we take into consideration the fact that the so-called trident is one of the signs found not only in the Cypro-Minoan script but also in Linear A (Daniel, 1941: 256; for a table of Cypro-Minoan signs see Masson, 1974: 12-15). In fact, even someone who is not an expert on the Cypro-Minoan script can pick out more signs that may very well belong to the Cypro-Minoan script. One of the marks that have not been identified by Sibella as Cypro-Minoan, is the one found both on a copper ingot and a tin ingot[1] (Pulak, 1998: 196). Yet this is a sign that belongs to the Cypro-Minoan script (CM 6, Masson, 1974: 13) and that survives in the Cypriot Syllabary of the Iron Age representing the sound *pa* (Masson, 1995: 66). The same mark is found stamped on an ingot from Cape Gelidonya (Bass, 1967: 72), inscribed on Cypriot ceramics (Daniel, 1941: 274), Mycenaean sherds from Ugarit (Hirschfeld, 2000: 188) and Cyprus (Daniel, 1941: 277), as well as, Cypriot ceramic loomweights (Smith, 1994: 230, 232). The problem is of course that the same sign is found in other scripts such as Linear A and Linear B (Daniel, 1941: 254). Nevertheless, Pulak does not even consider the slight possibility that the marks inscribed on the ingots may belong to the Cypro-Minoan script.

This is, perhaps, the result of the general reluctance one detects in both Pulak and Bass to accept even the possibility that the ship could be a Cypriot one. The following statement by Bass (1991: 75) is I believe characteristic:

> *If the Uluburun ship was Alasian, that would be strong evidence against the unproven identification as Cyprus, for there is no reason to believe that the ship was Cypriot, although much of its cargo was Cypriot: six tons of four handled copper ingots and several pithoi filled with unused Cypriot pottery (including Base Ring II Ware, White Shaved Ware, White Slip Ware, Bucchero Ware, unused lamps, and "wall brackets")*

Although both often stress how difficult it is to identify the nationality of a ship (Bass, 1991: 70; Pulak, 1997: 252), they strongly support that the ship was of a Syro-Palestinian or Levantine origin (Pulak, 1998: 218; Pulak, 1997: 252). This in spite of the fact that when looked at closely the evidence presented is not conclusive: the anchors of the ship belong to a type common to Cyprus and the Syro-Palestinian coast (Pulak, 1998: 218; Bass, 1991: 74), the weights belong to a standard employed both in Cyprus and the Syro-Palestinian coast (Pulak, 2000b: 259, 264), the wood (cedar) (Pulak, 1998: 213) that was used to make the boat could have come either from Cyprus or the Syro-Palestinian coast, the type of bush (thorny burnet) used for dunnage in the bottom of the hold and below the copper oxhide ingots (Pulak, 1998: 197) is common both in Cyprus and the Levant, etc. In the effort to identify the nationality of the boat they turn to the used, utilitarian ware found on the ship, as they believe that it would be characteristic of the region where the ship originated. However, the so-called "ship's galley ware" has not yet been identified and studied in detail (Pulak, 1998: 216). The only supporting clue that remained was that among the two types of lamps that the ship was carrying, Cypriot and Syro-Palestinian, only the latter seemed to have been used (Pulak, 1998: 218). I am not claiming that the ship is Cypriot, I am only noting that considering the difficulty in assigning nationality to a ship, to the ship's owner, to the ship´s crew, or to the person who perhaps chartered the ship to carry the cargo, I think we should leave this question open and suggest that both possibilities are equally probable. The nationality of the Uluburun ship, however, is not the subject of this paper. The trade of metals and the role of Cyprus is what concerns me and this is the reason why I have spent so much time looking at the marks on the copper and tin ingots.

We already knew that the copper oxhide ingots from various Late Bronze Age sites in the Eastern and Central Mediterranean were marked either with Cypro-Minoan stamped marks modelled on the still soft metal or inscribed after the metal had cooled down. The discoveries of tin ingots at Kfar Samir and at Uluburun have shown that the tin ingots are also inscribed and that many of the marks belong to the Cypro-Minoan script. Although we still cannot understand the significance of these primary and secondary marks they are still very important, because as suggested by Pulak (1998: 196) in the paragraph quoted above, the secondary marks probably relate to the people and the centre that was involved in the trade of these metals.

The strong possibility that the marks found not only on the copper but also on the tin ingots belong to the Cypro-Minoan script leads me to suggest, as indeed Maddin, Muhly and Stech (Maddin et al., 1977, 46; Muhly, 1978: 45; Wheeler, 1977: 24; Wheeler et al., 1979: 148) had suggested more than twenty five years ago, that this centre or centres (taking into consideration the number of LBA Cypriot urban centres on the coast, i.e. Enkomi, Kition, Hala Sultan Tekke, Maroni-Tsaroukkas etc) was located on Cyprus and that it was involved in the distribution of **both** metals around the Eastern Mediterranean.

Nevertheless, because I suspect that such a statement may not be easily accepted I will modify it slightly. I make this comment because even the possibility that a Cypriot international trade centre, such as Enkomi, may have been the Uluburun ship's last port of call is mentioned rather reluctantly, even though the largest volume of the cargo namely the ten tons of copper and the ten Cypriot pithoi filled with ceramics and comestibles originated on the island. Pulak (1997: 253; 1998: 215, 218) and Bass (1997: 155) propose instead that the ship was loaded at Ugarit, where the Cypriot copper, the Cypriot pithoi, the 149 Canaanite amphorae, the tin and the rest of the cargo could be gathered from the various points of origin. Even if we accept this, it does not exclude the possibility of Cypriots, based in Ugarit, being involved or being in charge of the metals trade. After all, Ugarit is the only site outside Cyprus, where we find documents written in the Cypro-Minoan script (Masson, 1995: 63; Masson, 1974).

To support my argument I would like to draw a parallel between the inscribed ingots and the inscribed pottery of this

[1] According to Pulak (1998: 196) the mark that is common to both the tin and the copper ingots is Number 4a in the table of incised marks (compiled by Sibella) that he reproduced in his paper. This mark, as stated above, is one that Sibella has not included in the group of Cypro-Minoan signs, although it is a well-known Cypro-Minoan sign. Sibella (1996: 11), on the other hand, writes that the mark common to the copper and tin ingots is No 7a in her table, and it is one that she did identify as Cypro-Minoan sign (Sibella, 1996: 10). It is impossible to know whether this discrepancy is due to a typographical error, a mistake or because the two authors are referring to two different ingots and therefore, two different marks. One way or another, and which ever of the two statements is correct, as argued above both the marks belong to the Cypro-Minoan and, therefore, can be used to support the arguments I present in the paper.

period. Hirschfeld (1993; 2000; 2001) has been studying and recording Cypro-Minoan signs and inscriptions on ceramic vessels. The marks are inscribed on handles or other obvious locations after firing, which sets them apart from potters marks, which are usually at the bottom of the vase and are made before firing. The Cypriots commonly inscribed different types of their pottery both before and after firing (Hirschfeld, 2001: 123). The practice of inscribing marks extended to imported pottery including Aegean Wares which are found in all the major Cypriot sites of the period, as well as, Near Eastern sites such as Ras Shamra, Minet el Beidha (in other words Ugarit and its port), Alalakh, Byblos, Beth Shan, and many more (Hirschfeld, 1993: 312; Hirschfeld, 2000). The fact that Cyprus has a long tradition of marking pottery, in contrast with the Aegean where the practice is very rare, in combination with the fact that large numbers of Aegean wares of a great variety of shapes marked with Cypro-Minoan signs are found on the island led Hirschfeld (1993: 313) to the conclusion that the incised marks point to a connection with Cyprus and to the following statement:

> *This final observation supports the hypothesis that marking Aegean pottery with signs incised after firing is a specifically Cypriot practice. This in turn, implies that vessels with these signs on them have either been routed via Cyprus at some stage, or that they have been handled by people familiar with the Cypriot marking system.*

I believe it would not be unreasonable for me to borrow Hirschfeld´s ideas and to make a similar statement. In other words, marking of **metal ingots**, whether they are copper oxhide ingots, copper bun ingots, tin oxhide ingots or tin plano-convex ingots, indeed even lead ingots, seems to be a specifically Cypriot practice which implies that the ingots have either been routed via Cyprus or that they have been handled by people familiar with the Cypriot marking system.

It is quite possible that these people were based at Ugarit where the tin would arrive over land from somewhere in the east following a trade route that we know from earlier texts. Let us not forget the text from Mari (Mari text A 1270), which mentions the presence of Cretan merchants who had their elder and an interpreter in Ugarit, where they received shipments of tin from Mari (Heltzer, 1989: 10). These texts date to the reign of Zimirlim, which corresponds to the first half of the eighteenth century (Heltzer, 1989: 25). It is possible that a similar situation existed in the Late Bronze Age at the time of the Uluburun shipwreck and later. Instead of Cretan merchants, Cypriot merchants would have been based at Ugarit, as is witnessed by the presence of texts written in the Cypro-Minoan script found there (Masson, 1974). These merchants would have been in charge of the import of Cypriot copper to Syria and Palestine. They also would have been in charge of receiving metallic tin coming from the East, which they brought back to Cyprus but also traded along with their copper to areas further west (Crete, the Aegean islands and mainland Greece) perhaps even to Egypt. This practice may be indicated in an Egyptian text (Papyrus Anastasi IV, 17.8-9) from the 19th Dynasty, which mentions ingots of copper and bars of tin being brought as presents to the pharaoh, on the neck of the children of Alashiya (Muhly, 1973: 245). According to Muhly (1981: 48): "*Now Cyprus certainly has no tin, so this reference can only mean that tin came to Egypt through Cyprus*".

The involvement of Cypriots is shown by the inscriptions of Cypro-Minoan signs on the ingots a practice which parallels the practice of marking Aegean pottery, the trade of which has often been suggested was also in the hands of Cypriots, because Aegean Wares are usually found in the Near East in association with Cypriot wares (Muhly, 1996: 52).

As pointed out again by Maddin, Muhly and Stech (1977: 46), the possibility that the Cypriots would trade tin, as well as copper need not surprise us. In order to produce bronze both copper and tin, are necessary. Tin, however, is not as readily available as copper and the identity of ancient tin sources creates even today heated discussions among scholars. By including tin along with a shipment of copper, as was done in all three cases of the ancient wrecks dating to the Late Bronze Age, i.e. Uluburun, Kfar Samir and Cape Gelidonya, the Cypriots ensured that their own product would be consumed. Similarly, even if we are dealing with a trade system that is based on gift-exchange between royals a gift that included both tin and copper would have been much more valuable, and much more welcome than one which included only copper.

CONCLUSION

In conclusion, let me recapitulate the basic ideas of this paper. I believe that the emerging elites used the advent of tin bronze on the island to establish their position, by controlling the import and distribution of tin in the island. I further suggest that having understood the importance of tin and the control of the tin supply, Cypriot merchants took an active role in the trade of tin from the Syro-Palestinian coast, where it arrived over land routes from the East, to areas across the sea such as the Aegean, and perhaps even Egypt. Let me take the argument one step further. When the complex system of trading collapsed and Ugarit among many other places was destroyed, this eastern source of tin was cut off. This, however, became an incentive for the search of tin elsewhere. Perhaps this is the main reason why the Cypriots travelled west reaching and establishing strong trade relations with Sardinia, as witnessed by the presence there of Cypriot oxhide ingots (Lo Schiavo, 1989; Lo Schiavo, 1998; Kassianidou, 2001), Cypriot pottery (Vagnetti and Lo Schiavo, 1989: 219-21), as well as, Cypriot smithing tools and tripod stands (Lo Schiavo, 1995: 49; Vagnetti and Lo Schiavo, 1989: 227-231). In Sardinia, they would find tin coming from the island (although Prof. Valera´s paper in this volume will answer the critical question of whether tin deposits on Sardinia had indeed been exploited in the Late Bronze Age) but also from richer sources in Iberia. Current archaeological work has shown that they were already exploited in the Late Bronze Age (Rodriguez Dias et al., 2001).

The Bronze Age metals´ trade has been a subject of paramount concern to many scholars; even so many questions remain unanswered and there is clearly a lot that remains to be learned. I do believe that what is critically needed is a detailed study of the marks on all metal ingots, not just the ones from Uluburun but also those from Cyprus, Crete, Sardinia, and elsewhere, by an expert in very much the same way as it was done for pottery. I hope, and it is a hope that I have expressed to Joanna Smith who together with Nicolle Hirschfeld has been working on creating a corpus of the Cypro-Minoan script (Smith and Hirschfeld, 1999), that someone will take up this project and shed some light to some of the issues raised in this paper.

Acknowledgements

I would like to thank Dr Fulvia Lo Schiavo and Dr Alessandra Giumlia Mair for inviting me to participate on this forum on the problem of ancient tin. I must admit that my initial reaction, one that I expressed immediately when the invitation was forwarded, was to state that I feared I would have nothing to say. After all, the one thing that we all know very well is that Cyprus has no tin. I am relieved that Dr Lo Schiavo did not accept this rush reaction and quickly reminded me that, of course there is tin in Cyprus, only it is in the form of bronze and, therefore, there would be something to say, even if it was just to explain when and how the tin came to the island. As is often the case with other papers written on the subject of Late Bronze Age trade in metals, this one has benefited greatly by the work of the team of Robert Maddin, Jim Muhly and Tamara Stech-Wheeler who as pointed out above presented the same idea almost 25 years ago. I would like to acknowledge their work and express my thanks to them. I would also like to thank Joanna Smith for her advice on issues regarding the Cypro-Minoan script, a subject I know little about. Special thanks to Michal Artzy for allowing me to reproduce the drawings of the inscribed ingots found in the area of Haifa. Last but not least, I would like to thank Maria Iacovou and George Papasavvas for reading through and discussing this paper with me and offering valuable comments and suggestions for improvement, and Phryni Hadjichristophi for the translation of the abstract into French. I hope that the final result will motivate someone to study systematically the primary and secondary marks on the metal ingots from all the sites in Central and Eastern Mediterranean, and shed some light on the burning issue of the metals´ trade.

Author's address

Vasiliki KASSIANIDOU
Archaeological Research Unit
Department of History and Archaeology
University of Cyprus
P.O. Box 20537
Nicosia, CY 1678 CYPRUS
Email: v.kassianidou@ucy.ac.cy

Bibliography

ARTZY, M., 1983, Arethusa of the Tin Ingot. *Bulletin of the American Schools of Oriental Research*, 250, p. 51-55.

BASS, G.F., 1997, Prolegomena to a study of maritime traffic in raw materials to the Aegean during the Fourteenth and thirteenth centuries BC. In, *TEXNH. Craftsmen, Craftswomen and Craftsmanship in the Aegean Bronze Age*. Proceedings of the 6th International Aegean Conference, Philadelphia, Temple University 18-21 April 1996, edited by R. Laffineur and P.P. Betancourt. AEGEAUM, 16, p. 153-170.

BASS, G.F, 1991, Evidence of Trade from Bronze Age Shipwrecks. In, Science and Archaeology: Bronze Age Trade in the Mediterranean, edited by N. H. Gale. *Studies in Mediterranean Archaeology*, 90. Göteborg: Paul Åström's Förlag, p. 69-82.

BASS, G. F., 1986, A Bronze Age Shipwreck at Ulu Burun (Kas): 1984 Campaign. *American Journal of Archaeology*, 90/3, p. 269-296.

BASS, G.F., 1967, Cape Gelidonya: A Bronze Age Shipwreck. *Transactions of the American Philological Society*, 57/8. Philadelphia: The American Philological Society.

BASS, G. F., C. PULAK, D. COLLON, and J. WEINSTEIN, 1989, The Bronze Age Shipwreck at Ulu Burun. 1986 Campaign. *American Journal of Archaeology*, 93, p. 1-29.

BEAR, L.M. 1963, The Mineral Resources and Mining Industry of Cyprus. *Ministry of Commerce and Industry. Geological Survey Department. Bulletin* No. 1.

BELGIORNO, M.R., 2000, Archaeological, metallurgical and historical evidence at Pyrgos (Limassol). *Report of the Department of Antiquities, Cyprus*, p.1-18

BELGIORNO, M.R., 1999, Preliminary report on Pyrgos excavations 1996, 1997. *Report of the Department of Antiquities, Cyprus*, p. 71-80

BUCHHOLZ, H.G., 1959 Keftiubarren und Erzhandel in zweiten vorchristlichen Jahrtausend. *Praehistorische Zeitschrift*, XXXVII/ 1: 1-40.

CATLING, H.W., 1971, A Cypriot bronze statuette in the Bomford collection. In, *Alasia I. Mission Archéologique d'Alasia* 4, edited by C. F.-A. Schaeffer. Paris: Klincksieck, p. 15-32.

CATLING, H.W., 1964, *Cypriot Bronzework in the Mycenaean World*. Oxford: Clarendon Press.

CATLING, H.W., 1962, Patterns of settlement in Bronze Age Cyprus. *Opuscula Atheniensia*, 4, p. 129-169.

CHARLES, J.A., 1980, The coming of copper and copper-base alloys and iron. In *The Coming of the Age of Iron*, edited by Th. A. Wertime and J.D. Muhly. New Haven and London: Yale University Press, p.151-183.

CHARLES, J.A., 1978, The development of the usage of tin and tin-bronze: some problems. In *The Search for Ancient Tin*, edited by A.D. Franklin, J.S. Olin, Th. A. Wertime. Washington, D.C: U.S. Government Printing Office, p. 25-32

CONSTANTINOU, G., 1982, Geological features and ancient exploitation of the cupriferous sulphide ore bodies of Cyprus. In *Early Metallurgy in Cyprus 4000 - 500 BC*, edited by J.D. Muhly, R. Maddin and V. Karageorghis. Nicosia: Pierides Foundation, p.13-23.

DANIEL, J.F. 1941, Prolegomena to the Cypro- Minoan script. *American Journal of Archaeology*, 45, p. 249-282

DIKAIOS, P., 1946, A new chapter in the long history of Cyprus: Wartime discoveries of the earliest Copper Age. *Illustrated London News*, 208/5576, p. 244-245.

DIMITROV, B., 1979, Underwater research along the south Bulgarian Black Sea coast in 1976 and 1977. *The International Journal of Nautical Archaeology and Underwater Exploration*, 8/1, p.70-79.

GALILI, E., SHMUELI, N. and M. ARTZY, 1986, Bronze Age ship's cargo of copper and tin. *The International Journal of*

Nautical Archaeology and Underwater Exploration 15/1: 25-37.

GASS, I.G., MACLEOD, C, J., MURTON, B.J., PANAYIOTOU, A., SIMONIAN, K.O. and C. XENOPHONTOS 1994, *The Geology of the Southern Troodos Transform Fault Zone*. Memoir No. 9. Nicosia: Geological Survey Department.

GIARDINO, CL. 2000, Prehistoric copper at Pyrgos. *Report of the Department of Antiquities, Cyprus*, p.19-32.

HELTZER, M., 1989, The trade of Crete and Cyprus with Syria and Mesopotamia and their Eastern Tin-Sources in the XVIII-XVII Century B.C. *Minos: Revista de filologia Egea* 24, p. 7-28.

HIRSCHFELD, N., 2001, Cypriots to the West? The evidence of their potmarks. In *Italy and Cyprus in Antiquity*, edited by L. Bonfante and V. Karageorghis. Nicosia: The Costakis and Leto Severis Foundation, p. 121-130.

HIRSCHFELD, N., 2000, Marked Late Bronze Age pottery from the kingdom of Ugarit. In, *Céramiques mycéniennes. Ras Shamra-Ougarit XIII* edited by M. Yon, V. Karageorghis and N. Hirschfeld. Nicosia: A.G. Leventis Foundation and Mission Française de Ras Shamra-Ougarit.

HIRSCHFELD, N., 1993, Incised Marks (Post-Firing) on Aegean Wares. In *Wace and Blegen: Pottery as Evidence for Trade in the Aegean Bronze Age 1939-1989*. Proceedings of the International Conference held at the American School of Classical Studies, Athens, December 2-3, 1989, edited by C. Zerner, P. Zerner, and J. Winder. Amsterdam: J.C. Gieben, p. 311-318.

KARAGEORGHIS, V., 1996, Some aspects of the maritime trade of Cyprus during the Late Bronze Age. In *The Development of the Cypriot Economy, from the Prehistoric Period to Today*, edited by V. Karageorghis and D. Michaelides. Nicosia: University of Cyprus, p. 61-67.

KARAGEORGHIS, V., 1976, *Kition. Mycenaean and Phoenician Discoveries in Cyprus*. London: Thames and Hudson.

KARAGEORGHIS V. and M. DEMAS, 1985, *Excavations at Kition V. The Pre-Phoenician Levels Areas I and II. Part 1*. Nicosia: Department of Antiquities, Cyprus.

KNAPP. A.B., 1996a, Power and ideology on prehistoric Cyprus. In *Religion and Power in the Ancient Greek World*. Proceedings of the Uppsala Symposium 1993, edited by P. Hellstöm and Br. Alroth. Boreas 24, Acta Uppsala: Universitatis Upsaliensis, p. 9-25.

KASSIANIDOU, V., 2001, Cypriot copper in Sardinia. Yet another case of bringing coals to Newcastle? In *Italy and Cyprus in Antiquity: 1500-400 B.C.* Proceedings of an International Symposium held at the Italian Academy for Advanced Studies in America at Columbia University, Nov. 16-18, 2000, edited by L. Bonfante and V. Karageorghis. Nicosia: Costakis and Leto Severis Foundation, p. 97-119.

KESWANI, P.S., 1993, Models of local exchange in Late Bronze Age Cyprus. *Bulletin of the American Schools of Oriental Research*, 292, p. 73-83.

KNAPP, A.B., 1997, *The Archaeology of Late Bronze Age Cypriot Society: The Study of Settlement, Survey and Landscape*. Department of Archaeology, University of Glasgow, Occasional Paper Series No.4. Glasgow: University of Glasgow.

KNAPP, A.B., 1996b, Introduction. In *Sources for the History of Cyprus. Volume II Near Eastern and Aegean Texts from the Third to the First Millennia BC*, edited by A. B. Knapp. Altamont, New York: Greece and Cyprus Research Center, p.1-13.

KNAPP, A.B., 1994, Emergence, development and decline on Bronze Age Cyprus. In *Development and Decline in the Mediterranean Bronze Age*, edited by C. Mathers and S. Stoddart. Sheffield Archaeological Monographs 8. Sheffield: John Collis Publications, p. 271-304.

KNAPP, A.B., 1986, Production, exchange and socio-political complexity on Bronze Age Cyprus. *Oxford Journal of Archaeology*, 5, p. 35-60.

KNAPP A.B and J.F. CHERRY, 1994, *Provenience Studies and Bronze Age Cyprus: Production, Exchange and Politico - Economic Change*. Monographs in World Archaeology 21. Madison: Prehistory Press.

LO SCHIAVO, F., 1998, Sardinian oxhide ingots 1998. In *Metallurgica Antiqua*. In honour of Hans-Gert Bachmann and R. Maddin, edited by Th. Rehren, A. Hauptmann, and J.D. Muhly. Der Anschnitt - Beiheft 8. Bochum: Selbstverlag des Deutschen Bergbau Museum, p. 99-112.

LO SCHIAVO, F., 1995, Cyprus and Sardinia in the Mediterranean. Trade routes toward the West. In *Cyprus and the Sea*, edited by V. Karageorghis and D. Michaelides. Nicosia: University of Cyprus and Cyprus Ports Authority, p. 45-60.

LO SCHIAVO, F., 1989, Early metallurgy in Sardinia copper oxhide ingots. In *Old World Archaeometallurgy*, edited by A. Hauptmann, E. Pernicka and G.A. Wagner. Der Anschnitt Beiheft 7. Bochum: Selbstverlag des Deutschen Bergbau-Museums, p. 33-38

MADDIN, R., 1989, The copper and tin ingots from the Kas shipwreck. In *Old World Archaeometallurgy*, edited by A. Hauptmann, E. Pernicka and G.A. Wagner. Der Anschnitt-Beiheft 7. Bochum: Selbstverlag des Deutschen Bergbau-Museums, p. 99-105.

MADDIN, R. WHEELER, T.S. and J.D. MUHLY, 1977, Tin in the ancient Near East. Old questions and new finds. *Expedition* 19 / 2, p. 35-47.

MASSON, E., 1974, *Cyprominoica. Répertoires documents de Ras Shamra-Essais d'interpretation. Studies in the Cypro-Minoan scripts 2.* Studies in Mediterranean Archaeology, XXXI: 2. Göteborg: Paul Åströms Förlag.

MASSON, O., 1995, Les écritures antiques à Chypre. In *Chypre Au Coeur des Civilisations Méditerranéennes*, edited by V. Karageorghis and A. Hermary. Dossiers d' Archeologie, 205, p. 62-67.

MERRILLEES, R.S., 1984, Ambelikou Aletri: A preliminary report. *Report of the Department of Antiquities, Cyprus*, p. 1-13.

MERRILLEES, R.S., 1972, Alasia. In *Proceedings of the First International Cyprological Congress (Nicosia 14-19 April 1969). Volume A*, edited by V. Karageorghis and A. Christodoulou. Nicosia: Society of Cypriot Studies, p. 111-119.

MISCH-BRANDL, O., GALILI, E. and S. WACHSMANN, 1985, *From the Depths of the Sea*. Jerusalem: the Israel Museum.

MUHLY, J.D., 1996, The significance of metals in the Late Bronze Age economy of Cyprus. In *The Development of the Cypriot Economy from the Prehistoric Period to the Present Day*, edited by V. Karageorghis and D. Michaelides. Nicosia: University of Cyprus and Bank of Cyprus, p. 45-60.

MUHLY, J.D., 1989, The organisation of the copper industry in Late Bronze Age Cyprus. In *Early Society in Cyprus*, edited by E. Peltenburg. Edinburgh, Edinburgh University Press in association with The National Museums of Scotland and the A.G. Leventis Foundation, p. 298-314.

MUHLY, J.D., 1986, The role of Cyprus in the economy of the Eastern Mediterranean. In *Acts of the International Symposium "Cyprus Between the Orient and the Occident"*, Nicosia 8-14 September 1985, edited by V. Karageorghis. Nicosia: Department of Antiquities, Cyprus, p. 45-60.

MUHLY, J.D., 1985, The Late Bronze Age in Cyprus: A 25 year retrospect. In *Archaeology in Cyprus 1960-1985*, edited by V. Karageorghis. Nicosia: A.G. Leventis Foundation, p. 20-47.

MUHLY, J.D., 1981, The Bronze Age setting. In *The Coming of the Age of Iron*, edited by T.A.Wertime, and J.D. Muhly. New Haven: Yale University Press, p. 25-67.

MUHLY, J.D. 1978, New evidence for sources of and trade in Bronze Age tin. In *The Search for Ancient Tin*, edited by A.D. Franklin, J.S. Olin, and Th. A. Wertime. Washington, D.C.: U.S. Government Printing Office, p. 43-48.

MUHLY, J.D., 1973, Copper and Tin. The Distribution of Mineral Resources and the Nature of the Metals Trade in the Bronze Age. *Transactions of the Connecticut Academy of Arts and Sciences*, 43. Hamden, Connecticut: Archon Books, p. 155-535.

MUHLY, J.D., 1972, The land of Alashiya. References to Alashiya in the texts of the second millennium BC and the history of Cyprus in the Late Bronze Age. In *Proceedings of the First International Cyprological Congress (Nicosia 14-19 April 1969)-Volume A'*, edited by V. Karageorghis and A. Christodoulou. Nicosia: Society of Cypriot Studies, p. 202-219

PAPASAVVAS, G., 2001, *Chalkinoi Ypostates apo tin Kypro kai tin Kriti (Bronze stands from Cyprus and Crete)*. Nicosia: The A.G. Leventis Foundation.

PULAK, C., 2001, The cargo of the Uluburun ship and evidence for trade with the Aegean and beyond. In *Italy and Cyprus in Antiquity*, edited by L. Bonfante and V. Karageorghis. Nicosia: The Costakis and Leto Severis Foundation, p. 13-61.

PULAK, C., 2000a, The copper and tin ingots from the Late Bronze Age shipwreck at Uluburun. In *Anatolian Metal I*, edited by Ü. Yalçin. Der Anschnitt, Beiheft 13, p. 137-157.

PULAK, C., 2000b, The balance weights from the Late Bronze Age shipwreck at Uluburun. In *Metals Make the World go Round*, edited by C. F. E. Pare. Oxford: Oxbow Books, p. 247-266.

PULAK, C., 1998, The Uluburun shipwreck: An overview, *International Journal of Nautical Archaeology*, 27, p. 188-224.

PULAK, C., 1997, The Uluburun shipwreck. In *Res Maritimae. Cyprus and the Eastern Mediterranean from Prehistory to Late Antiquity*, edited by S. Swiny, R.L. Hohlfelder and H. Wylde Swiny. CAARI Monograph Series. American Schools of Oriental Research Archaeological Reports No 4. Atlanta, Georgia: Scholars Press, p. 233-262.

PUSCH, E.B., 1995, High temperature industries in the Late Bronze Age capital of Piramesse (Qantir): II. A quasi-industrial bronze factory installations, tools and artifacts. In *Proceedings of the First International Conference on Egyptian Mining and Metallurgy and Conservation of Metallic Artifacts,* edited by F. A. Esmael. Cairo: Ministry of Culture, Supreme Council of Antiquities, p. 121-132.

RABAN, A. and E. GALILI, 1985, Recent maritime archaeological research in Israel- a preliminary report. *The International Journal of Nautical Archaeology and Underwater Exploration*, 14/4, p. 321-356.

RODRIGUEZ DIAZ, A., PAVON SOLDEVILLA, I., MEREDITH, C. and J. JUAN I TRESSERAS, 2001, *El Cerro de San Cristobal, Logrosan, Extremedura, Spain. The archaeometallurgical excavation of a Late Bronze Age tin-mining and metalworking site. First excavation season 1998.* BAR International Series 922. Oxford. BAR Publishing.

SCHAEFFER, C.F.-A, 1965, An ingot God from Cyprus. *Antiquity*, XXXIX, p. 56-7.

SIBELLA, P., 1996, The copper oxhide and bun ingots. *The Institute of Nautical Archaeology Quarterly*, 23/1, p. 9-11.

SMITH, J.S., 1994, *Seals for sealing in the Late Cypriot period.* Unpublished PhD dissertation submitted to the Faculty of Bryn Mawr College.

SMITH, J.S. and N. E. HIRSCHFELD, 1999, The Cypro-Minoan Corpus Project takes an archaeological approach. *Near Eastern Archaeology*, 62/2, p. 129-130.

STOS-GALE, Z.A. and N. GALE, 1994, Metals. In *Provenience Studies and Bronze Age Cyprus: Production, Exchange and Politico - Economic Change*, edited by, A.B. Knapp and J.F. Cherry. Monographs in World Archaeology 21. Madison: Prehistory Press, p. 92 – 121.

SWINY, S., 1982, Correlations between the composition and function of Bronze Age metal types in Cyprus. In *Early Metallurgy in Cyprus, 4000-500 B.C.*, edited by J.D. Muhly, R. Maddin, and V. Karageorghis. Nicosia: Pierides Foundation and Department of Antiquities, p. 69-80.

VAGNETTI, L. and F. LO SCHIAVO 1989 Late Bronze Age long distance trade in the Mediterranean the role of the Cypriots. In *Early Society in Cyprus*, edited by E. Peltenburg. Edinburgh: Edinburgh University Press in association with The National Museums of Scotland and the A.G. Leventis Foundation, p. 217-43

WEINSTEIN BALTHAZAR, J. 1990 *Copper and Bronze Working in Early through Middle Bronze Age Cyprus*. SIMA Pocketbook 84. Jonsered: Paul Åströms Förlag

WHEELER, T. S., 1977, The ancient tin trade in the Eastern Mediterranean and Near East. In *The Aegean Bronze Age: An Interdisciplinary Approach*. Temple University March 25, 1977 edited by P.P. Betancourt. Temple University Aegean Seminar, 2. Philadelphia: Department of Art History, Temple University, p. 23-26.

WHEELER-STECH, T., MUHLY, J. D. and R. MADDIN, 1979, Mediterranean trade in copper and tin in the Late Bronze Age Mediterranean. *Istituto Italiano di Numismatica, Annali* 26, p. 139-152.

THE PROBLEM OF EARLY TIN FROM THE POINT OF VIEW OF NURAGIC SARDINIA

Fulvia LO SCHIAVO

The paper focuses on the archaeological aspects of metallurgical research and on the archaeological conclusions that can be drawn from the results of metallurgical studies, with special attention to Nuragic Sardinia. The chronological setting (Fig. 1) is not the object of the present discussion, as there is now enough data on Mycenaean pottery, found in local context and also on Nuragic materials found outside of the island.

A. METALLIC TIN IN NURAGIC SARDINIA

The discovery of metallic tin[1] in archaeological complexes of Nuragic Sardinia was recognised only in three almost contemporary cases in the past: by F. Vivanet (Fiorelli 1878), E. Pais and F. Nissardi (Pais 1884; Baux & Gouin 1884; Fiorelli 1890) in the Nuragic sanctuary of Abini, Teti (Nu); by F. Nissardi (Fiorelli 1882; Nissardi 1884) in the hoard of Forraxi Nioi, Nuragus (Nu); by F. Vivanet (Fiorelli 1890) in the "foundry" of La Maddalena, Silanus/Lei (Nu). Nowadays (1999) there are new finds by M.A. Fadda and R. Valera (Fadda, in this volume; Valera, in this volume) in the sanctuary of S'Arcu 'e is Forros, Villagrande Strisàili (Nu) (Fig. 2).

Since these sites are well known, attention will be aimed at an essay of archaeological interpretation, with reference to the presence of tin.

Up to now, **Abini (Teti district, Nuoro province)** was referred to as a "hoard", and the same Ettore Pais, who published it in 1884 (Pais 1884), defined it "*ripostiglio*". On the contrary, the study of the oldest bibliography[2] (Spano 1865 p.7; Pinza 1901 col.163; Taramelli 1931), as well as the result of recent excavations, demonstrate beyond any possible doubt that it is a "typical" Nuragic sanctuary, with a Temple (it is difficult now to say if it was a Spring Temple or a Well Temple) and with other structures, built up with the characteristic refined "isodomic" masonry, within a wider enclosure.

On three different occasions, this site was mercilessly ravaged and looted. An enormous quantity of bronze offerings was collected; part of it is conserved in the National Archaeological Museum of Cagliari, assembled in three collections: Timon Collection (1865), Vivanet Collection (1878) and Gouin Collection (1882). They take the names from the collectors who bought the bronzes from the peasants (Timon, Vivanet) or excavated the pieces (Gouin) and gave them to the Museum. Further, the bronzes found by A. Taramelli in the 1929-30 excavations (Taramelli 1931) should also be added.

All old descriptions are so inadequate that it is impossible to figure out the original situation. In many cases there are references to groups of offerings in pots or buried in holes in the ground: it is plausible, by comparing the data with the situation of other Nuragic Sanctuaries, that near the Temple to the Water God/Goddess and in the sanctuary, there was **more than one hoard**.

Unfortunately we cannot exactly reconstruct the content of each deposit found in the past, and have to wait for the results of archaeological excavations under way, which might give a better understanding of the site (Fadda, Tuveri & Murru 1992 pp.250-251; Fadda 2000). We also have to consider that the sanctuary is located near the Nuragic village of "Interrojos" (Fiorelli) or of "Interroga/Interrogas" (Gouin, Pais), and this renders more complex the archaeological reconstruction and widens the chronological span, since EBA Bonnannaro pottery was also found here (Ferrarese Ceruti 1978 p.240 n.52 fig.229).

At Abini, Vivanet (Fiorelli 1878; Pinza 1901 col.155) reported the presence of more than 6 kg of tin ore fragments, together with copper plano-convex ingots ("bun ingots", Italian "*panelle*") and fragments of lead and iron. In 1904 Pigorini noted in the Cagliari Museum a fragment of ox-

[1] *"The use of the term "metallic" tin is redundant; there is no "non-metallic" tin. Tin is a metal and should be distinguished from other, as the case may be, "tin ore" or "tin oxide" etc. Tin is blue (or even sky blue) and lead is greyish. Both are heavy. Furthermore, as I have often said and published, tin would not have survived over the years when the ambient temperature fell below 13 C, as it would have during the Sardinian winter; it would transformed to the grey variety and have disintegrated as a powder and mixed with the earth.*" (R. Maddin, pers. comm.).
In this paper, the use of "metallic" tin is just a help for non-metallurgists, in order to define tin already smelted and ready to be alloyed with copper, possibly in the regular or irregular shape of an ingot. For a similar use, see Muhly 1985 pp.278, 279, sgg.

[2] The most difficult to interpret is the description of Léon Gouin who, allowed to conduct a regular excavation on the site, did not follow it directly, leaving the work to an inexperienced assistant - rising everybody critics (Pigorini 1865 p.77; Pinza 1901 col.156; Taramelli 1931 p.48) - who went on digging without a proper method, collecting bronze objects, now in Cagliari Museum in the Gouin Collection.
Gouin says: "*Dans une sorte de caveau souterrain entre trois nuragues, caveau formé de gros blocs, recouvert d'une matière bitumeuse et de terre, s'est trouvé un dêpot contenant des faisceux de longues épées liées par de bandes de cuivre, des statuettes, de beaux instruments de bronze, de poignard, de pain de cuivre et de l'étain oxydé (cassitérite).*" (Baux & Gouin 1884 p.195). ... "*A Teti les nuragues sont groupés, on en compte onze sur une espace de quatre hectare environ. Trois d'entre eux, peut-être (?) étaient reliés par une muraille, au milieu de l'enceinte s'est trouvé le caveau.*" (Baux & Gouin 1884 p.198).

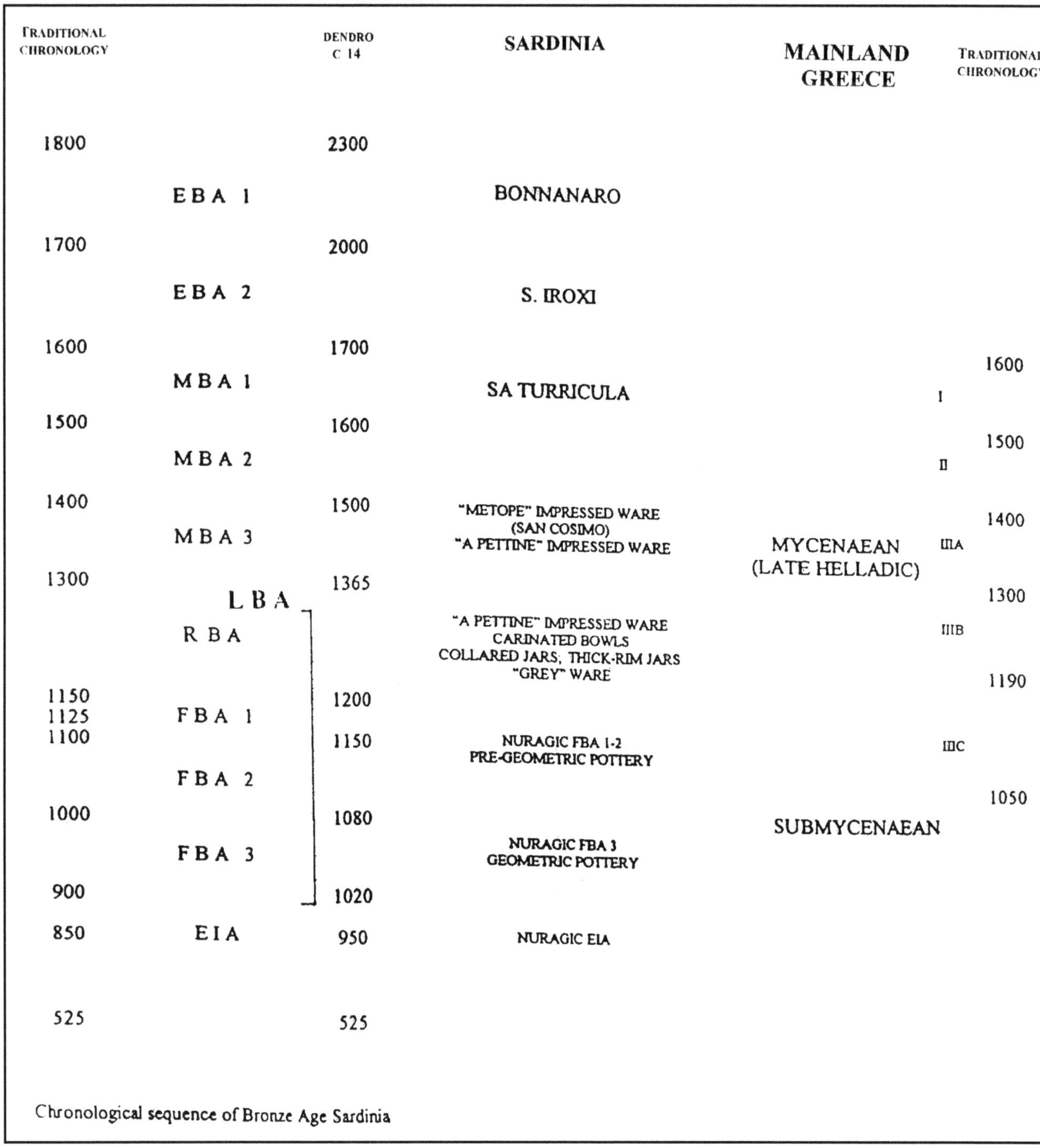

Figure 1. Chronological table of Bronze Age Sardinia.

hide ingot, which did not belong to the Serra Ilixi ingots (Pigorini 1904 p.92 n.2c; Balmuth & Tylecote 1976 p.198 fig.2; Lo Schiavo & Vagnetti 1980 p.381-2 n.6; Vagnetti & Lo Schiavo 1989 p.224 fig.28.3e). Further, many ox-hide ingot fragments in the deposits of Cagliari Museum bear the indication of "Teti (?)" as provenance (Lo Schiavo 1990 pp.24-5 n.7). There is no reason to reject this indication, but it is impossible to locate them more precisely in time and place.

All ox-hide ingots labelled "Teti (?)" were analysed by R. Maddin. He comments that, from the metallurgical point of view, they appear to be very similar to the ones from Ittireddu. Minor differences, concerning the trace elements, can be referred to the variations of the flux. The presence of so little iron indicates the result of a refining operation of an originally "impure" copper. Tin is absent (Maddin & Merkel 1990 pp.86-104). Lead isotope analyses were carried out on artefacts from Teti provided by Dr. H. Schickler, Stuttgart by N. Gale and Z. Stos-Gale (Gale & Stos-Gale 1992 tab.5).

Ettore Pais (1884, p.147) adds that "*ad Abini si sono trovati frammenti di cassiterite, biossido di stagno, come pure se ne sono trovati nella fonderia di Forraxi Nioi*" (at Abini, as well as in the Forraxi Nioi foundry, fragments of cassiterite, tin dioxide, were found). It was precisely the presence of these ore and metal fragments that convinced Pais, and before him Nissardi, the excavator of **Forraxi Nioi (Nuragus district, Nuoro province)**, that in both cases they were dealing not with votive deposits but with foundries. Concerning Abini, Pais and Nissardi observed that the

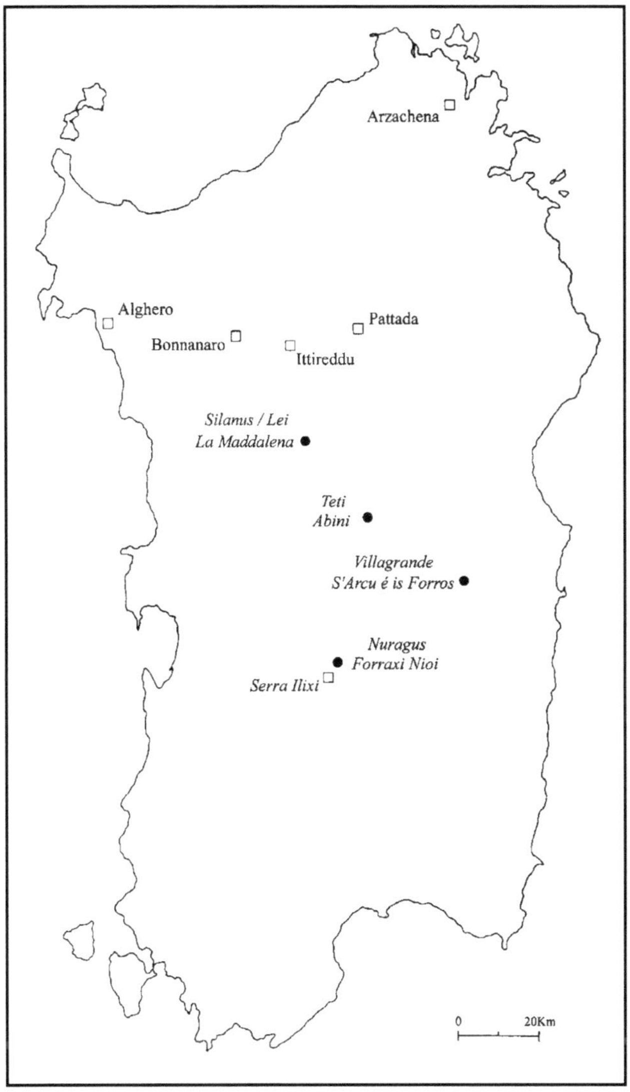

Figure 2. Geographic location of the sites mentioned in the text.

offerings, i.e. the bronze figurines, were pulled away from the "Tables of Offerings", so that the legs were broken and the feet were still stuck in the lead that fixed them to the stones. In their opinion, this profanation of the votive offerings could be explained only by a hasty assemblage of the bronzes, aimed at re-melting them, made in a great hurry because of an impending danger.

Pais wrote his work on Abini after the discovery and the three main episodes of looting and excavation of the site (Pais 1884). On the contrary, Filippo Nissardi personally excavated (though in only one week, from December 1-7, 1881) the site of Forraxi Nioi and concentrated on the study of the metallurgy of both complexes. Apart from the final printed edition, he left many hand-written notebooks and papers[3], from which it is possible to evaluate precisely the evolution of his understanding of the matter.

He generally speaks of "bronze", in the case of complete or fragmentary ingots and of all kinds of "green" metal, without any chemical analysis - as unfortunately happens also today. His first reference to tin was "*una materia grigiastra, pesante, in apparenza una scoria*" ("a greyish and heavy stuff, apparently a slag").

Only at the end of his work, after he studied the same kind of material, found in Abini, and after having had both analysed, he declares: *"quella sostanza grigiastra, pesante, che in seguito all'analisi riconobbi per minerale di stagno, cassiterite, con qualche pezzo di stannina, sottoposti a torrefazione, che è il primo trattamento del minerale*" ("after analysis I recognised as tin mineral that greyish and heavy material, cassiterite, with some pieces of stannine, treated by roasting, which is the first treatment process of the ore") (see also Baux & Gouin 1884 p.203; Pinza 1901 col.162).

By the stratified disposition of the pieces of ore and of fuel (lignite?), and of animal bones, that he thought were necessary to add phosphor to the smelting, he came to the conclusion that the container was in fact a crucible.

In 1983 and 1984 Ronald Tylecote described the remains of what was formerly examined by Cambi and by Vodret (Cambi 1959a p.427; Vodret 1959 p.257 n.3), and stated that what they interpreted as cassiterite treated in a crucible, was in fact 90% "corroded" tin and that, if this was to be interpreted as a crucible, it was a unique case of a crucible heated from the inside and not from the outside and therefore a melting process without slag, with a few comparable examples in Cyprus (Tylecote, Balmuth & Massoli Novelli 1983 p.70 note 40; 1984 p.134; Muhly 1985 p.286).

Tylecote was particularly interested in the presence of a copper smelting slag rich in iron, that is, fayalite (iron silicate) in a glassy matrix. In his conclusions, he said that this slag, in the case it was the first example of many others to be found, could be the solution to the problem of Nuragic metallurgy. Since it was impossible to check these materials recently, Tylecote's observations are precious. Lead isotope analyses were carried out by N. Gale and Z. Stos-Gale on artefacts from Forraxi Nioi, provided by Dr. H. Schickler, Stuttgart (Gale & Stos-Gale 1992 tab. 5).

From the archaeological point of view, Forraxi Nioi is far from being fully published. It has not been possible to locate the site of the discovery from the original description. Only few of the materials collected in the pottery buried in the ground were illustrated in the 1882 preliminary notice and some of them, extrapolated from the context, were studied and published[4]. The chronological span goes from ripe FBA (about XI c. BC) to the Iron Age (about VII c. BC.).

The third site of old discovery of metallic tin is **La Maddalena**, at the border of **Silanus and Lei districts, Nuoro province**. Groups of tools, weapons and ornaments, and pieces of metals, among which Vivanet mentions "*un pezzo di metallo bianco del peso di 700 grammi che*

[3] The warmest thanks are due to Maria Luisa Ferrarese Ceruti, who many years ago allowed me to check the Nissardi papers and kindlysupported a thorough archaeometallurgical study based on them.

[4] For many years a complete catalogue of Nuragic bronzes, with an essay of reconstruction of their contexts has been underway. For a preliminary overview of the bronzes collected and exhibited in the Sassari Museum, see Lo Schiavo 2000.

sottoposto ad analisi chimica, risulterebbe essere stagno puro" ("a piece of white metal weighing 700 g that a chemical analysis demonstrated to be pure tin") were found in different locations and different periods (Spano 1876, Fiorelli 1890). Nothing remains of this "white metal".

Only few of the bronze artefacts are still in the Museums of Cagliari and Sassari. A shaft-hole axe, two flanged axes, a chisel, a fragment of a second one, and a long spear-butt - were described and illustrated by Spano (Spano 1876 p.15 Tav.nn.6, 7, 9, 13, 14; Pinza 1901 coll.149-150). He gave them to the Sassari Museum and they were published one hundred years after his death (Lo Schiavo 1979 pp.81-82 nn.42-47 tav.VII, 1-6).

The fourth and recent discovery occurred in the **Nuragic Sanctuary of S'Arcu 'e is Forros**[5]**, Villagrande Strisàili district, Nuoro province**.

In 1985-86 clandestine diggings took place in a Nuragic village, dominated by a huge "Megaron Temple", and seriously damaged structures and stratigraphy. The date goes from the LBA to the beginning of the EIA.

During the first survey and in the following archaeological excavations, heaps of scrap metal and fragments of both copper ox-hide and plano-convex ingots, of lead ingots and of bronze artefacts were found (Fadda 1991; forthcoming 2; Vagnetti & Lo Schiavo 1989; Lo Schiavo 1990 p.26-27 n.11a).

In October 1999, a few fragments of metallic tin were also collected from the ground, confirming the firm belief that a foundry, serving the sanctuary, must have existed on the site; however, it is not yet possible to recognise the location of the furnaces (Valera R., Valera P. & Lo Schiavo F., in press; Fadda, in this volume; Valera, in this volume).

From the evaluation of the old and new archaeological discoveries the assumptions are:

1. It is highly probable that in the past much more metallic tin was found on Nuragic sites[6], generally considered hoards or foundries, namely where metallurgical operations could have taken place, and went unrecognised. Regrettably, this could have happened and be also happening today, mostly in rescue excavations. The problem is that only seldom is a specialist present. It was a lucky case in Villagrande Strisàili with Roberto Valera, who was able to recognise this metal, mistaken for lead because it was "greyish and heavy" ("*materia grigiastra e pesante*" Nissardi 1884) (see also R. Maddin's opinion on this matter[1]).
2. In the cases described above, metallic tin was found together with copper, lead and iron bars (nothing is said of silver). This allows the supposition that on these sites melting and alloying operations could have taken place.
3. Up to now, no stone moulds carved in steatite - a local refractory stone commonly used in Nuragic Sardinia - were found. However, since in the past raw pieces of baked clay were not preserved, as still can happen today in rescue excavations, where looters often destroy the site, turning it upside-down in their treasure-hunting, terracotta- and soft stone moulds could have existed and the use of the lost-wax process for the production of bronze objects and figurines is also a possibility.
4. In three of the four cases mentioned above, with the exception of Villagrande, still in course of evaluation, and in the other Nuragic complexes known up to now, there are no traces of smelting operations. No remains of smelting furnaces were found. However it is possible that melting and alloying operations were carried out in the sites, for the production of objects and figurines: see Santa Barbara, Bauladu (Atzeni C. et al. 1991; 1992a; 1992b; Atzeni C. et al. 1994; Gallin et al. 1994; Gallin & Tykot 1993), Santa Anastasia, Sardara (Ugas & Usai L. 1987) and Genna Maria, Villanovaforru (Atzeni C. et al. 1998).

B. THE PERCENTAGE OF TIN IN INGOTS AND ARTEFACTS

1. **Copper ingots**. Without entering into the debated question of the provenance of the copper ox-hide ingots, the analyses confirm that all ox-hide ingots, with very few exceptions, are of almost pure copper (Maddin & Merkel 1990; Begemann et al. 2001). A low level of tin is considered "*the normal level of impurities in smelted metal of this period ... that could have come from a contaminated ore deposit*" (Tylecote et al. 1983 p.84; Vodret 1935 p.111; Maddin & Merkel 1990 p.92-93; Begemann et al. 2001 p.55).

Scientists do not agree on the relationship between ox-hide and bun ingots: some of them note evident metallurgical differences (Garagnani & Martini 1998; Atzeni C. et al. 1998; Demurtas 1998), while others, analysing the same pieces, find "*no significant differences in trace element abundances*" (Maddin & Merkel 1990; Begemann et al. 2001 p.51).

However, they agree in excluding the possibility that the bun ingots could have been the first smelting product, used to produce, in a second melting operation, the ox-hide ingots. The archaeometallurgical and archaeological problem concerning the reason for producing two different shapes of copper ingots in Cyprus - as they are found in the Cape Gelidonja and Uluburun cargoes, as well as all over Sardinia - is still open (Pulak 2000, 2001).

2. **Bronze ingots**. In three or four cases there is such a high percentage of tin that it is possible to speak of "bronze ingots".

[5] The name of the site is reported as "S'Arcu 'e is Forros" by M.A. Fadda, as it is the actual common spelling, while on the IGM 1:25.000 is written "S'Arcu 'e is Forras".

[6] Gouin reports the discovery of a wreck near Carloforte island in southern Sardinia, consisting in 17 kg. of copper ingots of "unusual" oval shape, together with a consistent quantity of tin ingots, soon sent and sold in Marseille. Also associated were little elongated (25 cm., 150 gr.) brass (Cu 85,294, Zn 14,016, Sn 0,059) ingots (Baux & Gouin 1884 p.203). It is impossible to connect the finding to Nuragic period.

A fragment of an ox-hide or bun ingot is in the second hoard of Ittireddu (inv.n.62405, Sn 6.8 / 2.2%: Maddin & Merkel 1990; Sn 11: Begemann et al. 2001 p.51 n.5); another "measurable" content of tin (130 ppm) is in Ittireddu 62417 (Begemann et al. 2001 tab.1).

A little bun ingot from Camposanto, Olmedo (Sn 15.67%), was analysed by Vodret; he also mentions two "ingots", one from Monte Pelau-Bonnanaro and one from unknown provenance, with respectively 8,57 and 7,32 of tin (Vodret 1935, 1959).

The question has no clear solution, both from archaeometallurgical and archaeological points of view. In the case of the Ittireddu II hoard, there is a true close complex with 45 fragments, some of them of tiny dimensions. Of course it is impossible to distinguish the composition without a metallurgical analysis.

The five (three complete and two sections) little bun ingots are part of the few votive offerings left by the looters in 1926 in the Nuragic Sacred Well of Camposanto, Olmedo (Taramelli 1933) (Fig.3). They are unique for their tiny dimensions and low weight (inv. n.502, diam. cm.9, h. cm.1,4; gr.359), compared to the natural size - for example - bun ingots of Flumenelongu hoard (inv. n.10249, diam. cm.13,8 x 13, h. cm.3,5/4; gr.1752). They seem mostly miniature reproductions of real bun ingots. It was a widespread custom in the Nuragic period to have miniatures for any kind of bronze figure: men and women, animals, monuments, furniture, weapons and tools, which were subsequently deposited in the sanctuaries. Therefore it is possible that they were not used for melting, but as votive offerings.

On the whole, on these few and disputable cases it is not possible at the moment, to recognise a rule or a pattern.

3. **Tin in bronze artefacts**. There are many more analyses of ox-hide ingots than of bronze figurines, and there are only very few and occasional weapons and tools analyses. For those carried out in a less recent past; unfortunately, we do not have enough references to identify the objects (Vodret 1959 Tabb.3-6 (Tab.4); SAM 1960; Tylecote, Balmuth & Massoli Novelli 1983 tab.6 (Tab.5)). The "artefacts" analysed

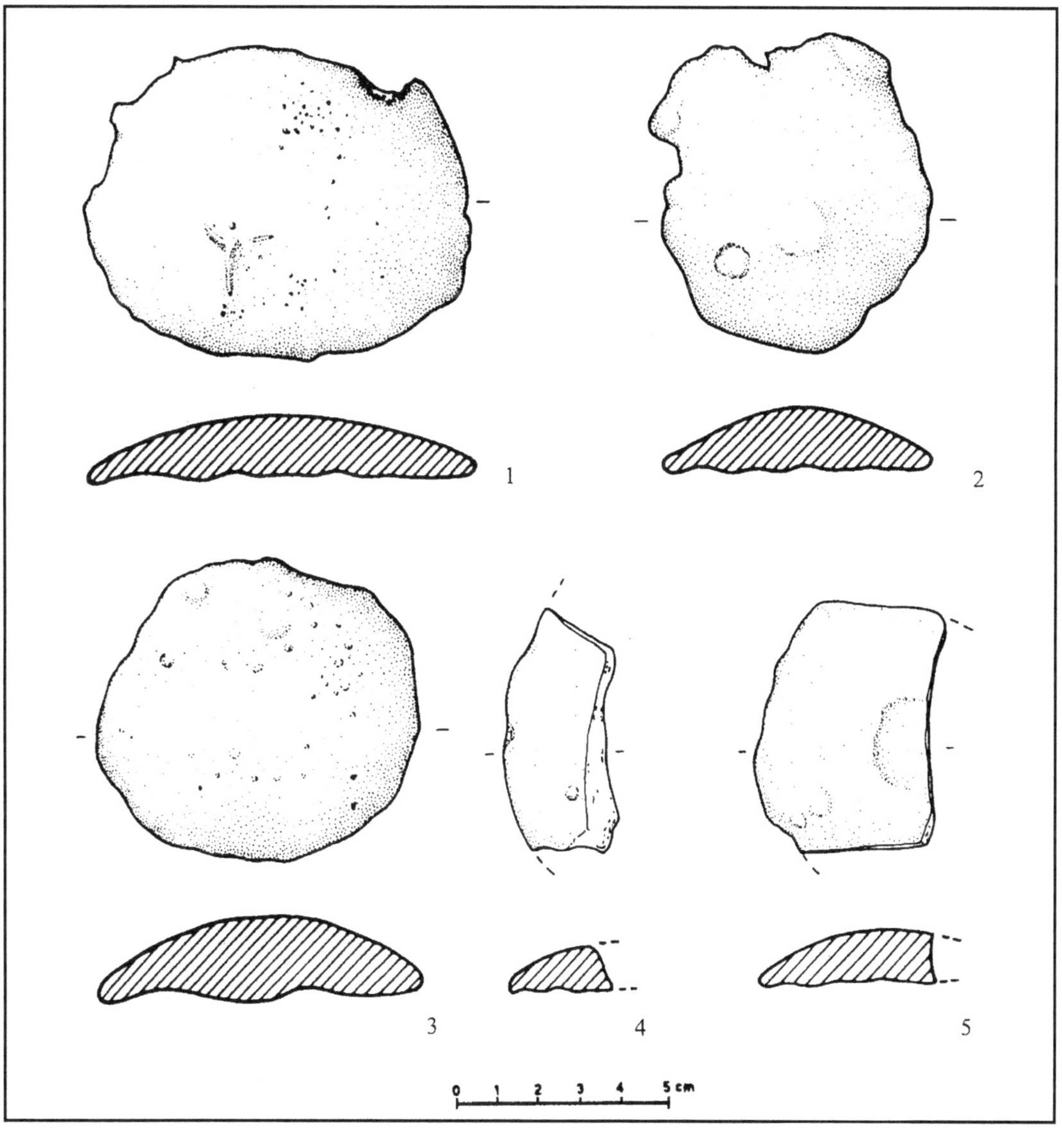

Figure 3. Little bronze ingots from the Nuragic Sacred Well of Camposanto, Olmedo (Sassari). 1- Inv.n.502; 2 – Inv.n.503; 3 – Inv.n.504; 4 – Inv.n.505; 5 – Inv.n.506.

in recent times are often only tiny fragments of rods and sheets and it is impossible to reconstruct and to evaluate archaeologically their original shape.

The highest percentage of tin is said to be in ornaments, followed by tools and finally by weapons; the percentage is not always "standard", but there is no doubt of the fact that it is a deliberate alloy (Sn from 4,9 to 11,6 %: Cambi 1959a; Sn from 7,6 to 16,27 %: Vodret 1959; Sn from 8.1 to 11%: Giardino 1987 fig.9.8, 1995 fig.84 based on Tylecote, Balmuth & Massoli Novelli 1983 and Craddock 1984).

Recent analyses of the hoards of Albucciu-Arzachena, Funtana-Ittireddu, Funtana Janna-Bonnanaro (Begemann et al. 2001) (Tab.6) and Flumenelongu-Alghero (Marabelli et al. in preparation) included a wider choice of tools and weapons. The results are different and show a greater variety of composition within the same category of objects (Sn from 6.8 to 14.3%, average 10,8%: Begemann et al. 2001 p.53) and many new interesting metallurgical aspects (Fig.4).

a) **Votive swords**. The lowest percentage of tin was found, in previous analyses (Vodret 1959 p.262), in the "*bandelle*", the metallic strips used to bind together groups of votive swords. The votive swords - as already remarked in the past (Baux & Gouin 1884 p.195, 204) - are very long (80/120 cm) and thin weapons, characterised by a strong midrib, asymmetrical on the two sides. They were stuck, point up, on the top of the roofs of the Nuragic Temples (Su Tempiesu-Orune, and Monte S. Antonio-Siligo), on the top of the "Ceremonial Wall" of Gremanu-Fonni Round Temple and on the "Tables of Offerings" (Ugas & Usai L. 1987; Fadda 1993; forthcoming 1; Lo Schiavo 1997).

They can be found broken in short pieces of about 25 cm, sometimes adapted as blades to the most typical Nuragic "gamma-hilted" daggers.

The tiniest fragments (2/5 cm) were found in Albucciu-Arzachena, Funtana-Ittireddu and Baccu Simeone-Villanovaforru hoards, associated, among other things, with ox-hide ingot fragments. At Albucciu-Arzachena and Funtana-Ittireddu the percentage of tin is around Sn 1% (Begemann et al. 2001 p.62), as if the bronzesmiths knew that this was not going to be a functional weapon, but a votive object, produced to be an offering, whose breakage was eventually foreseen and the fragments ritually re-adjusted.

The main point is the chronology: the Albucciu hoard is dated to the RBA (about XIII cent. B.C.) and the Baccu Simeone-Villanovaforru and Funtana-Ittireddu II (Lo Schiavo et al. in press) have the same chronology or are perhaps just a little later (end of RBA/beginning LBA), i.e. they are dated at the beginning of the most typical Nuragic production.

b) **Bronze figurines**. In general the percentage of tin of the bronze figurines varies (Sn 2.2 to 11.8). There is relatively little lead and contamination with iron, probably due to the use of iron tools, or to mixed chalcopyrite ore deposits (Balmuth & Tylecote 1976; Balmuth 1978; Tylecote, Balmuth & Massoli Novelli 1983, 1984).

The very few notable exceptions demonstrate how important it is to widen the data-base and to evaluate the different compositions; one of these exceptions is the bronze figurine of a long-braided and huge-handed man, from Abini (Lilliu 1966 pp.217-219 n.116; Atzeni C. et al. 1992a): Sn 0.27; Ag 6.5; Pb 0.29; Zn 0.09; Fe 0.01; Ni 0,03; Co <0.004. Another almost identical bronze figurine is apparently made of "standard" tin-bronze (Lilliu 1966 p.216 n. 115 from Urzulei).

C. THE LEVEL OF "KNOW-HOW" OF THE NURAGIC BRONZESMITHS

1. The most ancient and typical Nuragic product of the RBA is votive swords: they are deliberately without or with very low tin; the lowest tin level is also found in the metallic strips used to bind together votive swords.

2. The ox-hide, plano-convex and truncated-conical (Bonnanaro) ingots so far analysed are made of pure copper, evidently second refining (Merkel 1986; Maddin & Merkel 1990). Some of them show a lead-isotope fingerprint corresponding to the local Sardinian ores (Begemann et al. 2001 p.57)[7]. Some of the hoards so far analysed are RBA or beginning of the FBA.

3. The bronze artefacts show a "standard" bronze composition and generally a correlation between the use of the object and its tin percentage; the "deviations" can be the result of remelting scrap.

The statistical data-base we possess up to now, is not large enough to allow the distinction of definite patterns. Nevertheless, there are still exceptions which deserve careful study: one is the "mixed bag" from Albucciu-Arzachena, where the little fragments are not scraps, but perhaps show the tinkering with various kind of ores or else experimental essays of melting and alloying (Begemann et al. 2001 p.63).

D. OPEN PROBLEMS

Nuragic Sardinia seems to have developed a metallurgical technology at the level of Mediterranean Bronze Age. Still, there are inexplicable questions, difficult to answer also elsewhere in the Mediterranean:

1. Why did they produce and trade, on an internal and external network **both** ox-hide and bun ingots?

2. Did Bronze Age workshops practise a "standard" empirical alloying, based on the colour, or did they employ a weighing and measuring system, based on precise ponderal scale? And if yes, which ponderal system?

3. Were **bronze** ingots commonly in use? Why are they so rarely found?

---

[7] See in Begemann et al. 2001 the updated bibliography on Sardinian copper ingots analyses.

Oggetti	Località	Cu	Sn	Pb	Fe	Ni	Zn	SiO2
Bandelle	Abini (Teti)	96.56	1.15	0.14	0.03	0.01	2.24	
Bandelle	Abini (Teti)	95.89	1.62		1.75		0.175	
Pugnale votivo	Abini (Teti)	85.6	10.85	0.06	0.15		2.87	
Spada votiva	Abini (Teti)	86.86	9.26	0.03	0.01	0.03	3.02	
Spada votiva	Abini (Teti)	96.53	2.21	0.46	0.51		0.22	0.04
Puntale lancia	Nuragus	86.61	8.67	0.43	0.27		1.89	0.3
Puntale lancia	Nuragus	87.08	10.04	0.11	0.08		2.73	
Puntale lancia	Abini (Teti)	91.11	7.82	0.6	0.09	0.04	0.11	
Puntale lancia	Abini (Teti)	93.22	5.7	0.14	0.18	0.07	0.12	
Spada combattimento	Nuragus	87.48	10.27	0.32	0.1		2.1	
Spada combattimento	Abini (Teti)	94	4.8	0.87	tracce	0.05	0.22	
Spada combattimento	Abini (Teti)	94.1	5.31	0.06	0.3		0.15	0.03
Spada combattimento	Abini (Teti)	85.63	12.77	0.02	0.09		1.37	
Spada combattimento	ignota	91.21	5.3	0.87	0.11	tracce	2.63	0.82
Spada lama larga	Maracalagonis	90.36	0.32	0.08	0.01	tracce	8.27	0.68
Lama pugnale (framm.)	Olmedo	93.22	5.77	0.07	0.19		0.002	0.4
Frammento pugnaletto	Serra Ilixi	82.5	11.28		0.53	tracce	5.02	
Frammento pugnaletto	Isili	82.5	11.28		0.53	tracce	5.02	
Spada combattimento larga	Maracalagonis	90.36	0.32	0.08	0.01	tracce	8.27	0.68
Puntale lancia	Nuragus	86.61	8.67	0.43	0.27		1.89	0.3
Punta lancia	Nuragus	87.08	10.04	0.11	0.08		2.73	
Spada combattimento	Nuragus	87.48	10.27	0.32	0.1		2.1	
Lama pugnale	Olmedo	93.22	5.77	0.07	0.19		0.002	0.4
Bandelle	Abini (Teti)	96.56	1.15	0.14	0.03	0.01	2.24	
Bandelle	Abini (Teti)	95.89	1.62		1.76		0.18	
Punta lancia	Abini (Teti)	91.11	7.82	0.6	0.09	0.04	0.11	
Punta lancia	Abini (Teti)	93.22	5.7	0.14	0.18	0.07	0.12	
Spada combattimento	Abini (Teti)	94	4.8	0.87	tracce		0.22	0.05
Spada combattimento	Abini (Teti)	85.63	12.77	0.02	0.09		1'37	
Spada combattimento	Abini (Teti)	94.1	5.31	0.06	0.3		0.15	0.03
Spada votiva	Abini (Teti)	86.82	9.26	0.03	0.01	0.03	3.02	
Spada votiva	Abini (Teti)	96.53	2.21	0.46	0.51		0.22	0.04
Pugnale votivo	Abini (Teti)	85.6	10.85	0.06	0.15		2'87	
Ascia bipenne	Lotzorai	84.29	11.59	0.43	0.18		2.84	
Ascia bipenne	incerta	89.13	4.15	5.6	0.26		tracce	0.75
Ascia piatta	incerta	94.01	3.9	1.56	0.04			0.78
Accetta	Lei	87.65	6.24	1.16	0.31		2.69	2.75
Accetta a margini rialzati	Terralba	83.36	9.61	0.23	0.09		3.71	3.16
Accetta a margini rialzati grossa	Terralba	80.24	10.47	0.82	0.11		4.29	4.62
Accetta a margini rialzati contorta	Terralba	90.11	5.13	1.6	0.01		3.01	
Accetta a margini rialzati norm.	Abini (Teti)	85.26	9.36		0.001		5.64	
Accetta a margini rialzati norm.	Nule	85.44	9.17		0.002		5.25	
Accetta a margini rialzati norm.	Ozieri	83.78	8.52	3.5	0.01		3.65	1.01
Accetta a margini rialzati norm.	Sarrok	97.15	0.38		0.001		2.22	0.19
Accetta a margini rialzati norm.	Quartu S.Elena	87.11	9.26		0.001		2.8	1.61
Accetta a margini rialzati norm.	Tertenia	90.57	7.6	0.45	tracce		1.53	
Accetta a margini rialzati norm.	Gairo	96.02	0.9	0.34	tracce		3.21	
Accetta a margini rialzati norm.	Lotzorai	83.07	13.07	0.002	tracce		3.57	
Frammento lima	incerta	85.12	13.05	0.76	0.02		1.55	
Frammento sega	incerta	82.58	10.87	2.06	tracce		2.08	2.65
Accetta a margini rialzati	Gairo	96.02	0.9	0.35	tracce		3.21	
Accetta	Lei	87.65	6.24	1.16	0.31		2.69	2.75
Accetta a margini rialzati	Lotzorai	83.07	13.07	tracce	tracce		3.57	
Ascia bipenne	Lotzorai	84.29	11.59	0.43	0.18		2.84	
Accetta a margini rialzati	Nule	85.44	9.17		0.002		5.25	
Accetta a margini rialzati	Ozieri	83.78	8.52	3.5	0.01		3.65	1.01
Accetta a margini rialzati	Quartu S.Elena	87.11	9.26		tracce		2.8	1.61
Accetta a margini rialzati	Sarrok	97.15	0.38		tracce		2.22	0.19
Accetta a margini rialzati	Terralba	83.36	9.61	0.23	0.09		3.71	3.16
Accetta a margini rialzati grossa	Terralba	80.24	10.47	0.82	0.11		4.29	4.62
Accetta a margini rialzati contorta	Terralba	90,11	5.13	1.6	0.01		3.01	
Accetta a margini rialzati	Tertenia	90.57	7.6	0.45	0.001		1.53	
Accetta a margini rialzati	Abini (Teti)	85.26	9.36		tracce		5.64	

Figure 4. Tin in weapons and in tools, according to the different metallurgical analyses: Vodret 1959, tab.3 p.261, tab.4 p.263, tab.5 p.264-5 (seulement armes, bandes métalliques et outils / only weapons, metallic strips and tools).

5a

Oggetti	No.	Località	Sn	Pb	As	Sb	Ag	Ni	Bi	Co	Fe
Rivet	7942	Abini	10	0.67	0.83	tr	0.18	0.03			tr
Rivet	7943	Abini	10	0,89	0.08		0.01	0.02			strong
Dagger	7944	Abini	9	0.03	0,54	tr	0,03	0.03			tr
Dagger	7945	Abini	5	1	0.68	tr	0.2				tr
Dagger	7946	Abini	> 10	0.08	0,17	tr	0,02	0.01	0,009		tr
Dagger	7948	Abini	> 10	0.12	0.74		< 0,01	0.35			some
Dagger	7956	Abini	1,25	0.03	0.09		tr	0.02			strong
Dagger	7957	Abini	> 10	0,09	0.49		tr	0.02			tr
Dagger	7958	Abini	9.1	0,05	0.96		0,01	0.36		0,06	some
Dagger	7961	F. Nioi	6,7	1.45	0.27	0,54	0.2	0.25	0.011		some
Dagger	7964	F. Nioi	7,5	4,9	0,61	0,3	0,17	0.03	0.02		some
Axe	7960	F. Nioi	10	2,6	1,05		0,49	0.03			stong

5b

Oggetti	British M. No.	Località	Cu	Sn	Pb	As	Sb	Fe	Ni	Co	Bi	Zn	Ag
Dagger	1926.5 - 11.6	S.M.Paulis	88	8,7	1,3	0,12	0,025	0,8	0,035	0,035	0,003	0,03	0,05
Dagger blade	1926.5 - 11.8	S.M.Paulis	89	9,3	1,1	0,2	0,2	0,15	0,4	0,015			0,24
Spearhead	1926.5 - 11.7	S.M.Paulis	87	11	0,8	0,15	0,2	1,1	0,03	0,045		0,03	0,05
Ferrule	1926.5 - 11.9	S.M.Paulis	91	4,5	0,8	0,43	0,03	1,8	0,03	0,035		0,02	0,09
Double axe	1926.5 - 11.18	S.M.Paulis	88,5	10,8	0,7	0,3	0,01	0,4	0,02	0,015			0,06

5c

Oggetti	No.	Località	CU	Sn	As	Zn	Pb	Fe	Ni	Ag	Co	Mn
Votive sword	20947	Albucciu	99.05	o.97	0.18	0.59	0.11	0.1	0.32	0.007	0.008	
Votive sword	62385	Ittireddu	97.67	0.8	0.18	0.23		0.25	0.22	0.12	0.01	0.009
Axe	SS-B	Flumenelongu	82.6	1.31			0.06		0.14	0.29		
Axe	SS-C	Flumenelongu	85.47	10.94	0.09		1.39		0.19	0.62		
Axe	SS-D	Flumenelongu	90.71	5.83	0.23	0.002	0.93	0.028	0.018	0.003		
Bar	SS-A2	Flumenelongu	75.91	0.16	0.03	0.004	16.82		0.28	0.002		
*	SS-A4	Flumenelongu	65.78	5.6	0.64	0.002	25.39		0.019	0.26		
Axe	SS-I	Funtana Janna	85,75	11.6		0.008	0.013	0.003	0.02	0.26		
Axe	SS-J	Funtana Janna	85.28	9.85		0.008	0.013	0.003	0.02	0.026		
Axe	SS-K	Funtana Janna	85,23	12.3	0.06	0.002	0.04	0,009	0,013	0.034		
Axe	SS-L	Funtana Janna	81.13	10.8		0.09	0.64	0,52	0,017	0.02		
Axe	SS-M	Funtana Janna	82,75	12.5	0.21	0.006	0,91	0.013	0.017	0.11		
Axe	SS-N	Funtana Janna	84.72	12.8		0.009	0.25	0.035	0.012	0.047		
Spearhead	Ol-4	Sa Sedda 'e Sos Carros	85.83	11.54	0.26	0.004	0.35	0.025	0.031	0.08		
Spearhead	OL5	Sa Sedda 'e Sos Carros	89,17	7.46	0.17	0.002	0,44	0.075	0.048	0.16		
Vot. sword	OL-6	Sa Sedda 'e Sos Carros	92.82	5.05	0.36	0.27	0.73	0,7	0,032	0.03		
Rod	OL-8	Sa Sedda 'e Sos Carros	94,24	5.74	0.33	0,009	0,65	0,25	0.031	0.01		
Axe blade	OL-9	Sa Sedda 'e Sos Carros	89.83	6.98	0.25	0.01	1	0.11	0.37	0.08		
Vot. sword	OL-16	Sa Sedda 'e Sos Carros	94,46	7.46	0.51	0.019	5.21	0.17	0.019	0.11		
Rod	OL-18	Sa Sedda 'e Sos Carros	87.58	10.99	0.03	0.14	0.38	0,012	0,026	0.08		
Rod	OL-19	Sa Sedda 'e Sos Carros	93,6	3.03	0.16	0.2	1.43	0.17	0.027	0.09		
Rod	OL-20	Sa Sedda 'e Sos Carros	88,66	8.23	0.26	0,009	0.9	0.16	0.013	0.05		
Rod	OL-21	Sa Sedda 'e Sos Carros	89.28	5.35	0.12	0.046	2,76	i.41	0.024	0.02		

* Handle fragment ?

Figure 5. a) Tin in weapons and in tools, according to the different metallurgical analyses: SAM 1960 in Tylecote, Balmuth, Massoli Novelli, 1983, tab.6 p.68 (seulement armes, rivets et outils / only weapons, rivets and tools). b) Tylecote, Balmuth, Massoli Novelli, 1983, tab.11 p.73, aussi / also in Craddock 1976 p.109; Craddock, Tite, 1984, pp.20-21 (seulement armes et outils / only weapons and tools). c) Lo Schiavo, Maddin, Merkel, Muhly, Stech, 1990, pp.51, 85, 145, 153, 167.

4. Were **tin** ingots commonly in use? Were they rarely found because of the instability of pure tin? Is this a sufficient explanation for finding everywhere so many ox-hide and no tin ingots?

5. Where and how took smelting operations place?

In Nuragic Sardinia "*metallurgy does not seem to have created social changes but rather responded to it*" (Muhly & Stech 1990 p.209). A next step was to ascertain that: "*with the present data at hand, it is clear that during Nuragic times, artefacts made of bronze were not just imported into Sardinia but also manufactured locally*" (Begemann et al. p.73).

SARDINIAN ARTIFACTS

	Sample	Cu %	Sn ppm	Pb ppm	As ppm	Sb ppm	Co ppm	Ni ppm	Ag ppm	Au ppm	Fe ppm	Zn ppm	Se ppm	Te ppm
Implements														
Arzachena														
chisel	20976	89	143000	690	3800	135	530	289	300	14.8	<550	31	61	85
Bonnanaro														
double axe	10712	86	110000	2400	2530	390	97	206	890	3.5	220	70	40	<20
double axe	10713	90	116000	950	750	1610	22	650	470	0.63	520	<20	135	<25
socketed axe	10714	93	101000	710	1090	530	16	520	430	7.8	<300	<15	42	<20
double axe	10715	92	123000	3650	840	134	65	283	300	1.34	980	57	13	<10
axe blade	10716	92	105000	280	2170	390	16	370	350	10.3	160	9	38	<15
axe blade	10717*	92	103000	250	2120	370	16	380	340	10.2	<250	15	35	<15
axe blade	10718*	90	86000	160	1790	310	16	370	293	9.1	138	11	30	<10
Oliena														
spatula	58981	91	139000	430	5300	161	233	320	207	17.8	<700	16	61	102
Pattada														
double axe	1	90	103000	2560	2770	97	221	295	610	3.3	2390	198	27	<15
double axe	2	87	119000	1350	2900	82	215	227	460	4.0	7700	263	28	<20
axe adze	3	86	99000	2410	8300	188	137	247	580	3.5	550	256	24	<20
flanged axe	4	86	133000	2140	2570	131	277	340	430	4.2	7400	207	39	<20
flanged axe	5	89	121000	1580	3200	124	265	285	490	6.2	2680	133	39	<40
chisel	6	89	85000	1960	3400	118	470	390	570	5.9	5800	166	31	24
chisel	7	84	110000	4000	2590	133	172	297	580	5.1	2280	116	32	<20
chisel	8	90	112000	1000	5500	91	320	202	800	3.6	4900	380	39	<30
chisel	9	89	68000	3780	3700	141	149	277	710	5.1	8200	239	40	<30
handle of a chisel	10	83	109000	2040	3400	116	254	291	850	3.1	10700	298	40	<20
dagger	11	89	94000	1680	2400	139	170	249	320	17.2	1370	111	58	<20
dagger	12	90	74000	4780	3300	154	380	360	960	8.8	5600	179	35	<30
dagger	13	85	112000	3370	1700	84	232	281	370	3.6	3800	196	25	<15
Sword fragments														
Arzachena														
	20974A	99	<200	2730	9200	97	2	<7	13600	2.4	<1000	222	31	<70
	20974B	92	16200	50	2170	83	61	340	30	5.0	740	390	79	50
	20974C	98	17400	20	1390	42	28	191	20	3.4	1720	420	115	42
	20974D	100	9300	510	2240	144	201	310	114	5.3	1110	620	78	45
	20974E	97	9600	450	1920	57	123	298	30	4.6	790	117	89	49
	20974F	101	10100	230	1950	60	79	420	41	4.9	1500	137	94	49
	20974G	100	3300	90	2870	79	176	290	43	6.7	1360	237	76	58
	20974H	105	4700	40	2000	52	86	261	37	5.2	1450	150	92	63
	20974I	106	8500	150	1920	57	100	380	32	4.8	2550	350	114	51
	20974K	95	11200	60	2300	55	279	380	32	5.4	2360	283	60	41
	20974L	99	21700	390	2480	95	66	270	26	4.0	1270	118	114	42
	20974M	97	7800	40	1650	47	113	330	29	4.3	2460	241	83	46
Ittireddu														
	62385	101	11500	450	2530	118	100	340	246	5.7	1540	210	86	58
	62385A	83	75000	7400	1910	141	74	263	235	5.9	230	32	44	23
	62385D	100	4800	370	1470	38	37	210	26	3.8	1840	550	112	48
Non-descript metal														
Arzachena														
	20975A	102	<60	60	21	7	1230	<75	288	0.36	13700	<60	97	<40
	20975B	94	<70	4090	53000	420	31	165	2520	1.36	11900	1560	21	<60
	20975C	87	<110	5130	39000	1940	74	520	4400	0.98	12300	1660	24	<80
	20975D	104	<30	2350	18100	50	6	79	630	0.69	160	299	13	<10
	20975E	101	<250	900	4400	86	1	<1	32000	0.97	<2000	640	<1	<10
	20975G	103	<30	10	480	8	9	114	211	1.04	610	9	73	<15
	20975H	102	<150	3100	6300	118	2	<70	16700	1.14	<1000	590	23	<70
	20975R	99	<75	10	53000	450	<1	<35	1150	0.26	<500	20	2	<50

* Same object.

Figure 6. Tin in weapons and in tools, according to the different metallurgical analyses: Begemann, Schmitt-Strecker, Pernicka, Lo Schiavo, 2001, tab.3, pp.60-61.

As archaeological research go deeper and as our knowledge of tin sources and trade widens (Muhly 1973, 1982, 1985; Gillis 1990, 1991; Gillis, Clayton, forthcoming; Begemann et al. 1999; Wen Yi et al. 1999; various contribution on tin in the Mediterranean, in this volume), the more fundamental appears the role of this large island, rich in resources and in landing places, and just on the route of western tin, in a moment in which the eastern tin ores could have been difficult, if not impossible, to get (Lo Schiavo 2001; Kassianidou 2001 p.110).

Still, Mycenaean presence and Cypriot metallurgical influence in Nuragic Sardinia are undeniable and still the provenance of metallic tin found in Villagrande Strisàili still remains unknown.

Many more analyses and archaeological evaluation are needed, to help solving the problems, but the archaeological elements collected so far are highly stimulating.

Author's address

Fulvia LO SCHIAVO
ISMEA – CNR
Via Giano della Bella 18
00162 Roma, ITALY
Email: f.loschiavo@ime.rm.cnr.it

Bibliography

ATZENI C., BADAS U., CINCOTTI A.M., DEMURTAS G., MASSIDDA L., & SANNA U., 1988, Compendio delle caratteristiche chimiche, metallografiche e isotopiche dei reperti metallurgici di Baccu Simeone (Villanovaforru), Quaderni della Soprintendenza Archeologica per le province di Cagliari ed Oristano 15, pp.150-167.

ATZENI, C., MASSIDDA, L. SANNA, U., & VIRDIS, P.F., 1991, Notes on Lead Metallurgy in Sardinia during the Nuragic period. Journal of Historical Metallurgy Society 24, pp. 97-105.

ATZENI C., MASSIDDA L., SANNA U., & VIRDIS P.F., 1992a, Some metallurgical remarks on the Sardinian bronzetti, In Sardinia in the Mediterranean: a Footprint in the Sea, Studies in Sardinian Archaeology presented to M. Balmuth, ed. by R.H. Tykot & T.K. Andrews. Sheffield, pp.347-354.

ATZENI C., MASSIDDA L., SANNA U., & VIRDIS P.F, 1992b, Bronze metal working at nuragic site of Santa Barbara. Journal of Historical Metallurgy Society 26, pp.31-35.

ATZENI C., MASSIDDA, L. SANNA, U., & SISIU, G., 1994, The Use of Ceramic Materials in Metallurgy in Sardinia in the Early Iron Age. In World Ceramic Congress. The Ceramic Heritage, Firenze, pp.279-286.

BALMUTH M.S., 1978, Sardinian bronzetti in American Museums. Studi Sardi XXIV, 1975-77, pp.145-156.

BALMUTH M.S., & TYLECOTE R.F., 1976, Ancient Copper and Bronze in Sardinia: excavation and Analysis. Journal of Field Archaeology 3, pp.195-201.

BAUX A., & GOUIN L., 1884, Essai sur les nuragues et les bronzes de Sardaigne. matériaux pur l'histoire primitive et naturelle de l'homme XVIII, 3, 1, pp.187-201.

BEGEMANN F., KALLAS K., SCHMITT-STRECKER S., PERNICKA E., 1999, Tracing ancient tin via isotope analyses. In: The Beginnings of Metallurgy. Proceedengs of the International Conference, (Bochum 1995). Ed. by A. Hauptmann, E. Pernicka, T. Rehren, Ü. Yalçin. Der Anschnitt, 9, pp.277-284.

BEGEMANN F., SCHMITT-STRECKER S., PERNICKA E., & LO SCHIAVO F., 2001, Chemical composition and Lead isotopy of Copper and Bronze from Nuragic Sardinia. European Journal of Archaeology 4, 1, pp.43-85.

CAMBI L., 1959a, Ricerche chimico-metallurgiche su leghe cupriche di oggetti ornamentali preistorici e protostorici dell'Italia centrale e settentrionale. Studi Etruschi 27, pp.191-198.

CAMBI L., 1959b, Problemi della metallurgia etrusca, Studi Etruschi 27, pp.415-432.

DEMURTAS G., 1998, Metodologie d'indagine microstrutturale applicate ad alcuni aspetti della metallurgia del rame e delle sue leghe (archeometallurgia). Dottorato di Ricerca in Ingegneria Metallurgica, XI Ciclo, Politecnico di Torino, Anni Accad. 1995/96 - 1997/98 (Tutor Prof. U. Sanna).

CRADDOCK P.T., 1976, Copper alloys used by the Greek, Etruscan and Roman civilizations, 1, Journal of Archaeological Science, III, 103-123.

CRADDOCK P.T., & TITE M.S., 1984, Report on the composition of bronzes from the hoard found at Santa Maria in paùlis, Sardinia. In: Macnamara, Ridgway & Ridgway 1984, pp.19-21.

FADDA M.A., 1991, Villagrande Strisàili (Nuoro) Località S'Arcu e' is Forros. Tempio a megaron. Bollettino d'Archeologia 10, pp.108-110.

FADDA M.A., 1993, Fonni (Nuoro). Complesso nuragico di Madau o Gremanu, Bollettino d'Archeologia 19-21, 176-181.

FADDA M.A., 2000, Teti: nella Terra Santa dei Nuragici. Archeologia Viva XIX, n.80 n.s., mar.-apr., pp.98-103.

FADDA M.A., à paraitre / forthcoming 1, Fonni (Nuoro). Complesso di templi nuragici. Bollettino d'Archeologia 34-36.

FADDA M.A., à paraitre / forthcoming 2, Villagrande Strisàili (Nuoro) Località S'Arcu e' is Forros. L'abitato nuragico intorno al tempio a megaron. Bollettino d'Archeologia 34-36.

FADDA M.A., dans ce volume / in this volume, Nuragic sanctuary of S'Arcu 'e is Forros (Villagrande Strisàili, Nuoro). Colloquium "The Problems of Early Tin". XIV UISPP Congress, Liège, 2-8 september 2001, edited by A. Giumlia-Mair & F. Lo Schiavo.

FADDA M.A., TUVERI C., & MURRU G., 1992, Le tecniche edilizie del periodo nuragico nell'architettura delle acque presenti nel territorio della Barbagia. In Sardinia in the Mediterranean: a Footprint in the Sea, Sardinian Archaeology presented to M. Balmuth, ed. by R.H. Tykot & T.K. Andrews. Sheffield, pp.250-261.

FERRARESE CERUTI M.L., 1978, Sisaia, una deposizione in grotta della cultura di Bonnannaro. Quaderni 6, Soprintendenza Archeologica per le province di Sassari e Nuoro, Sassari.

*FIORELLI G., 1878, (communiqué par / comunication by Vivanet F.), Teti. Notizie per gli Scavi, pp.244-250.

*FIORELLI G., 1882a, (communiqué par / comunication by Nissardi F.), Nuragus. Notizie degli Scavi, pp.305-311.

*FIORELLI G., 1882b, (communiqué par / comunication by Nissardi F.), Teti. Notizie per gli Scavi, pp.378-379.

*FIORELLI G., 1890, (communiqué par / comunication by Vivanet F.), Lei. Avanzi di una fonderia dell'età del bronzo scoperti nel territorio del comune. Notizie degli Scavi, pp.334-336.

GALLIN L.J. & TYKOT R.H., 1993, Metallurgy at Nuraghe Santa Barbara (Bauladu), Sardinia. Journal of Field Archaeology 20, 3, pp.335-345.

GALLIN L.J., TYKOT R.H., ATZENI C., VIRDIS P., & SISIU G., 1994, Attività metallurgiche al nuraghe S. Barbara di Bauladu (Oristano) (prima parte). Quaderni della Soprintendenza Archeologica per le province di Cagliari ed Oristano 11, pp.141-153.

* Fiorelli G(iuseppe) was the General Director of the Museums and Ancient Excavations ("Musei e Scavi di Antichità" or "Antichità e Belle Arti") of the Ministry of Public Education in those years, and the official person allowed to sign the archaeological reports sent to him by all over Italy, in order to be published in the official revue "Notizie degli Scavi di Antichità". Since he was not archaeologist, he often reported within inverted comas the reports, as due to the Superintendents of the different regions of Italy. This praxis was so well known and normal that in many cases the name of the Superintendent is not even quoted.

GARAGNANI G.L. & MARTINI C., 1998, Microstructural investigations of ox-hide and plano-convex copper ingots from Sardinia. In: Science and Technology for the Safeguard of Cultural Heritage in the Mediterranean Basin. proceedings of the Ist International Congress (Catania, 27 nov.-2 dic., 1995), CNR, pp.461-465.

GIARDINO C., 1987, Sfruttamento minerario e metallurgia nella Sardegna protostorica. In: Nuragic Sardinia and the Mycenaean World. Studies in Sardinian Archaeology III, ed. by Miriam S. Balmuth. BAR International Series 387, pp.189-222.

GIARDINO C., 1995, Il Mediterraneo Occidentale fra XIV ed VIII secolo a.C. Cerchie minerarie e metallurgiche - The West Mediterranean between the 14th and 8th centuries B.C. Mining and metallurgical spheres. BAR International Series 612.

GILLIS C., 1990, All that glitters is not gold, Bulletin, Medelhavsmuseet (The Museum of Mediterranean and Near Eastern Antiquities), 25, pp.24-32.

GILLIS C., 1991, Tin in the Aegean Bronze Age, Hidra 8, pp.1-31.

GILLIS C., CLAYTON R., à paraitre / forthcoming, Further Developments in the Use of Tin Isotopes for Sourcing Tin. In: METRON. Measuring the Aegean Bronze Age, 9th International Aegean Conference, Yale University, Dept. Of Near Eastern Languages & Civilisations, 18-21 April, 2002.

KASSIANIDOU V., 2001, Cypriot copper in Sardinia: another case of bringing coal to Newcastle? In: Italy and Cyprus in Antiquity. (New York, Italian Academy, Columbia University, November 16-18, 2000), ed. By L. Bonfante & V. Karageorghis. Nicosia 2001, pp.97-119.

LILLIU G., 1966, Sculture della Sardegna nuragica. Verona. La Zattera.

LO SCHIAVO F., 1979, Storia e consistenza della Collezione Spano al Museo Sanna di Sassari. In Contributi su Giovanni Spano. Sassari, pp.63-89.

LO SCHIAVO F., 1990, Copper ox-hide and plano-convex ingots in Sardinia. In: LO SCHIAVO et al.1990, pp.15-40.

LO SCHIAVO F., 1997, Bronze Age Metals: their distribution and manufacture in Nuragic Sardinia. In: Mensch und Umwelt in der Bronzezeit Europas. (Berlin, 17-19 März 1997), ed. by B. Hänsel, Kiel, pp.193-216.

LO SCHIAVO F., 2000, Bronzi e Bronzetti del Museo "G. A. Sanna" di Sassari. ed. by F. Lo Schiavo & G.M. Demartis, Sassari. (Imago Media, Piedimonte Matese).

LO SCHIAVO F., 2001, Late Cypriot bronzework and bronzeworkers in Sardinia, Italy and elsewhere in the West. In: Italy and Cyprus in Antiquity. (New York, Italian Academy, Columbia University, November 16-18, 2000), ed. By L. Bonfante & V. Karageorghis. Nicosia 2001, pp.131-152.

LO SCHIAVO F., MADDIN R., MUHLY J.D., MERKEL J., & STECH T., 1990, Analisi metallurgiche e statistiche sui lingotti di rame della Sardegna - Metallographic and Statistical Analyses of Copper Ingots from Sardinia. Quaderni 17, Soprintendenza Archeologica per le province di Sassari e Nuoro. Ed. by F. Lo Schiavo & R. Maddin. Sassari. (Il Torchietto, Ozieri).

LO SCHIAVO F. ANTONA A., BAFICO S., CAMPUS F., COSSU T., FONZO O., FORCI A., GARIBALDI P., ISETTI E., LANZA S., LEONELLI V., PERRA M., PUDDU M.G., RELLI R., ROSSI G., SANGES M., USAI A., & USAI L., in press. In La Sardegna nell'età del bronzo recente: Articolazioni cronologiche e differenziazioni locali - La Metallurgia. In L'Età del Bronzo recente in Italia. Lido di Camaiore, 26-29 ottobre 2000.

LO SCHIAVO F. & VAGNETTI L. 1980, Micenei in Sardegna? Rendiconti Accademia dei Lincei, classe scienze morali, serie XXXV, 5-6, 1980, pp.372-391.

MADDIN R., MERKEL J., 1990, Metallographic and statistical analyses. In LO SCHIAVO et al. 1990, pp.41-199.

MACNAMARA E., RIDGWAY D. & RIDGWAY F.S., 1984, The Bronze Hoard from S. Maria in Paùlis, Sardinia. British Museum Occasional Papers n.45. Dept. Greek and Roman Antiquities. London.

MARABELLI M., CESAREO R., GIGANTE G., GIARDINO C., PINARELLI L., & CANEVA C., en preparation / in preparation. Progetto CNR: Estrazione e produzione metallurgica nell'Italia preromana. Archeometria e sperimentazione. Ripostigli di S.Imbenia, Sassari (Villaggio e nuraghe). (Relazione trasmessa da M. Marabelli, il 13.08.1997, prot.n.8342/22).

MERKEL J.,1986, Ancient smelting and casting of copper for "ox-hide" ingots. In Sardinia in the Mediterranean. Studies in Sardinian Archaeology II. Ann Arbor, pp.251-264.

MUHLY J.D., 1973, Copper and Tin. The distribution of metal resources and the nature of metals trade in the Bronze Age. Transactions of the Connecticut Academy of Arts and Sciences 43, pp.155-535.

MUHLY J.D., 1985, Sources of Tin and the beginning of Bronze metallurgy. American Journal of Archaeology 89, pp.275-291.

MUHLY J.D., STECH T., 1990, Final observations. In: LO SCHIAVO et al. 1990, 201-221.

NISSARDI F., 1884, Intorno ai due ripostigli di Abini e di Forraxi Nioi presso Nuragus. Lettera al Prof. E. Pais, Direttore del BAS (Risposta ad E. Pais, Il ripostiglio ...). Bullettino Archeologico Sardo I, 2, suppl. pp.3-29.

NISSARDI F., Giornale di scavo relativo allo scavo in località Forraxi Nioi. In Pellini A. 1973-74, Doc. 36, pp.125-138.

PAIS E., 1884, Il ripostiglio di bronzi di Abini presso Teti. Bullettino Archeologico Sardo I, 2, serie II, pp.67-181.

PELLINI A., 1974, Il Nissardi e la preistoria sarda. Tesi di Laurea, Università degli Studi, Cagliari, Facoltà di Lettere e Filosofia, Anno Accademico 1973-74. (Rel. Prof. G. Lilliu).

PIGORINI L., 1895, Pani antichissimi di rame da fondere della Sardegna, Bullettino di Paletnologia Italiana 21, 1, p.243.

PIGORINI L., 1904, Pani di rame provenienti dall'Egeo scoperti a Serra Ilixi in provincia di Cagliari. Bullettino di Paletnologia Italiana 31, 1, pp.91-107.

PINZA G., 1901, Monumenti primitivi della Sardegna. Monumenti Antichi dei Lincei XI. (U. Hoepli, Milano).

PULAK C., 2000, The Copper and Tin Ingots from the Late Bronze Age Shipwreck at Uluburun. In: Anatolian Metal I, Beiträge des Symposiums, Bochum Nov. 1998, Der Anschnitt, Beiheft 13, pp.137-157.

PULAK C., 2001, The Cargo of the Uluburun Ship and Evidence for Trade with Aegean and Beyond, In: Italy and Cyprus in Antiquity. (New York, Italian Academy, Columbia University, November 16-18, 2000), ed. By L. Bonfante & V. Karageorghis. Nicosia 2001, pp.13-60.

SAM 1960: Junghans S., Sangmeister E. & Schröder M. 1960, Metallanalysen Kupferzeitlicher and Früheisenzeitlicher Bodenfunde aus Europa. Studien auf den Anfangen der Metallurgie. I. Berlin.

SERRA RIDGWAY, F.R., 1986, Nuragic Bronzes in the British Museum. In Sardinia in the Mediterranean. Studies in Sardinian Archaeology II. Ann Arbor, pp.85-101.

SPANO G., 1865, Memoria sopra alcuni idoletti di bronzo trovati nel villaggio di Teti e Scoperte Archeologiche fattesi nell'isola in tutto l'anno 1865. Cagliari. (Tip. Arcivescovile, Cagliari).

SPANO G., 1873, Memoria sopra l'antica cattedrale di Galtellì e Scoperte archeologiche fattesi nell'Isola in tutto l'anno 1872. Cagliari. (Tip. A. Alagna, Cagliari).

SPANO G., 1876, Scoperte archeologiche fattesi nell'Isola in tutto l'anno 1876. Cagliari.

STOS-GALE Z. & GALE N., 1992, New Light on the Provenience of the Copper Ox-hide Ingots Found on Sardinia. In Sardinia in the Mediterranean: a Footprint in the Sea. Studies in Sardinian Archaeology presented to M. Balmuth. Sheffield, pp.317-346.

TARAMELLI A., 1931, Teti. I. Esplorazione del santuario nuragico di Abini. Notizie degli Scavi, pp.45-63.

TARAMELLI A., 1933, Tempietto protosardo del Camposanto di Olmedo (Sassari). Bullettino di Paletnologia Italiana LIII, pp.118-

TYLECOTE R.F., BALMUTH M.S., & MASSOLI NOVELLI R., 1983, Copper and Bronze metallurgy in Sardinia. Journal of the Historical Metallurgy Society 17, 2, pp.63-78.

TYLECOTE R.F., BALMUTH M.S., & MASSOLI NOVELLI R., 1984, Copper and Bronze metallurgy in Sardinia. In Studies in Sardinian Archaeology. Ann Arbor, pp.115-162.

UGAS G. & USAI L., 1987, Nuovi scavi nel santuario nuragico di S. Anastasia di Sardara. In La Sardegna nel Mediterraneo tra il II e il I millenio a.C. Atti del II Convegno di Studi, Selargius-Cagliari 27-30 nov. 1986. Cagliari, pp.167-218.

VAGNETTI L., LO SCHIAVO F., 1989, Late Bronze Age long distance trade: the Role of the Cypriots. In Early Society in Cyprus. Edinburgh: 217-243

VALERA R. dans ce volume / in this volume, Tin in the Mediterranean area: history and geology. Colloquium The Problems of Early Tin. IV UISPP Congress, Liège, 2-8 september 2001 (edited by A. Giumlia-Mair & F. Lo Schiavo).

VALERA R., VALERA P., & LO SCHIAVO F., in press, Lead in Nuragic Sardinia: ores, isotopy and archaeology. In Beginnings of Metallurgy in the Old World. Euroseminar, 18-20. November in Freiberg (Erzgebirge), Germany, (edited by E. Pernicka).

VODRET F., 1935, Ricerche chimiche sui bronzi preistorici della Sardegna. Atti del V Congresso Nazionale di Chimica. Roma, pp.105-120.

VODRET F. 1959, Sui bronzi preistorici dell'epoca nuragica. Rendiconti del Seminario della Facoltà di Scienze dell'Università di Cagliari 29, 3-4, pp.248-272.

ZWICKER U.,VIRDIS P., CERUTI M. L., 1980, Investigations on copper ore, prehistoric copper slag and copper ingots from Sardinia. In Scientific Studies in Early Mining and Extractive Metallurgy, edited by P.T. Craddock, British Museum Occasional Paper 20, London, pp.135-163.

WEN YI, BUDD P., McGILL R.A.R., YOUNG S.M.M., HALLIDAY A.N., HAGGERTY R., SCAIFEW B., POLLARD A.M., 1999, Tin isotope studies of experimental and prehistoric bronzes. In: The Beginnings of Metallurgy. Proceedings of the International Conference, (Bochum 1995). Ed. by A. Hauptmann, E. Pernicka, T. Rehren, Ü. Yalçin. Der Anschnitt, 9, pp.285-290.

VILLAGRANDE STRISAILI-NUORO – THE DISCOVERY OF TIN NEAR THE MEGARON TEMPLE OF S'ARCU 'E IS FORROS

Maria Ausilia FADDA

In autumn 2000, during the excavation campaign in the sanctuary village of S'Arcu 'e is Forros, near the town Villagrande Strisaili, several tin fragments were found together with fragments of copper and lead ingots. Because of the importance of the discovery it is important to describe the whole archaeological area, as the different buildings are closely related to the cultual function of the village and, most probably, to the metallurgical activity which took place for the production and perhaps for the trade of bronze objects, offered by pilgrims in the megaron temple.

The megaron temple of S'Arcu 'e is Forros belongs to a large archaeological complex, on a hill at 1000m above the sea level, under Mt. Allue in Fogu at 1300 m. The hill is enclosed on the West side by the river Bacu Alleri and on the East side by the river Abbatrula, both affluents of the river Flumendosa. A Giant's tomb, belonging to the village, was built on a small elevated space with a sheer drop to the river, while the nuraghe Ortacesu, built on a nearby hill, controlled the river on the other side.

The village consists of several huts, built on the upper side of the hill and on the hill side, toward the river on the East. On the hill top the structure of a megaron temple was excavated. The name is given to these buildings because of their planimetry, which is similar to that of the Mycenaean buildings; however the chronological setting of these special type of buildings for cult in Sardinia is very different from the older Mycenaean buildings, which were mainly for civilian use. (Photo 1)

The temple shows a rectangular shape, with a total length of 17 m and a non uniform width, which goes from 5.50 m to a maximum of 6.50, because of the evident irregularity of the perimetral walls with different thickness and go from 1.10m to a maximum of 1.50 m.

THE TEMPLE

The temple has a complex planimetry with a rectilineal front which now shows seven layers of polygonal medium-sized stones, more precisely worked near the door posts. A trapezoidal corridor (trapezoidal (2.10 m length X 1.50 m width) broader towards the inside leads to the first rectangular room A (3.10 m length X 3.60 m width). A small open mouthed vessel with concave bottom was found in room A, in the corner on the left side at h.2.55m, under a rather hard beaten floor, which in some places showed carbon layers from a fire during the first building phase of the temple in the early period of the recent BA. A fragment of votive sword and part of an askos-shaped vessel with geometrical decorations on the shoulder come from the same excavation area.

A corridor leads from the first room to the remaining three rooms belonging to the temple.

The first rectangular space (1.10 m length X 1.07 m width) opens into room B (3.70 m length X 1.80 m width) which has an irregular rectangular shape. Few ceramic fragments of very altered fabric were collected in this room. A fragment of votive sword and some unidentifiable bronze fragments were found in the corridor (1.20 m length X 1.40 m width) leading to room C. The corridor still has an architraved opening with a little trapezoidal window above. The clearly overhanging perimetral walls enclose a rectangular room (2.30-2.50 m length X 3.50 m width), in which several schist plates and other flat stones, certainly belonging to the roof of the temple were found in the earth layer disturbed by grave robbers. Several limestone and fossiliferous sandstone plates were found in room C. Some are simply shaped and smoothed on the sides, other show regular circular holes. Among the plates there was also a T-shaped hewn fossiliferous limestone which shows, on the visible flat side, a large breast-shaped ashlar. Several rectangular flat based sandstone slabs, with several holes, come from the same room C. Some of them are ordered in regular lines, other are not. Remains of the lead, which had been used for fixing votive bronzes into the holes are still visible on some of the slabs. Here there was also a broken quadrangular limestone basin, with a small circular depression in the middle and well worked surfaces. Numerous ceramic sherds of globular vessels and small carenated cups come from the same room C. In the debris there are also bronze sheets, pin fragments and pieces of runners, used as fixing rods in the bases of the offerings in the cult rooms.

A quadrangular corridor (1.10 m length X 1.10 m width) with overhanging walls, which form an ogival door, leads to room D, with an irregular rectangular shape (2.60-2 m length X 3.40 m width). In the bottom wall there is a small trapezoidal niche, which results to be axial to all corridors of the four rooms. In room D there was an enormous granite slab, thinner at both ends. Among the ceramic sherds there are fragments of cooking pots, small ollae with cylindrical neck, steeply carenated cups of different size with flaring concave neck and small bowls with perforated handles.

A rectangular stone slab (0.57 m length X 0.24 m width X 0.20 height), with large holes on the upper part and a votive

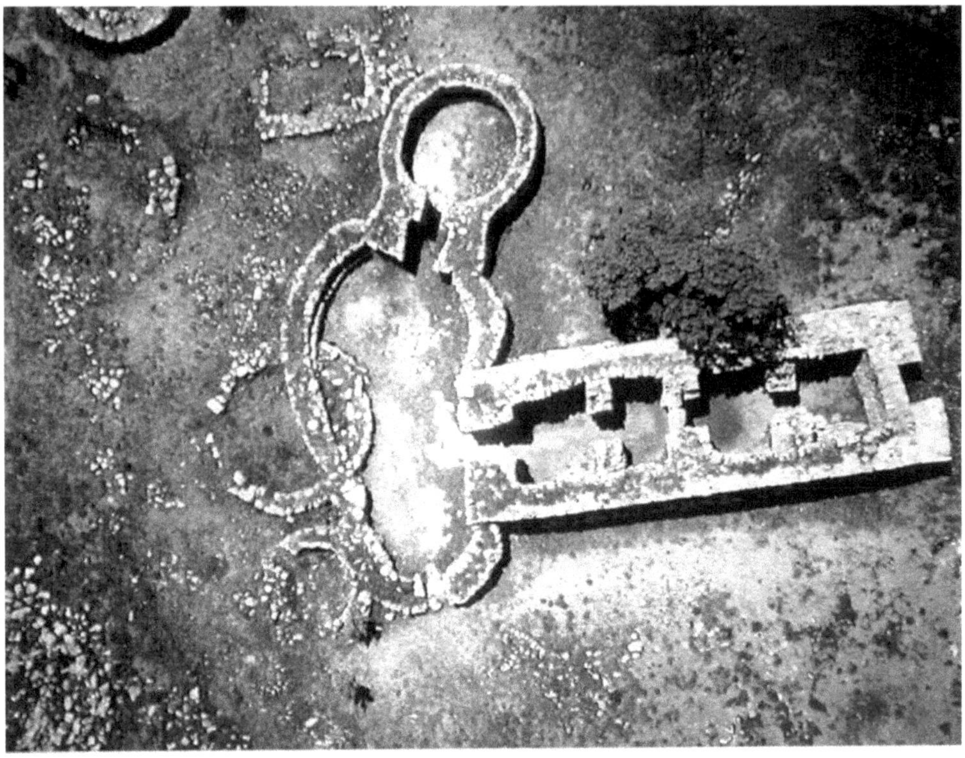

Photo 1: VILLAGRANDE STRISAILI (Nuoro)
Aerial view of the megaron temple and of the temenos before it.

Photo 2: VILLAGRANDE STRISAILI (Nuoro)
Aerial view of the building with two circular rooms, used for metallurgical activities.

sword still fixed in one of them, was leaning on the right side of the front. The study of the building front has shown that there were two building phases, with an original plan with two lateral elongated walls (in antis), which were later filled by a wall and formed a rectilineal front.

In the surroundings of the temple there are numerous traces of metallurgical activities connected with the temple. Many fragments of ingots of Aegaean type, bun ingots, three axes, a sanguisuga fibula, fragments of bronze statuettes, a squatting Orientalizing bronze lion, many copper remains collected in a bronze sheet vessel (diam 0.40 m) with many rivetted repairs, a 11.700 kg disc-shaped lead ingot(diam 0.47 m) with 2 holes and fragments of a Cypriot tripod come from the nearby nuraghe and from the huts.

The temple section on the N-W-N-E sides, still covered by a heap of debris and materials from a clandestine excavation in the four rooms and in the area before the temple, were explored in the second digging campaign in Octobre-November 1990. The archaeological layers which had not previously been touched because of a very steep terrain, which had formed a thick layer of sliding earth, was also explored.

The excavation has clearly shown the plan if the monument with elongated lateral walls, which form a double in antis building. The wall appendices on the back side in antis are irregularly connected to the wall, with a construction texture different from that of the perimetral walls and show a superior thickness, which is further accentuated at the end. The in antis extension on the entrance side looks however more regular in the stone layers, which in this case do not show differences from the perimetral wall.

Among the stone debris of the walls there are many flat stones of different thickness, which appear to belong to the walls of the last rooms of the temple, in which the walls are steeply overhanging the rooms A,B and C. The excavation evidenced the skirting board at the wall foot of the building, which follows the length of the side walls. The skirting board, made of large summarily worked stone blocks, lined in brick-on-the-edge course, guaranteed the stability of the building and could be used as a bench or as a support.

Several layers of very dark earth, mixed with different stones – river pebbles of different shape, used perhaps as pestles or whetting stones, were found in the trench on the back side.

Some fragments of sandstone with differently shaped holes on the side possibly belong to the bases for offerings, found in great quantity also inside the temple. The presence of many fragments of votive swords and lead fragments are to be seen in connection with the same bases.

From the same excavation trench come also the upper part of a nuraghe model made of sandstone (height 0.182 m , diam. 0.15 m) with seven oval holes forming a circle and one hole in the middle. The model has exact comparisons with other models found in Sardinia and originally carried on its top seven votive objects, fixed with lead. The presence of the bases and of the model in the outer area shows that the temple offerings were put also into the vestibule in antis. The archaeological depot contained, particularly in the deeper layers, a large quantity of ceramic sherds, dated to the later periods of the MBA (15th-14 c.BC), which correspond to the first period of the first phase of the settlement and is therefore older than the construction of the buildings for religious use. At the base of the temple external walls there is a uniform thick carbon layer, most probably due to a fire of the wooden structures completing the projection of the stone slabs roof, layered on wooden beams with double slope, which released the weight on the thick perimetral walls. It is also possible that the fire was also the reason for the changes made to the building in the parts in antis, to give more stability to the roof. The exploration of the last room D of the temple, which was disturbed by clandestine diggers, but still shows the original stone pavement was completed in the same campaign

In room C the excavation has shown a thick cultural layer rich in ceramic sherds and 2 pots, one of which was whole. The two containers were set into the badly and only in part preserved beaten clay floor in two corners of the cella.

The three containers found in room A in the first campaign and in room C in the last, are ovoid ollae with open shape, with rounded lips and indistinct bottom or with flat bottom with rounded margins. The ollae handles are flat ribbons applied on the widest spot of the container wall and of reverse elbow type, very broad at the attaching point and with concavo-convex section.

The arrangement of the two containers set into the corners of the temple rooms suggests that they were employed as containers for a water reserve, used for purification rituals. The hypothesis is supported by the presence of fragmented sandstone basins and by a channel at the basis of the wall in room B and at the point of highest slope for facilitating the water discharge.

The ollae found in the monument belong to a typology of Nuragic material, which is very common all over Sardinia and can be dated to the recent and latest phase of the LBA (12th-9th c.BC).

The third excavation campaign brought to light on the front side a large elliptical walled enclosure, which leans on the lateral walls of the in antis vestibule. The walls surrounding the enclosure show at the basis a bench, similar to that found on the outer walls of the monument and similarly employed as support for the offerings. The entrance of the elliptical area, on the left side, shows a singular threshold, made of two chunks of sandstone with two small depressions in which are small vessels of bronze sheet, fixed with lead and two rectangular holes, which most probably held originally the wooden door posts of the entrance. The enclosure walls are set on the axis to the entrance on an earlier hut, which contained ceramics dated to the MBA (15th-14 c.BC). On the left side of the enclosure there is a building consisting of a rectangular vestibule, which leads to a circular room, most probably used as depot for the temple. The whole area, explored in the last campaign, showed a clay floor mixed with earth and coal. The remains of burnt paving were visible especially before the entrance to the temple, where many fragments of votive bases , small, simple daggers, pins, an axe-shaped pendant and more bronze fragments were found. Together with the bronze items there were fragments of carenated bowls, with plastic decorations, semispherical bowls and collared ollae, to be dated to the RBA and FBA (13th-9th c. BC). The entire and fragmented bases for the offerings, indicate that the whole enclosure area before the megaron was used for to collect votive offerings, with a ritual which seems to be rather similar to that of the better known well-temples in the territory of the Nuoro province.

After several excavation campaigns, which brought to light the temple structures of the megaron and the temenos, work in the village area was taken up again, on the sloping side of the hill, which is enclosed between two rivers. The new excavation was decided as rescue of some hut remains, which had been badly damaged by a caterpillar, stolen by looters from a yard, put up for the construction of the new road to Lanusei.

The excavation revealed four huts, a rectangular sacellum and a polilobate building, still not examined inside.

The room next to the enclosure wall of the megaron temple has an outer circumference of 4.5 m and an inner of 3.10 m The height of the hut can be discerned only on the side leaning to the temple wall, while the outer side shows only one stone line on the basis and the archaeological layer. The ceramics found here, are open-shaped ollae with a thick lip with annular section, truncated conical tegami made of coarse material with many inclusions, open carenated bowls and many pestles, basins and smoothing tools with evident traces of use and a broken small votive bronze dagger. The inside of the hut still shows part of the beaten clay floor and traces of coal from a central and a lateral hearth. The archaeological layer shows two cultural phases. The more archaic one belongs to the first period of the settlement and is contemporary with the large hut, which had been partly enclosed and destroyed by the construction of the walled enclosure of the temple in the RBA.

In the surroundings several fragments of votive sandstone bases, similar to those found inside the megaron, a finely worked bronze pin-head and small fragments of a non-identifiable bronze sheet container were found. A rectangular room (ext.perimeter 6 X 4 m, internal p. 4.5 X 2.3 m), missing several parts of the walls, was found inside the temple enclosure.

The excavation, down to 1.42 m , showed a thick burnt level at the basis, a second clay layer(0.90 m) with rubble and beaten clay next to three hearths. Two hearths were in the central part of the sacellum, while the third was at the entrance in the left corner. The presence of burnt layers also in the outer part of the temple suggests that the rectangular construction was built on an earlier layer, corresponding to the burnt layer previously documented in the deeper stratigraphy of the temple and the temenos on the front. A third layer, consisting of a new burnt layer and belonging to the last building phase, contained a large semispherical bowl, sustained by a large number of minute pebbles at the wall foot. The ceramic sherds in the deepest layer are fragments of truncated conical tegami, open shaped ollae with thickened, angular sectioned lip and ollae with flat ribbon handles, to be dated to the earlier phases of the RBA.

The more recent layers of the FBA and of the EIA contained very fragmented and non-indicative material, rod handles, fragments of shouldered bowls and small open-shaped ollae with rounded lips. The question of the use of the small rectangular room is still open, but the presence of three hearths suggests a domestic use more than a room for cult.

In the near vicinities of the rectangular room the hut B still conserves its entire perimetral walls and part of the height (ext. diam. 6.30 m, int.diam.4.30 m) and an entrance on the East side, with a large stone slab as threshold. Inside there are a bench and a short wall enclosing a storeroom. The floor and the bench remains, as well as the central hearth are covered by a layer of reddish clay mixed with coal, which had been used also for plastering the walls of the room. At the base of the bench there were, obviously intentionally heaped up, hundreds of river pebbles with traces of use at the points (apparently pestles or millstones), long whetting stones with eroded central part and a hole at the top. There were also smoothing stones for burnishing, some of which with incisions, drilled pendants and an unfinished ram bust.

The whole archaeological layer did not contain any common archaeological material and this suggests that this room was employed exclusively for the storage of stone artefacts. The presence of a thick burnt layer is documented all over the surface of hut B, beginning from the entrance and might indicate that the hut was temporarily abandoned because of a fire and successively reused for the storage of non-perishable goods. The outer area of room B also shows signs of a fire on a thick clay layer. In the same direction, also the huts C and D, built on an irregular rock surface, still visible in the rooms with small depressions and protuberances, were explored.

The wall height is conserved only in small parts, because of the destruction due to the passage of grave robbers with the caterpillar. The irregular shape of the rock conserved some entire ceramics. Hut C has an outer diameter of 5.90 m and an internal diameter of 4.80 m. The bad conservation of the wall does not allow to precise the orientation of the entrance. The archaeological layer, almost miraculously still intact on the side with more pronounced slope, contained one-handled cups, truncated conical tegami, ollae with simple and thickened lips, bowls with handles and plastic decoration, broad handles with impressed oblique pattern, spindles and fragmented millstones. The floor remains show that it was made of beaten clay. Hut D has an external diameter of 6 m and an internal diameter of 4,35 m with an entrance to the East. The remaining walls are set on an irregular rock surface, levelled inside with gravel and clay.

The archaeological layers are well preserved and contained open-shaped ollae with two handles, open-shaped cups with one handle, small ollae with continuous neck. A vessel with ring-bottom and surface with parallel, but irregular holes, which might have been used as a torch. In natural rock cracks there were some tegami fragments, with a combed decoration, to be dated to the earliest period of the settlement. Despite the destructive passage of the caterpillar, the area between the two huts still shows the remains of more hearths. One of them has two circular holes, left by the original firedog in the beaten clay.

The data do not allow the calculation of the original size of the huts and of the other structures. It is however quite clear that the village was destroyed by fire and several buildings were later reconstructed in place of the previous huts, without respecting the earlier position of the walls. This is suggested by the presence of several hearths under the remaining walls or in rather illogical spots inside the huts.

In the same campaign another large hut, leaning on a bilobate structure and with considerable height, was identified. The new rooms show on the outside a large amount of flat stone slabs, similar to those found around the megaron and which were employed for the double sloping roof of the temple. The remaining wall of the bilobate structure still preserves in the wall texture a line of stone slabs, forming a sort of shelf around the circular construction. The amount of slabs is much higher on the N.W side. These might belong to a covered communication passage, sustained by wooden posts.

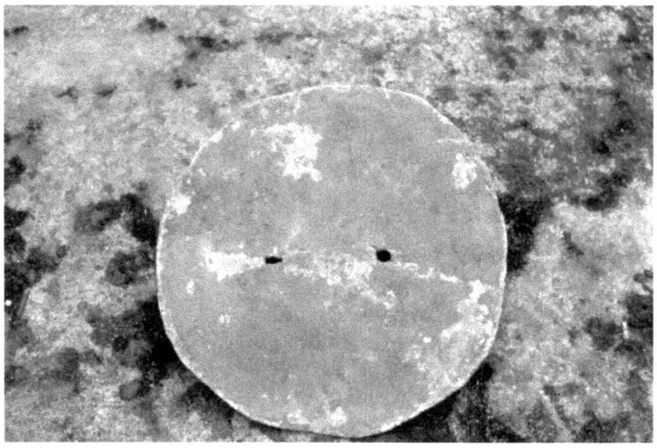

Slide 1: Circular lead ingot with two holes, found in the megaron temple.

Slide 2: General view of the building for metallurgical activities.

THE BUILDING

The building shows evident anomalies and consists of two rooms (Photo 2). The total length is 10.50 m the width 5.50 and the max. remaining height 2.50 m on the Southern part, with a width of 1.80 m.

The second room on the N-side has a max. height of 1.80m and a base of 2.40 m. The wall thickness is of 2-2.30 m at the basis and both structure show, particularly inside, a noticeable overhang. On the W-side, at the base of two rooms there are two openings, which might be interpreted as channel and correspond to the direction of the most common winds in Sardinia, which mainly come from the West. The openings dimensions are 50 X 30 cm and the width corresponds to the thickness of the wall (2.10 m). The opening on the Northern side still preserves a long slab on the basis. The slab was set into the building wall during the construction and this fact excludes the hypothesis of a later transformation of the structure for a different use (fig. slides 2-3). The main interpretation problems of the building as a metallurgical structure are in particular the dimensions of the two "furnaces", which are far too large compared to other structures found in other contexts. It is however possible that small furnaces were constructed inside the existing structures and that they were systematically destroyed and re-built for new metallurgical activities. Inside the two circular structures, which were repeatedly ruined by illegal excavations, there were some ceramic sherds of Nuragic time, remains of clay mixed with coal and ashes, tiny copper prills, bronze fragments and copper and lead drips.

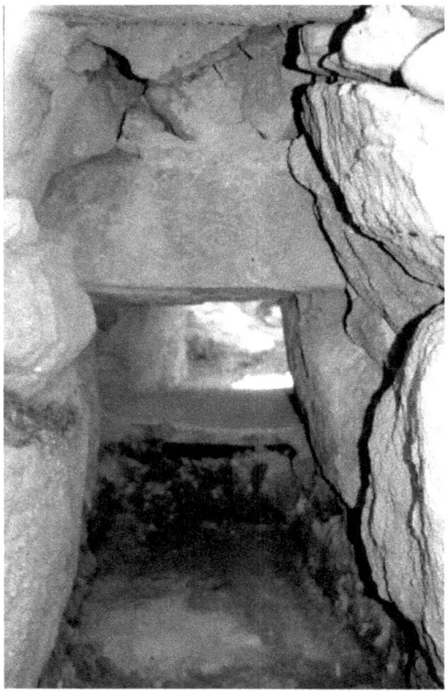

Slide 3: Channel at the basis of the wall of the metallurgical building.

The data collected during excavation inside the buildings are still not enough for a clear interpretation of the real function of the two rooms, however on the outer area there are clear traces of intense metallurgical activities. From the space between the megaron and this building come several

Slide 4: Some of the tin fragments collected in the area around the megaron temple and the building for metallurgical activities.

fragments of bun ingots and parts of ox-hide ingots, the already mentioned disc-shaped lead ingot and some elongated bun ingots and a large amount of copper remains. The most important discovery is that of several tin fragments, which were collected on the surface, together with copper and lead fragments in the area surrounding the megaron (slide 4). The chemical analyses, which , published in this volume by Roberto Valera, demonstrate an intense activity and experimenting of alloys of different composition, by reducing the use of tin as much as possible. The tin had to be imported and was substituted in the alloy by the local lead, which could be easily extracted from the rich ores of the island. It is just possible that the Nuragic artisans managed to produce several highly leaded alloys for objects with different function and use.

It is quite clear that the Nuragic bronze smiths tried to reduce the tin use as much as possible, but the question of the source of the tin supply is still open. The Nuragic population had a potentially privileged geographic position on the maritime trade routes from the East to the West and viceversa.

It is still not clear what kind of trade exchange, direct or indirect, and which trade goods might have been exchanged by the Nuragic traders with the more probable intermediaries, such as the Mycenaeans, the Cypriots and, later on the Proto-Villanovans. From the most recent finds, it seems that the intermediation of the Villanovans might have been significant for the import into Sardinia of large quantities of amber, which could be bartered, at least partly, for costly products, such as bronze and ceramics, documented in several contexts of the Tyrrhenian coast.

The problem of tin import might have been solved by the Nuragic populations with the payment of a tax for the permission of landing on the shores of the island, which in the BA and in the EIA, was an obligatory bridge-head for all trade routes from the East and the West.

All hypotheses formulated in this paper are going to be further investigated and supported by future excavations, which are apparently going to last a long time and will hopefully bring new interesting results.

Author's address

Maria Ausilia FADDA
Soprintendenza Archeologica
Via Ballero, 30
08100, Nuoro, ITALIE

Bibliography

FADDA M.A. 1985. *Il Tempio a Megaron di S'Arcu e' Is Forros in "Nuovo Bollettino Archeologico Sardo.* Delfino Editore.

FADDA M.A. 1989. *Il Tempio a megaron di S'Arcu e' is Forros in Early Society in Cyprus.* Edimburgh University Press

FADDA M.A. Il tempio nuragico di Sa Carcaredda *Bollettino di Archeologia del Ministero per i Beni Culturali e Ambientali,* n°6 Poligrafico dello Stato.

FADDA M.A. Il tempio a Megaron di S'Arcu e' is Forros (Villagrande Strisaili). *Bollettino di Archeologia del Ministero per i Beni Culturali e Ambientali* n°6 Poligrafico dello Stato.

FADDA M.A. 1992. Il tempio a megaron di S'Arcu e is Forros (Villagrande Strisaili). *Bollettino di Archeologia* 13/14/15. Poligrafico dello Stato.

FADDA M.A. 1993. Raccolta e tesaurizzazione delle offerte negli edifici cultuali della Sardegna nuragica. *Atti Fifty Conference of Italian Archeology Oxford.*

FADDA M.A. 1993. *Archeologia Prenuragica e Nuragica in Ogliastra.* EDISAR.

FADDA M.A. 1996. Antichi Sardi Purificati. *Rivista di Archeologia Viva* n 57 maggio giugno.

FADDA M.A. 1998. Nuovi templi a Megaron della Sardegna Nuragica. *Atti XII International congress of prehistoric and protohistoric sciences*, vol 4 , Forli.

FADDA M.A. 1998. Il patrimonio Archeologico dell'Ogliastra. *Gli antichi padri Quaderni arzanesi* n°. 5.

FADDA M.A. 2000. Un esempio di architettura religiosa in Ogliastra. In *L'Identità di una provincia.* Edizione Comunità Montana n 11 Ogliastra.

FADDA M.A. 2001. *Nuove acquisizioni del magalitismo nella provincia di Nuoro Aspetti del megalitismo preistorico* Grafica del Parteolla.

LA METALLURGIE DU BRONZE EN SICILE (ITALIE)

Rosa Maria ALBANESE PROCELLI

Résumé: Afin d'évaluer l'utilisation de l'étain dans la métallurgie du bronze en Sicile à la période protohistorique, on considère les gisements métallifères de l'île, les différentes sources documentaires relatives à cette production (analyses métallographiques; objets utilisés dans la chaîne opératoire des procédés métallurgiques) et les échanges avec les régions susceptibles d'être les sources d'approvisionnement en métaux pour la Sicile.

Abstract: This paper focuses on the problems of the use of tin in Sicilian bronze metallurgy on the basis of the archaeological evidence and metallurgical research.

Il est problématique de proposer un cadre global de l'utilisation de l'étain dans la métallurgie protohistorique de la Sicile en considérant l'état actuel de la documentation, qui est dépourvue d'un dossier systématique d'analyses métallurgiques. Nos considérations doivent nécessairement s'appuyer sur les très faibles données connues jusqu'ici, ce qui les rend très hypothétiques. Il serait nécessaire de comprendre comment le progrès technologique s'avance en lien strict avec les changements intervenus à l'intérieur des institutions sociales et politiques d'une communauté, dans lesquelles la technologie, aussi bien que d'autres branches de l'économie, est strictement «imbriquée».

1. GISEMENTS MÉTALLIFÈRES

En Sicile des gisements de minéraux métalliques sont localisés dans la région du Nord-Est (Monti Peloritani, district de Fiumedinisi). Ces gisements ont été exploités dès la période protohistorique et probablement déjà à partir du Chalcolithique récent[1].

Ils comprennent des minéraux de plomb, de zinc, de cuivre, d'arsenic, d'antimoine et de fer, et des minéralisations à sulfures mixtes. Des analyses chimiques d'échantillons minéraux de la vallée de Fiumedinisi ont révélé la présence de cuivre et de fer (en quantité élévée dans la composition des tétraédrithes) et l'inclusion d'antimoine dans la galène, qui est du type argentifère. Dans certains échantillons, l'étain est présent parmi les éléments en traces entre 6,87 et 296 p.p.m.[2]. D'autres analyses chimiques de scories de l'âge romain, provenant de la région de Messine, ont démontré leur provenance de minéraux riches en sulfures d'arsenic, de cuivre et de nickel (ce qui s'accorde bien avec la composition des gisements des Monti Peloritani) et la présence d'étain entre 2 et 4 p.p.m.[3].

Il s'agit évidemment de quantités trop faibles pour une utilisation dans les alliages de bronze. L'étain était importé en Sicile: les régions de provenance les plus proches étaient la Toscane (*Colline Metallifere*) et la Sardaigne.

2. ANALYSES MÉTALLURGIQUES

Les données relatives à des analyses métallurgiques sont encore peu nombreuses.

En ce qui concerne le Chalcolithique, on connait seulement l'analyse d'une hache plate de Licodia Eubea (Catane) (fig. 2.11), composée de cuivre (98,1 %) avec des traces faibles d'étain (0,78 %) et d'arsenic (0,03 %)[4].

Une hache à rebords du Bronze Ancien (fig. 2.10), retrouvée près de Catane et analysée par spectrophotométrie par Barker-Slater et par fluorescence X par C. Giardino, est composée de cuivre sans étain, avec des quantités faibles de plomb (0,165), d'arsenic (0,027), de nickel (0,056), de zinc (0,075) e de fer (0,021) et de quantités plus consistantes d'antimoine (1.25) et d'argent (1.21), qui sont dûes au cuivre primaire[5].

Des analyses récentes d'objets du Bronze Ancien et Moyen ont été effectuées par XRF à l'Université de Catane. Il s'agit de trois poignards provenant de contextes du Bronze Ancien de Palagonia (Catane)[6]. Ils sont composés de cuivre sans étain, mais avec des petites quantités d'arsenic et d'antimoine: ce qui correspond à la technologie connue en Italie dans la même période.

Le premier poignard à lame triangulaire de la tombe Nord 5 (inv. 4341, fig. 2.1) a deux rivets majeurs, d'origine, et deux

1 GIARDINO 1997, p. 406.

2 OTERI ET ALII 1986; BALDANZA - TRISCARI 1987; CIMINO - TRISCARI 1990.

3 BONANNO - BARONE - TRISCARI 1998.

4 ALBANESE PROCELLI 1993, p. 31, fig. 1; GIARDINO 1997, p. 406, fig. 1: 3; cf. GIARDINO 1995, p. 189, note 15 (avec références aux objets siciliens analysés par Junghans-Sangmeister-Schroeder au Musée de Mainz).

5 BARKER 1971; SLATER 1971, fig. 8a, inv. 36470; GIARDINO 1997, p. 406.

6 MANISCALCO 2000. Les objets de la fig. 1b/c sont des polissoirs.

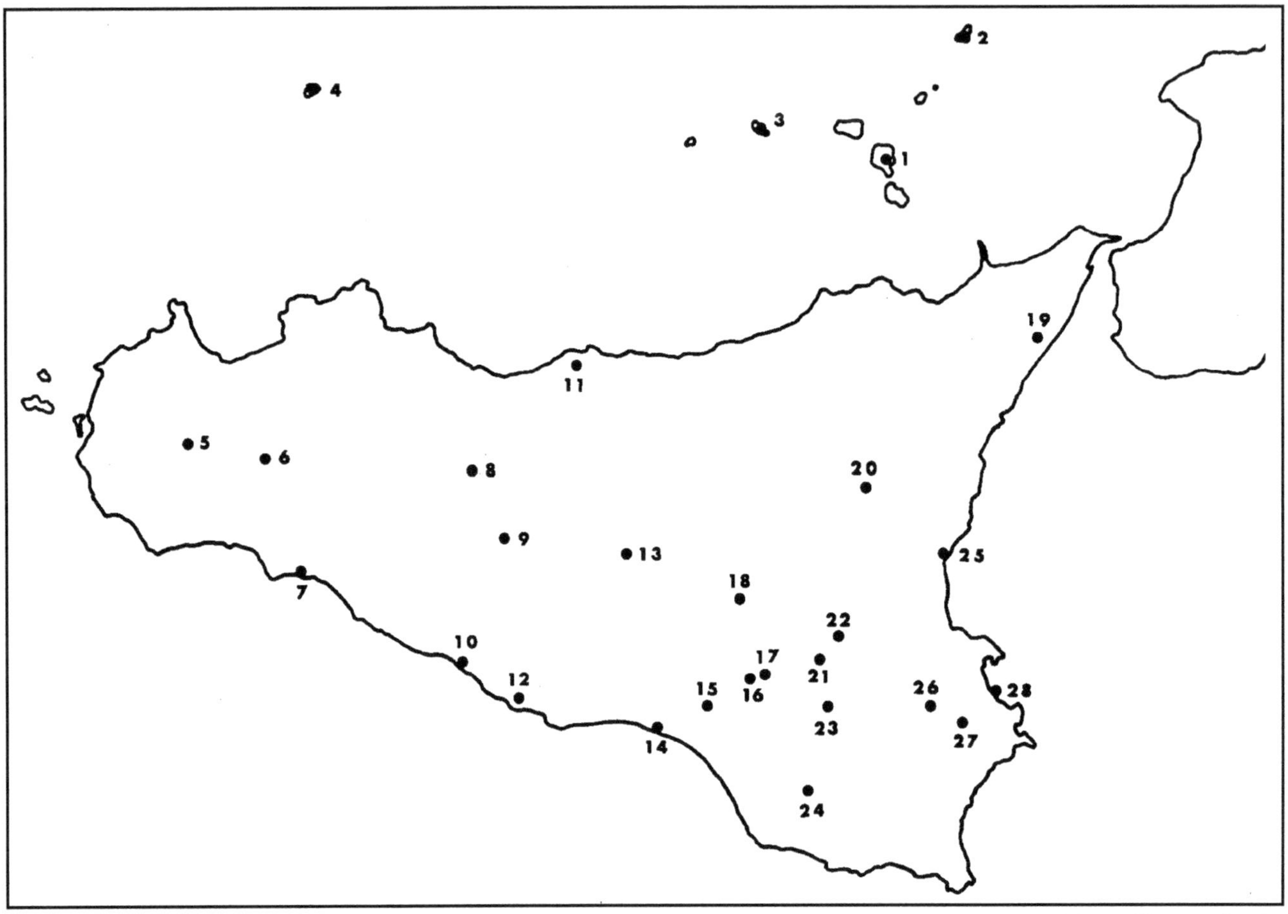

Figure 1. Carte des lieux cités dans le texte. 1. Lipari. 2. Panarea. 3. Filicudi. 4. Ustica. 5. Mokarta. 6. Monte Finestrelle. 7. Sciacca. 8. Colle Madore. 9. Polizzello. 10. Cannatello. 11. Chiusilla di Isnello. 12. Palma Montechiaro. 13. Sabucina. 14. Manfria. 15. Monte Dessueri. 16. Monte S. Mauro. 17. Caltagirone. 18. Morgantina. 19. Fiumedinisi. 20. Adrano (et Mendolito). 21. Mineo. 22. Palagonia. 23. Licodia Eubea. 24. Monte Racello. 25. Catania. 26. Pantalica. 27. Grotta della Chiusazza. 28. Thapsos.

rivets plus petits, résultats d'une restauration ancienne. Cette forme a des précédents en Sicile dans un exemplaire à lame triangulaire de la Chiusilla d'Isnello, daté du Chalcolithique. Le poignard de Palagonia est composé de cuivre avec arsenic (7/10 %) mais sans étain. On sait qu'une quantité élevée d'arsenic rend le cuivre plus résistant et donne des caractéristiques estéthiques très appréciées. Cela ferait penser qu'il s'agit d'un alliage intentionnel. Mais l'on sait que R.J. Forbes pensait que la volatilité et la toxicité de l'arsenic n'aurait pas permis l'utilisation intentionelle de ce métalloïde. A son avis il y a «many dark points» dans l'histoire des trois métaux (étain, plomb et antimoine), puisque la différence entre eux ne serait pas appréciable par des gens qui n'avaient pas de notions chimiques[7].

Les deux autres poignards de Palagonia (inv. 5950 e 5951: fig. 2.2-3) proviennent de la tombe Sud 1. Par leur base mi-ogivale ils sont similaires aux types Guardistallo et Massa Marittima de la péninsule italienne, dont la diffusion dans les régions centrales est attribuée aux phases initiales et avancées du Bronze Ancien[8]. Les poignards de Palagonia sont composés de cuivre sans étain, avec une quantité consistante d'antimoine.

Une hache à rebords d'Adrano-*grotta* Maccarrone de la période avancée du Bronze Ancien est composée de cuivre sans étain, avec des quantités faibles d'antimoine et d'autres éléments présents dans le cuivre d'origine[9] (fig. 2.5).

La section de deux poignards des tombes 4 et 5 de la nécropole de Monte Racello, produits en moule bivalve (inv. 17454 et 17480, fig. 2.7-8), pourrait indiquer une datation au début du Bronze moyen. La composition de ces poignards se diffère. Cela est très important aux fins de la perception du manque de standardisation et du haut degré de l'expérimentation dans les procédés opératoires métallurgiques à la fin du Bronze Ancien et au début du Bronze Moyen, si l'on considère qu'il s'agit d'objets contemporains, de la même provenance et similaires du point de vue typologique. L'un des deux exemplaires (inv. 17454, tombe 4) a des quantités considérables d'arsenic (7/10 %), comme dans le cas du poignard le plus ancien de Palagonia (inv. 4341). Cela pourrait prouver la persistance des traditions métallurgiques pendant de longues périodes. Par contre l'autre poignard de la tombe 5 (inv. 17480) n'a aucune trace d'arsenic[10]. L'utilisation de cuivre avec des quantités variables d'arsenic et d'antimoine correspond à nos connaissances sur la métallurgie

[7] FORBES 1972, p. 171, 177.

[8] BIANCO PERONI 1994, p. 16, 19.

[9] MANISCALCO 2000, p. 164.

[10] MANISCALCO 2000, p. 164.

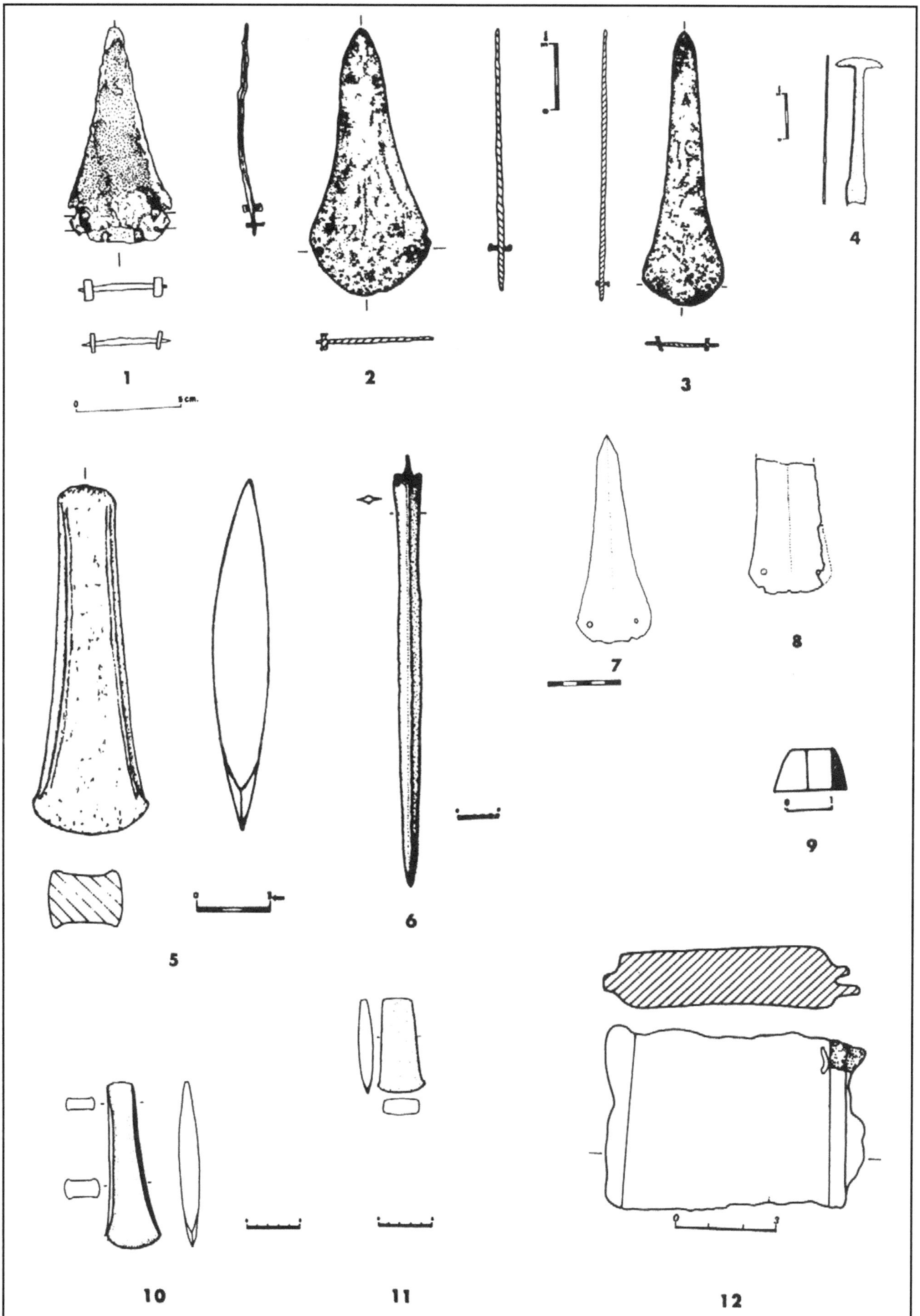

Figure 2. 1-3. Palagonia. 4. Grotta Chiusazza. 5. Adrano-Maccarrone. 6. Caltagirone. 7-8. Monte Racello. 9. Pantalica, tombe Sud-Est 81. 10. Catane. 11. Licodia Eubea. 12. Polizzello (1-3 et 5 d'après Maniscalco 2000; 4, 6, 10, 11 d'après Giardino 1997; 7-8 d'après Leighton 1999).

contemporaine en Italie péninsulaire (faciès de Remedello et Rinaldone)[11].

Ce n'est qu'à la fin du Bronze Moyen et au début du Bronze Récent qu'une vraie métallurgie du bronze, caractérisée par un bon alliage de cuivre et d'étain, se pratique en Sicile. L'analyse d'une épée de Caltagirone-Montagna (fig. 2.6) du type Thapsos-Pertosa[12], indique un bon alliage du bronze, contenant plus que 12 % d'étain, ce qui donne à cette arme de bonnes qualités de dureté et de résistance aux chocs[13].

Des analyses qualitatives par XRF ont été récemment effectuées sur des armes en bronze du Bronze Récent et Final (XIII-Xe s. av. J.-C.) de la nécropole de M. Dessueri (Caltanissetta). Il s'agit de poignards, qui semblent caractérisés par des alliages différents: un poignard du type Pertosa de la tombe Canalotti 12 est composé de cuivre et d'étain; un autre de la tombe Fastucheria 5 est caractérisé par l'argent présent dans l'alliage de cuivre et d'étain et par un rivet en argent et en or; un troisième de la tombe Canalotti 24 est en cuivre uniquement[14].

D'autres analyses récentes ont été effectuées par Caneva - Giardino sur des objets du dépôt de Polizzello (Caltanissetta), qui comprend des objets du Bronze Final et du Premier Âge du Fer. Ces analyses ont permis de constater la présence d'inclusions riches en soufre et en fer, qui indiquent une utilisation possible des minéraux de cuivre du district des Monti Peloritani, où le minéral principal est constitué de chalcopyrite[15].

3. LINGOTS

En Sicile on connait des lingots de cuivre (cest-à-dire de minéral provenant d'extraction) et des lingots de bronze, résultant de la refonte d'objets en bronze. De ce type sont par exemple au VIIIe/VIIe s. av. J.-C. les lingots du dépôt du Mendolito près de Adrano.

Les fragments des lingots *ox-hide* de l'âge du Bronze Moyen et Récent de Lipari, Thapsos, Cannatello sont évidemment en cuivre. Celui de Cannatello, analysé par Mosso, est en cuivre presque pur[16]: cela indiquerait que Chypre serait une des sources de ce minéral pour la Sicile en cette période.

Il serait de grand intérêt de connaître la composition d'un fragment de lingot retrouvé à Ustica-Faraglioni dans un contexte du Bronze Moyen-Récent. Dans ce village cinq moules spécifiquement destinés à la production de haches plates ont été retrouvés[17].

En Sicile, la seule indication relative à des lingots en étain concernerait le dépôt de Polizzello (Caltanissetta) de la première moitié du VIIIe s. av. J.-C.: mais cette indication, contestée par E. Gabrici, n'est pas vérifiable aujourd'hui[18].

L'analyse d'un lingot discoïdal en cuivre primaire du dépôt de Polizzello a permis de supposer des liens avec des gisements à sulfures mixtes, du type de ceux qui sont présents dans la région des Monts Peloritani en Sicile[19]. Mais le cuivre, aussi bien que d'autres minéraux comme l'étain, pouvait parvenir en Sicile de régions différentes de la Méditerranée. Un fragment de lingot «a barra» du Premier Âge du Fer provenant de Polizzello (fig. 2.12) appartient à un type qui se retrouve dans les régions italiennes du Latium et de la Toscane. Cela confirmerait que c'est de ces regions-là qu'une partie de l'approvisionnement en cuivre et en autres minéraux (peut-être le fer et aussi l'étain) arrive en Sicile au Premier Âge du Fer.

Il ne faut cependant pas penser que l'échange de l'étain se faisait exclusivement en lingots. Des documents de la seconde moitié du deuxième millènaire, de l'archive de Amarna en Egypte, indiquent que le commerce de l'étain se faisait surtout «in crude bronze»[20].

4. DOCUMENTS RELATIFS À LA PRODUCTION MÉTALLURGIQUE

4a. Creusets

La documentation la plus ancienne, relative à la production des objets en cuivre, concerne un fragment de creuset, provenant d'une couche du Chalcolithique Ancien (faciès de Diana-Spatarella) à Lipari[21].

Parmi les matériaux du village de Manfria près de Gela, datés de la période du Bronze ancien, des creusets sont identifiables[22].

À un creuset de l'habitat de Morgantina (Aidone) appartient peut-être un fragment de paroi convexe en «limestone», ayant des traces de métal à l'intérieur[23].

---

[11] BARKER 1971.

[12] Un exemplaire de ce type a eté identifié par L. Vagnetti dans l'épave de Ulu Burun-Kash: VAGNETTI 2000.

[13] Analyse effectuée par A. Mosso: cf. GIARDINO 1997, p. 408, fig. 3:1.

[14] PANVINI sous presse. Les analyses ont été effectuées par le laboratoire LANDIS de Catane.

[15] GIARDINO 1995, p. 185.

[16] Cf. GIARDINO 1996, p. 129.

[17] Lingot: HOLLOWAY & LUKESH 2001, p. 20, fig. 1.10, *trench* 22. Moules: HOLLOWAY & LUKESH 1995, p. 81; HOLLOWAY & LUKESH 2001, p. 24, fig. 2.23.

[18] GIARDINO 1995, p. 26.

[19] GIARDINO 1997, p. 408.

[20] FORBES 1972, p. 157.

[21] BERNABO' BREA - CAVALIER 1980, p. 490, pl. CV, 5, inv. 8993.

[22] Communication de E. Procelli.

[23] LEIGHTON 1993, p. 190, n°. 328, pl. 112.

4b. Moules

Le corpus des moules connus jusqu'ici en Sicile et dans les îles Eoliennes comprend environ 70 exemplaires en pierre et deux en bronze (dépôts de Lipari et Monte S. Mauro), datés du Bronze Ancien au Premier Âge du Fer. Ils proviennent de sites de la Sicile (Cannatello, Colle Madore, Mineo-Camuti, Mokarta-Cresta di Gallo, Monte Finestrelle, Monte S. Mauro, Morgantina, Palma Montechiaro-Piano Vento et Galia, Pantalica, Sabucina) e des îles environnantes (Filicudi-Capo Graziano; Lipari; Panartea-Milazzese; Ustica-Faraglioni)[24]. Ils ont été retrouvés généralement en habitats, mais très souvent ils sont sans contextes.

En ce qui concerne les valves lithiques, la majeure partie (53,62 %) est attribuable au Bronze Final. Les moules relatifs au Bronze Moyen et Récent, au Premier Âge du Fer et au Bronze Ancien atteignent des pourcentages moins élevés (respectivement 26,08, 13,04 et 5,79 %).

La typologie des différents objets produits au fil des siècles est un indicateur précieux du progrès technologique, des contacts et des échanges et aussi des réponses données par les communautés aux exigences locales.

En général les moules connus en Sicile sont plutôt déstinés à la production d'outils (haches plates; ciseaux, alênes) à l'époque du Bronze Ancien. À celle du Bronze Moyen et Récent appartiennent des moules déstinés à la production d'armes (épées), de bijoux (bracelets) et d'outils (haches plates et à collet).

À l'époque du Bronze final les catégories fonctionelles sont relatives aux armes (pointes de lance et de javelot, épées et poignards) et à l'outillage, en particulier les haches (à appendices, à douille, à collet) et les ciseaux. On produit aussi des pendentifs, comme des micro-modèles de hachettes à collet, ayant une valeur magique.

Au Premier Âge du Fer une partie considérable de la production est réservée aux anneaux (Mineo-Camuti, Morgantina).

Si l'on considère, en suivant Leroi-Gourhan[25], que le progrès technologique se mesure en termes de rapport entre l'effort fourni et le résultat obtenu, les solutions techniques adoptées dans les moules de Sabucina permettent de constater qu'à la période du Bronze Final les métallurgistes avaient acquis un'expérience considérable. Elle vise à obtenir des conditions et des résultats satisfaisants et une économie de matériaux (pierre et métal), de temps et d'énergies dans les différents moments de la chaîne opératoire, depuis la phase de la préparation des moules jusqu'à celles de la coulée et des procédés de finissage des produits successifs.

4c. Polissoirs

Un polissoir en «quarzarenite» se retrouve dans le groupe des moules du Bronze final de Sabucina[26].

Deux polissoirs en grès («arenaria») du village du Bronze Ancien de Palagonia sont probablement déstinés au finissage d'objets en silex ou en os[27].

5. CONDITIONS SOCIALES DU TRAVAIL

L'organisation du travail de cette production du métal est le signe de la relation qui existe entre la société globale et les métallurgistes, qui généralement font l'objet d'une particularisation.

Au niveau archéologique il est difficile d'établir une distinction nette entre ateliers permanents et ateliers temporaires.

Les contextes de certains moules, retrouvés à l'intérieur de cabanes à Lipari et à Capo Graziano - Filicudi, plaiderait en faveur de l'integration de l'activité métallurgique dans certains villages, déjà à l'âge du Bronze Ancien.

En cette période, caractérisée par un faible degré de spécialisation, on peut supposer que les métallurgistes opèrent en collaboration avec les potiers, comme l'indiquerait à Lipari la cabane *delta* XII, qui contenait un moule et une décharge de vases ratés à la cuisson.

La coopération avec d'autres artisans (ceux du travail de l'os ou de la corne, par exemple, pour l'emmanchement des objets) devait être fondamentale.

6. CONTACTS ET ÉCHANGES

En considérant que les importations d'objets en métal ou leurs relations typologiques peuvent être considerées comme des indicateurs de contacts avec les différentes régions impliquées dans les échanges, les changements technologiques intervenus au fil des siècles dans la métallurgie du bronze en Sicile pourraient permettre de comprendre les lieux de provenances possibles des sources du cuivre et de l'étain.

Si en Sicile le début de la métallurgie du cuivre est précoce et remonte au début du Chalcolithique (fin du IVe-début du IIIe millénaire av. J.-C.), comme le montre le creuset retrouvé a Lipari dans un contexte du faciès Diana-Spatarella (*supra*),

[24] Sicile: Cannatello: ORSI - RIZZO 1897. Colle Madore: VASSALLO 1999, pp. 85-89. Mineo-Camuti: MANISCALCO 2000, p. 159, fig. 1a. Mokarta: MANNINO-SPATAFORA 1995, p. 130, pl. XIV, 3. M. Finestrelle: TUSA 1997, p. 235, V. 191. Monte S. Mauro: ORSI 1910, col. 734, fig. 2. Morgantina: LEIGHTON 1993. Palma Montechiaro: CASTELLANA 1997, p. 49, fig. 21-22. Pantalica e Sabucina: ALBANESE PROCELLI 2000. Îles Eoliennes: Capo Graziano: ALBANESE PROCELLI 1991. Lipari e Panarea: BERNABO' BREA 1958, p. 101, 124. Ustica: à la note 17..

[25] LEROI-GOURHAN 1994.

[26] ALBANESE e MAZZOLENI 2000.

[27] MANISCALCO 2000, p. 159, fig. 1b, c.

la diffusion de cette pratique est encore peu répandue pendant le Chalcolithique et l'âge du Bronze Ancien. En cette dernière période, il semble que les traditions technologiques du Chalcolithique se prolongent, puisque la composition des objets est encore en cuivre. En revanche, les différences dans les alliages ne correspondent pas nécessairement aux différences typologiques, ce qui semble indiquer un manque de standardisation dans la production à l'âge du Bronze Ancien.

Au Chalcolithique Récent, la présence en Sicile (à Sciacca et dans la cave de la Chiusazza: fig. 2.4) de deux poignards du type Montebradoni, un site près de Volterre[28], permet de supposer des liens, qui pourraient impliquer aussi l'approvisionnement en cuivre en la région minière de Toscane.

À l'âge du Bronze Ancient et Moyen, les contacts entre la Sicile, l'Egée et la Mediterranée orientale sont systématiques et se répercutent dans le progrès de la métallurgie et dans les apports technologiques, comme le montre un moule de Filicudi-Capo Graziano, qui est en Sicile l'exemplaire le plus ancien destiné à produire des haches à collet[29].

Dans le cas du poignard le plus ancien de Palagonia (inv. 4341) le système de fixage de la poignée par deux rangs de rivets, placés en position carrée à la base de la lame, rappelle celui des plus anciens poignards égéens (type IIIc Renfrew)[30]. C'est par contre un système inconnu en Italie: cette caractéristique pourrait être une indication très précieuse de contacts dans le domaine de la métallurgie entre la Sicile et l'Egée. A des prototypes du Bronze Ancien 1-3 de la Grèce continentale, de la Crète et des Cyclades[31] se rapportent aussi les poignards de M. Racello, datables du début du Bronze Moyen.

En Sicile, de vrais alliages en cuivre et en étain ne sont pas utilisés avant l'Âge du Bronze Moyen avancé ni du Bronze Récent: ce sont des périodes durant lesquelles les «stimulus», dûs aux contacts de longue durée avec les artisans égéens, marquent un progrès considérable même dans ce secteur de la technologie[32].

Mais il faut aussi considérer les contacts avec les régions occidentales de la Méditerranée. Ils connaissent leur apogée au Bronze Final et au Premièr Âge du Fer. Il y a d'un côté la présence de bronzes de types occidentaux et atlantiques en Sicile (haches à talons et anneaux, rasoirs), d'autre côté des bronzes de types siciliens en Ibérie et dans l'Occident: haches à appendices, fibules coudées et à «antennes» [33].

Au Bronze Récent et Final et au Premier Âge du Fer la céramique nuragique retrouvée dans une certaine quantité à Lipari, et exceptionnellement en Sicile, pourrait indiquer la présence d'individus provenant de Sardaigne, en qualité de agents des trafics. Cette hypothèse pourrait s'appuyer aussi sur la présence d'une perle conique en plomb dans le mobilier de la tombe Sud-Est 81 de Pantalica, qui comprend aussi une petite *Schnabelkanne* nuragique[34] (fig. 2.9).

Au point de vue des liens de la Sicile et du Nord-Ouest de l'Europe, C. Giardino a souligné les rapports très stricts qui existent entre les rasoirs siciliens du Bronze Récent (faciès de Pantalica) et les deux exemplaires d'Angleterre et de Hollande[35]: le dernier présente deux rivets, ce qui ne se retrouve jamais dans les produits siciliens, caractérisés en général par un seul rivet.

Au Premier Âge du Fer des importations d'objets en bronze produits dans la région tyrrhénienne centrale de l'Italie (haches à ailerons et à douille de Pantalica, Polizzello, Monte S. Mauro) pourraient être considérées comme des indices d'échanges visés à l'approvisionnement de métaux de la région toscane, qui comprendraient du cuivre, de l'étain et probablement en cette période du fer[36].

Addresse de l'auteur

Rosa Maria ALBANESE PROCELLI
Dipartimento SAFIST
Facoltà di Scienze e Filosofia, Università di Catania
95124 Catania – ITALIE

Bibliographie

ALBANESE PROCELLI, R.M., 1991, Appendice IV. Matrice di fusione dalla capanna XIV della Montagnola di Capo Graziano. In BERNABO' BREA, L. - CAVALIER, M., Meligunìs Lipára VI. Palermo: Accademia Nazionale di Scienze, Lettere e Arti, p. 309-314.

ALBANESE PROCELLI, R.M., 1993, Ripostigli di bronzi della Sicilia nel Museo Archeologico di Siracusa. Palermo: Accademia Nazionale di Scienze, Lettere e Arti.

ALBANESE PROCELLI, R.M., 2000, Bronze metallurgy in protohistoric Sicily: the stone moulds. In Italy in its Mediterranean Setting. Studies in Honour of E. Macnamara (Accordia Research Institute). London, p. 75-90. With an Appendix by P. Mazzoleni, p. 90.

BALDANZA, B.-TRISCARI, M., 1987, Le miniere dei Monti Peloritani. Bibl. Arch. Stor. Siciliano. VIII. Industria poligrafica della Sicilia, Messina.

BARKER, G., 1971, The first metallurgy in Italy in the light of the metal analyses from the Pigorini Museum. BPI 80, p. 183-208; SLATER, E., 1971. Appendix II, *ibid.*, p. 209-210.

BERNABO' BREA, B.-CAVALIER, M., 1980, Meligunìs Lipára IV. Palermo: Flaccovio.

28 BIANCO PERONI 1994, p. 12-3.

29 ALBANESE PROCELLI 1991.

30 RENFREW 1973, p. 321, fig. 16.5; cf. BARKER 1971, p. 188.

31 RENFREW 1973, p. 321, fig. 16.5, type IVb.

32 Cfr. GIARDINO 1997, p. 408.

33 LO SCHIAVO 1990.

34 Lipari: BERNABO' BREA - CAVALIER 1980. Pantalica: ORSI 1912, pl. IX, 66. Un fragment de *brocchetta askoide* nuragique, daté du Premier Âge du Fer, a été individualisé à Mothié: LO SCHIAVO sous presse.

35 GIARDINO 1995, p. 225, fig. 109 D, 3-4.

36 Cf. ALBANESE PROCELLI 1993, p. 81 et 233.

BIANCO PERONI, V., 1994, I pugnali nell'Italia Continentale (PBF VI, 10). Stuttgart: F. Steiner.

BONANNO, C.-BARONE, G.-TRISCARI, M., 1999, Archeometallurgia di scorie metalliche di età romana a Messina. In Le Scienze della Terra e l'Archeometria, edited by C. D'AMICO - C. TAMPELLINI. Bologna: Patron, p. 137-140.

CASTELLANA, G., 1997, La grotta Ticchiara ed il castellucciano agrigentino. Palermo: Assessorato ai Beni Culturali e Ambientali e Pubblica Istruzione.

CIMINO, G.-TRISCARI, M., 1990, First chemical and mineralogical results achieved by investigations on «ancient alloys» from the Peloritan mountains (Sicily). L'industria mineraria 6, p. 29-33.

FORBES, R.J., 1972, Studies in Ancient Technology, IX. Leiden: E.J. Brill.

GIARDINO, C., 1995, Il Mediterraneo Occidentale fra XIV ed VIII secolo a.C. Cerchie minerarie e metallurgiche. BAR International Series 612. Oxford: BAR Publishing.

GIARDINO, C., 1996, Miniere e tecniche metallurgiche nella Sicilia protostorica: nuove linee di ricerca. In Early Societies in Sicily (Accordia Spec. St. on Italy 5), edited by R. Leighton. London: Accordia Institute, p. 129-138.

GIARDINO, C., 1997, La metallotecnica nella Sicilia pre-protostorica. In Prima Sicilia, edited by S. TUSA. Palermo: Ediprint, p. 404-414.

HOLLOWAY, R.R. & LUKESH, S.S., 1995, Ustica I. Excavations of 1990 and 1991, Providence - Louvain-La-Neuve: Brown University.

HOLLOWAY, R.R. & LUKESH, S.S., 2001, Ustica II. Excavations of 1994 and 1999. Providence, Rhode Island: Brown University.

LEIGHTON, R., 1999, Sicily before History, London: Duckwort.

LEROI-GOURHAN, A., 1994, Ambiente e tecniche, Milano, Jaca Book (éd. orig.: Milieu et technique, Paris 1943-45, Albin Michel).

LO SCHIAVO, F., 1990, La Sardegna sulle rotte dell'Occidente. I. La protostoria. In Atti XXIX Conv. St. Magna Grecia, Taranto 1989. Taranto: Ist. St. e Arch. Magna Grecia, p. 100-33.

LO SCHIAVO, F., sous presse. Un frammento di brocchetta askoide nuragica da Mozia. V Congresso Int. St. Fenici e Punici, Palermo/Marsala, 2-8 ott. 2000.

MANISCALCO, L., 2000, Osservazioni sulla produzione metallurgica in Sicilia nell'Antica Età del Bronzo. SicA 98, p. 159-166.

MANNINO, G.-SPATAFORA, F., 1995, Mokarta. La necropoli di Cresta del Gallo. Palermo: Assessorato Beni Culturali, Ambientali e Pubblica Istruzione.

ORSI, P. & RIZZO G.E., 1897, Nuovi materiali siculi del territorio di Girgenti. BPI 23, p. 113-122.

ORSI, P., 1910, Di una anonima città siculo-greca a Monte S. Mauro presso Caltagirone. MonAnt XXI, p. 729-850.

OTERI, F. & ALII, 1986, Confronto tra composizione chimica e modalità giaciturali di alcuni minerali metalliferi dei Monti Peloritani (Sicilia N.E.). Rendiconti Soc. It. di Mineralogia e Petrologia 41, 1, p. 41-52.

PANVINI, R., sous presse, Manufatti in avorio dalla necropoli di M. Dessueri. Tavola Rotonda «L'utilizzazione dell'avorio nell'Età del Bronzo in Italia (coordinatori: L. VAGNETTI, M. BETTELLI, I. DAMIANI), I Congr. Int. «La Terra degli Elefanti», Roma, 16-20 ott. 2001.

RENFREW, C., 1973, The Emergence of Civilisation. London: Methuen and Co.

TUSA, S., Ed., 1977, Prima Sicilia. Palermo: Assessorato Beni Culturali, Ambientali e Pubblica Istruzione.

VAGNETTI, L., 2000, I Micenei fra Mediterraneo orientale ed occidentale dopo la fine dei palazzi. Atti XXXIX Conv. St. Magna Grecia, Taranto, p. 63-89.

VASSALLO, S. (Ed.), 1999, Colle Madore. Un caso di ellenizzazione in terra sicana. Palermo: Assessorato Beni Culturali, Ambientali e Pubblica Istruzione.

IVème Section – Problèmes de métrologie et le commerce de métaux

IVth Section – Metrology and metal trade problems

INVESTIGATING WEIGHT SYSTEMS IN NURAGIC SARDINIA

Marisa RUIZ-GÁLVEZ

WEIGHT SYSTEMS IN NURAGIC SARDINIA?

Investigating weight systems in Bronze Age Europe is a hazardous task, in view of the open scepticism of most part of the scholars. Several reasons help to explain it.

Firstly, the scarcity of balance weight findings. Secondly, because metric systems are usually linked to complex, bureaucratic State Systems, which, in Bronze Age Europe outside the Aegean, is simply unthinkable. Thirdly, because most of processual archaeologists prefer to see the developments in prehistoric Europe as independent of whatever kind of contacts with the Mediterranean/Near East.

Nevertheless, human societies world-wide, have developed local systems of calculation, either to count periods of time, distances, food rations or equivalencies among goods (Ifrah 1997; Ruiz-Gálvez 1998a:66-69).

Moreover, history attests the adoption by some primitive societies of foreign means of payment for external transactions. This is the case of certain societies from sub-Saharan Africa. Coins were not totally unknown there, thanks to the contacts with caravans of Arab merchants. However, the systematic introduction of coins as a mean of payment began with the arrival of the Portuguese there in fifteenth century AD. Metal coins were very appreciated as a prestige good, and used exclusively for payments between Portuguese traders and local leaders, but not for inner exchange. (Rivallain 1994).

I hope that this example helps to convey my point, i.e., that some non State Bronze Age societies of Europe, could have been engaged in long distance Mediterranean trade systems, and therefore had adopted for their external transactions, the weight standard dominant at that moment (Ruiz-Gálvez 1998a: chapter 7 and 2000:275). The recognition at Ulu Burum of, at least three different weight standards, three set of scales and, very specially, of two wooden boards for writing ciphers and making calculations, suggests that traders on board were travelling around a wide area, and doing equivalencies among different goods, different measure systems, and perhaps, different trade and exchange systems (Pulak 2000: 247, 262-4).

So I think that, when analysing non State Bronze Age societies, the task should be lesser to track down clusters of weighed metal items and/or ingots, and to compare them with units, fractions or multiples of every known system of weight, than to inquire whether there are archaeological evidences of strong interactions with a certain trade area, that could make justifiable the local adoption of a foreign weight standard.

My case study is Nuragic Sardinia, where strong contacts with Mycenaean, Cypriots and later on, Phoenician traders are well attested and where a huge production of metallurgy is also well recorded.

Sardinia is the second largest island in the Mediterranean and constitutes an authentic port of call between East and West, as well from a nautical point of view as from an archaeological one.

Evidences of Mycenaean and Postmycenaean trade from the Middle Bronze Age 3 on (Lo Schiavo 2001), are not scarce there (Ferrarese Ceruti 1997: chapters 14, 15, 24, 25, 27, 30, 31, 32, 34; Lo Schiavo et alii 1985; Vagnetti 1993[a] & b, Vagnetti & Lo Schiavo 1989), being the fragments of a LHIIIA2 alabastron recovered on the courtyard and the central tower of nuraghe Arrubiu and the LHIIIA2/B sherds from Su Muru Manu (Tharros), the oldest Mycenaean finding to date (Bernardini 1989; Lo Schiavo & Vagnetti 1993). From Later Bronze Age on (*Bronzo Recente*), visible changes are taking place within the Nuragic Society: changes in architecture with the construction of complex, multitower nuraghi; the first evidence of supraterritorial well sanctuaries; the sudden appearance of ashlar, isodomic masonry; changes in territorial pattern, with special interest in the control of strategic gateway points, as outlets, fords, mountain passages and natural ways; population growing...etc (Lo Schiavo 1997:150,154-156).

Both, Italian and foreign scholars working on Nuragic Sardinia seem at least to agree, that such developments were achieved by the Nuragic Society on one's own (Lo Schiavo 1997: 151;Webster 1996:108). Nevertheless, and although acknowledging the absence of whatever kind of evidence of Aegean colonies in the Central Mediterranean, Vagnetti (1993a & 1998:69-71), makes the point, that the kind of specialised know-how, which implies the production of Aegean wheel made pottery, could not be acquired by local craftsmen by their own, and without a close contact with the Aegean masters (ibidem 1998:69. See also Bernardini 1991 & 1993).

Then not only technological know how, as for instance pottery wheel, but also lost wax technique, and probably the cultivation of olive trees[1] were introduced in Sardinia in

[1] Analysis of wall impregnation of pithoi from Broglio de Trebisacce, (Calabria, Southern Italy). Large, imported pithoi are reported in Sicily and Sardinia, that might be connected with the transport of olive oil (Vagnetti 1998:72).

Mycenaean and Postmycenaean times (Vagnetti 1998:Lo Schiavo et alii 1985; Lo Schiavo 2001). Other more subtle forms of know how might have been introduced too, as for instance, metric knowledge (isodomic technique) and, perhaps, a more accurate mean of time measurement[2]. Also certain forms of conspicuous consumption, as sets for eating and drinking (cauldrons and bronze vessels), or tripods and frankincense burners[3] (Lo Schiavo 1982; Lo Schiavo et alii 1985; Lo Schiavo & Ussai 1995; Vagnetti & Lo Schiavo 1989) as well as codes of aesthetic and of public conveyance of the leader image, were also probably introduced there. To such symbolic code belong mirrors, combs, razors, cosmetic tweezers and fibulae. The first four items are probably connected with the hair and, specially, beard care as a symbol of respect and dignity. The last one might be related to the introduction of new precious textiles, probably of geometric design and occasionally striped of purple. Such textiles are described both at the Iliad and the Odyssey and are likely mirrored on the geometric patterns of the painted, stroke burnished and incised pottery of Early Iron Age Greece (geometric pottery), Early Iron Age Cyprus (Cypro-Geometric), Early Iron Age Sardinia (*ceramica stralucida*. Campus & Leonelli 1999: 168.3; 180.1; 234.4-5; 243.4; 246.1-6), Late Bronze Age SE and SW France (Gómez 1990) and Late Bronze Age Spain and Portugal (Sherratt & Sherratt 1991 & 1993; Ruiz-Gálvez: 1998a:chapter 6;1998b: 100-13; & 2000a:13; Cáceres 1997).

The class of technical and ideological knowledge described above is rather complex, and its acquisition implies a close relation between the one who transmits it, and the one who learns it. Single foreign craftsmen could have installed themselves and be integrated, perhaps by intermarriage, within a native society. Archaeology would be unable of detecting his presence, when he would had been integrated in his adoptive society. In other cases, close contact could have been the result of a *koine*, as a multiethnic crew sailing under a State flag, i.e. in Mycenaean times, or in a free-lance, half trade half piracy way, in Postmycenaean times. This last type of trade is described by Homer in the Odyssey (IX.39-47) and is archaeologically recorded at the warrior/trader grave of Lefkandi (Popham & Lemos 1995). Ancient sources refer to Sardinian, Corsican and Etruscan people as traders and pirates (Serv. An. X.172; Strab. VI.2.2) (Torelli 1996:62,68; Lilliu 1982:133, 213-14)[4]. Nuragic bronze boats datable at the Final Bronze Age and Early Iron Age provide a symbolic evidence of the source of power and prestige of Sardinian leaders (Lo Schiavo 2000).

Summing up, there are proofs enough of close contacts between the Nuragic Society and foreign traders from Middle Bronze Age 3 onwards. These make justifiable my query about the adoption of one or several Mediterranean systems of weight in Later (*Bronzo Recente*) and/or Final Bronze Age (*Bronzo Finale*) Nuragic Sardinia.

ANALYSING THE EVIDENCE

Although infrequent, balance weights are not totally unknown in Sardinia. Ugas (1985:45,47-48 and footnotes 53 and 54; Ugas & Usai 1986:189) refers to four steatite balance weights found at Santu Brai, weighing respectively 25.17gr., 26.80 gr., 64.87 gr. And 63.70 gr. Balance weights 1 an 2 show a five dots mark. Number 3 has an incised X mark. The surface of balance weight 4 is quite eroded and therefore it is no more possible to find out if there was a mark. The above referred author mentions also some other balance weights from Nuragic sites, as i.e. a bronze balance weight from Teti-Abini weighing 2.128 Kg., a steatite balance weight from Forraxi Nioi of weight 23.84 gr. Eventually and of an uncertain chronology, a sample of 18 lead balance weights from the Gùspini area, with clusters around 3.33 gr., twice this unit, 6.7 gr., and some multiples of 333.4 gr. Later on, Zaccagnini (1991:344) identified Santu Brai weights with two well-known Mediterranean systems of weight: the Aegean of 6.5/6.8* gr. to which belong Santu Brai 3 and 4 (6.48 x 10), and the Microasiatic unit of 11.75* gr. represented by Santu Brai 1 and 2 (5 x 1/2 microasiatic shekels). With some caution, due to the uncertainty of their archaeological context, the 18 Gúspini's lead weights were ascribed by Zaccagnini (Ibidem:344), to the Aegean unit. The other two stone and bronze weights from Forraxi Nioi and Teti-Abini were not commented upon by Zaccagnini, but could also be ascribed under my view, to the Microasiatic unit. Thus, the first one is almost twice as much 2 microasiatic shekels of 11.75* gr. The second, although underweight, may also be related to a multiple of that same unit (i.e. 117.5 x 18 or 4. 1/2 mina of ca. 470 gr.[5]

2 Most part of sacred wells show a circular hollow on the top of the domed ceiling, apparently covered by a removable slag. I wonder, whether such hollows were meant to provide that the midsummer or midwinter sun shined through it, down to the middle of the sacred well and the corridor staircase. Rites of passage are closely linked to the rhythms of land. Both belong to that kind of ideological knowledge, traditionally controlled by shamans, priests and sacred leaders.(Bloch & Parry 1982:1-44;Cardini 1983; Earle 1997:chapter 5; Helms 1988:chapter 5, pp106-10; van Gennep 1986:13; Ruiz-Gálvez 1998:56-69; Shalins 1988)

3 Frankincense and myrrh are drugs associated to rituals of passage, as those concerning the access to priesthood or to the royal office, burials, and the communication with the divinity (Sherratt 1991). Therefore, the gifts of the Three Wise Men to the Child Jesus were gold (symbol of royalty), frankincense (symbol of Divinity) and myrrh (symbol of the mortal condition of every human being). That is why they were a royal monopoly. After the Mediterranean *collapse* of 1200 BC, frankincense began to appear more widely in Eastern, Central and, even Western Mediterranean. The opening of a new frankincense route, alternative to the Egyptian one, connecting Arabia to the Levantine ports, such as Tel Nami, opposite to the island of Cyprus, as well as the changing nature of trade routes, from a state-controlled trade until 1400-1300 BC to a more opportunistic, entrepreneurship trade from 1200 BC may be responsible for the *generalisation* of elite ways of display in wider areas of the Mediterranean and Europe, (Artzy 1994 & 1998; Harrison & Mederos 1996; Knapp 1990; Sherrat, S. 1994 & 1998; Sherratt S. & A. 1991 & 1993; Ruiz-Gálvez 1998a:272-89; Ruiz-Gálvez 1998b: 106-11; Ruiz-Gálvez 2000 a & b)

4 Piracy was an honourable way of acquiring prestige and booty provided that raids were directed against foreigners instead of friends or members of the own descent group.

5 One western mina of ca. 470 gr. was divisible into 70 shekels of 6.7* gr.; into 60 shekels of 7.9* gr. (the Eblaitic or Phoenician shekel); into 50 shekels of 9.4* gr. (the Egyptian qedet); or into 40 shekels of 11.75*gr. (the microasiatic shekel) (Zaccagnini 1986:416 & 1991:344-5; Parise 1993).

Table 1.

type	Hoard name	Inv. nº	Weight (gr.)	Micro-asiatic unit 11.75	canonical weight (gr.)	mean weight	Standard dev.
flanged axes	Ossi	9977/352	347,08	30	352.5	349,38	2,3
flanged axes	Ossi	9966/353	351,68	30	352.5	349,38	2,3
flanged axes	Ossi	9969/355	398,67	34	399.5	406,97	8,8
flanged axes	Ossi	9971/350	398,95	34	399.5	406,97	8,8
flanged axes	Ossi	9980/351	402	34	399.5	406,97	8,8
flanged axes	Ossi	9973/344	416,08	34	399.5	406,97	8,8
flanged axes	Ossi	9970/344	419,15	34	399.5	406,97	8,8
flanged axes	Ossi	9974/349	445,73	38	446,5	451,06	4,3
flanged axes	Torralba	10878	446	38	446,5	451,06	4,3
flanged axes	Ossi	9763/343	453,67	38	446,5	451,06	4,3
flanged axes	Ossi	9979/346	453,96	38	446,5	451,06	4,3
flanged axes	Ossi	341	455,97	38	446,5	451,06	4,3
flanged axes	Ossi	9967/353	467,93	40	470	468,98	1,5
flanged axes	Ossi	9965/340	470,03	40	470	468,98	1,5
flanged axes	Ossi	9968/345	520,91	44	517		
flanged axes	Forraxi Nioi	16769	569,84	48	564		
flanged axes	Ossi	9978/348	598,22	52	611		
flanged axes	Unknown place	5436	674,64	58	681,5	682,54	5,9
flanged axes	Ilbono	34721	679,03	58	681,5	682,54	5,9
flanged axes	Ilbono	34717	680,16	58	681,5	682,54	5,9
flanged axes	Nule	41431/G	688,62	58	681,5	682,54	5,9
flanged axes	Nule	41431/C	690,26	58	681,5	682,54	5,9
flanged axes	Nule	41431/D	694,32	60	705	700	4,6
flanged axes	Nule	41431/L	694,85	60	705	700	4,6
flanged axes	Ilbono	34719	695,46	60	705	700	4,6
flanged axes	Ilbono	34720	696,61	60	705	700	4,6
flanged axes	Nule	41431/N	697,45	60	705	700	4,6
flanged axes	Nule	41431/P	697,86	60	705	700	4,6
flanged axes	Ilbono	34718	698,15	60	705	700	4,6
flanged axes	Nule	41431/M	699,45	60	705	700	4,6
flanged axes	Nule	41431/O	699,86	60	705	700	4,6
flanged axes	Nule	41431/H	704,77	60	705	700	4,6
flanged axes	Nule	41431/B	706,07	60	705	700	4,6
flanged axes	Nule	41431/A	706,88	60	705	700	4,6
flanged axes	Nule	41431/E	708,39	60	705	700	4,6
flanged axes	Nule	41431/F	714,15	62	728,5	716,43	1,7
flanged axes	Ilbono	34715	717,15	62	728,5	716,43	1,7
flanged axes	Nule	41431/I	718,00	62	728,5	716,43	1,7

We do not know with certainty whether these balance weights were made by Sardinian craftsmen, or just used by them. But, with the exception of the Gúspini sample, all of them come from Nuragic contexts, either well sanctuaries as Forraxi Nioi or Teti, or villages as Santu Brai. Of special interest is the fact that two of the three findings of balance weights are sanctuaries. Their religious and supraterritorial character make of them neutral market places (Courtin 1984; Ruiz-Gálvez 1998: chapter 2). That fits as well with the finding of fifteenth lead ingots, some marked, at the complex of the Santa Anastasia de Sardara sacred well (Ugas 1985; Ugas & Usai 1986), apparently in Early Iron Age contexts. Some of them were ascribed by Zaccagnini (1991) to the Aegean unit and the others, to the Hittite or Microasiatic one.

Other kind of evidences for the existence of systems of weight in Nuragic Sardinia are metal items, specially ingots and axes, because their standardised form, allow to store them by size and shape. In my search for possible regularities in the weight of Nuragic bronzes I have exclusively made use of unfragmented items, whose weights are published or that have been weighed specifically for my investigation[6].

I have first analysed a group of flanged axes from diverse occurrences and chronology (*Table 1*). Flanged axes have been studied by Lo Schiavo (1989-90) from a typological

[6] I am deeply indebted to Dr Lo Schiavo, Dr Vicenzo Santoni Chief Director of the *Sopraintendenza Archeologica per le provincie di Cagliari e Oristano* an to the staff of the Archaeological Museum of Cagliari for providing me with a list of weights and fractions of the Ilbono, Nule, Ossi and Forraxi Nioi flanged axes, and of ingots and items of the Olmedo, Bonnanaro and Arrubiu hoards. Weights of Torralba and Flumenelongu hoard items are taken from Lo Schiavo's papers (Lo Schiavo 1976 & 1988).

Table 2.

Hoard	Inv. nº	Item	Weight (gr.)	Aegean unit 6.5/6.8	Canonical weight (gr.)	Phoenician unit 7.5/7.9	Canonical weight (gr.)	Micro-asiatic unit 11.75	Canonical weight (gr.)
Bonnanaro	10714	*Portuguese socketed axe*	780	12 (6.5x10x12)	780	10 (7.8x10 x10)	780	---	
Bonnanaro	10711	*Plano-convex copper ingot*	2060 (±2 1/2 times 780)	32 (65x32)	2080	26 (78x 26)	2028	175	2056,25
Bonnanaro	10709	*Plano-convex copper ingot*	6640 (±8 1/2 times 780)	102 (65x102)	6630	85 (78x 85)	6630	-----	
Bonnanaro	10710	*Plano-convex copper ingot*	9700 (±12 1/2 times 780)	150 (65x150)	9750	125 (78x 125)	9750		

point of view. According to her (ibidem: 266), Nule-Ilbono axe-type should be dated to the Middle Bronze Age, although the appearance of flanged axes in hoards as Ossi[7], Torralba and Forraxi Nioi indicates, that some evolved variants of that type, went on in Later and Final Bronze Age (*Bronzo Recente e Bronzo Finale*), (Ibidem:265-6).

Anyhow in spite of their apparent chronological differences, all of them seem to fit into the same system of weight: the Hittite or Microasiatic weight, of which they are multiples in a binary system (Petruso 1979). This standard was very common in Southern Anatolia, Cyprus and the Levant, in the 2nd millennium BC. (Courtois 1983; Lassen 2000; Parise 1993; Zaccagnini 1986 & 1991). Nevertheless, the Microasiatic unit seems not to be widespread in Sardinia as well as in the Western Mediterranean and Atlantic area, until the Postmycenaean times, probably connected with new forms of free lance trade (Ruiz-Gálvez 1998: chapter 7 and 2000a & b). Therefore a Middle Bronze Age date for the Nule-Ilbono axe type, seems to me a bit early.

The Bonnanaro hoard has been recently analysed and it seems that, as well the ingots as the Portuguese socketed axe and the Cypriot double axes have been cast locally (Begemann et all. 2001). Two possible weight standards could have been at work there (*Table 2*), the Aegean and the Phoenician, the two of them fitting quite well with Bonnanaro's weights. Yet, the sample is very scarce, and therefore not representative. Nevertheless, there is some kind of relation among the diverse ingots and tools, then (*Table 2)*, nº 10711 is two and a half times nº 10714; nº 10709 is eight and a half times nº 10714; and nº 10710 is twelve and a half times nº 10714. Bonnanaro's Cypriot double axes are datable to the Late Helladic IIIC, and the Portuguese double looped socketed axe is very well known in Late Bronze Age III Portuguese hillfords and hoards (10th-9th centuries BC) (Vilaça 1995;Ruiz-Gálvez 1998: chapter 6). Therefore a date of Late Bronze Age/Iron Age transition for the hoarding of Bonnanaro's items could be acceptable. Then, either the Aegean or the Phoenician units might have sense. The first one because is a widespread standard in Sardinia (Zaccagnini 1991). The second one, which is also attested at the Ulu Burum cargo (Pulak 2000), due to the Phoenician Diaspora to the West at the beginning of the First Millennium BC (Aubet 1994;Ruiz-Gálvez 1998:chapter 7 and annex 3; Torres 1998).

Torralba and Flumenelongu hoards (Lo Schiavo 1976 and 1988) were found at huts connected with the nuraghe Santu Antine and the nuraghe Flumenelongu respectively. The two of them are datable at the end of the Final Bronze Age (Lo Schiavo 1976:15; Ibidem 1988:221). Some objects from Flumenelongu such as the bracelets or the double looped flat palstave have close counterparts in Spanish hoards of the Late Bronze Age II and III (11th-9th BC.) (Galán & Ruiz-Gálvez 1996:153-6; Ruiz-Gálvez 1998: chapters 6 & 7; Ruiz-Gálvez 2000a:271-5). Two of the Torralba weights are very close to two times and four times a western mina (*Table 3 and 4*). In both cases the weight unit at work could be either the Aegean or the Microasiatic one. Nevertheless the second unit seems to be better fitted, then the hypothetical Aegean standard shows too many fractions (a quarter, three quarter, one third, two thirds) to be practical. Even more, the Microasiatic unit is present at other Final Bronze Age/ transition to the Early Iron Age contexts in Sardinia, as we saw above[8].

Lastly, copper and lead ingots from Olmedo and Arrubiu (Tables 5 and 6) could either belong to the Aegean or to the

7 Ossi hoard was found at the Sa Mandra 'e Sa Guia nuraghe and settlement. The occurrence there of Final Bronze Age *a pettine* pottery, and Early Iron Age geometric pottery (*a cerchielli*) (Ferrarese Ceruti 1997 pp322 & 422), as well as the C14 date series R-1094a (3050±50 BP) = 1430-1120 B.C; R.1096 (2810±50 BP) = 1130-830 BC; R-1097 (2800±50) = 1130-820 BC; R-1092a (2740±50 BP) = 1000-800 BC; R-1093a (2690±50 BP) = 930-790 BC; R-1098 (2670±50 BP) = 930-780 BC; R-347 (2600±70) = 800-510 BC; R-1065a (2590±50) = 750-450 BC; R-346 (2460±70) = 660-370 BC confirms a Postmycenaean chronology for the Ossi hoard. (after Tykot 1994. I have updated the calibration program to the Oxcal 3.5. All data calibrated at two s).

8 Balance weights and lead ingots from Santu Brai, Forraxi Nioi, Teti and Santa Anastasia di Sardara. I remind my readers that the flanged axe from the Forraxi Nioi fit also with the 11.75* gr. standard.

Table 3.

Hoard	Inv. nº	item	Weight (gr.)	Aegean Unit 6.5/6.8	Canonical weight (gr.)	Micro-asiatic unit 11,75	Canonical weight (gr.)
Torralba	10878	*Flanged axe*	446	65 (6.8x10x6 1/2)	442	38	446,5
Torralba	10864	*Plano-convex copper ingot*	570	85	578	48	564
Torralba	10874	*Plano-convex copper ingot*	738	110 (6.8x10x11)	748	62	728,5
Torralba	10873	*Plano-convex copper ingot*	750	110	748	64	752
Torralba	10879	*Vertical shaft hole axe*	882	130	884	75	881,2
Torralba	10863	*Plano-convex copper ingot*	970 (± a double mina)	142	965,6	82	963,5
Torralba	10861	*Plano-convex copper ingot*	990	146	992,8	84	987
Torralba	10867	*Plano-convex copper ingot*	1040	154	1047,2	88	1034
Torralba	10872	*Plano-convex copper ingot*	1459	214	1455,2	124	1457
Torralba	10875	*Plano-convex copper ingot*	1680	248	1686,4	144	1692
Torralba	10862	*Plano-convex copper ingot*	1710	252	1713,6	146	1715,5
Torralba	10866	*Plano-convex copper ingot*	1866 (ca. four minas)	274	1863,2	158	1856,5
Torralba	10871	*Plano-convex copper ingot*	2070	306	2080,8	176	2068
Torralba	10868	*Plano-convex copper ingot*	2210	326	2216,8	188	2209
Torralba	10869	*Plano-convex copper ingot*	2210	326	2216,8	188	2209
Torralba	10870	*Plano-convex copper ingot*	2740	400	2720	234	2749,5

Microasiatic unit. Again, the sample is too small to be significant. Anyway, both units have apparently coexisted in Sardinia until the end of the Bronze Age and the beginning of the Iron Age, as it happens in the Santa Anastasia di Sardara lead ingots (Zaccagnini 1991).

RETHINKING WEIGHT AND TRADE SYSTEMS

Two standard of weight appear as a constant in the sample analysed above: the Aegean and the Microasiatic one. The first one is easy to justify, if we think of a Mycenaean trade expanding to the Central Mediterranean and including Sardinia on it (Vagnetti 1993 a & b; Ferrarese Cerruti 1997).

But, what about the Microasiatic unit? Although both, the Aegean and the Microasiatic units have apparently coexisted in Sardinia, the last one seems to have been at work only very late, in Final Bronze Age/beginnings of the Early Iron Age contexts. Thanks to the work of Lo Schiavo, Vagnetti and other scholars, we know of the strong presence of Cypriots traders and craftsmen in Sardinia at Mycenaean and Postmycenaean times. But, at a great surprise, the Cypriot unit of 9.3* gr. (Petruso 1984) is not attested in Sardinia. Quite on the contrary, the Microasiatic unit was common in Anatolia, the Levant and Cyprus, the so called *Sea People* main area. Some scholars as Susan Sherratt (1994; 1998 2000 and Sherratt & Sherratt 1991 & 1993; Arzy 1994 & 1998; Negbi 1991; Bauer 1998) point out, that the political changes and turmoils taking place in the Eastern Mediterranean in the 13th. and beginning of the 12th. centuries BC. could be less the consequence of mass migrations or invasions than the subversion of the State systems of trade. Thus small trader groups, previously working under State control as officials, started to act on their own behalf, in a private, opportunistic trade or even piracy, against the interests of their ancient "*bosses*". Such kind of free lance trade could help to explain that commodities, techniques and skills, traditionally under palace control, as the iron smelting, the lost wax technique, the frankincense and the burners, the vertical loom and the purple etc, began to arrive after de 12th BC. to the Central and Western Mediterranean and to Continental Europe.

Scrap metal trade, including *European* and Central Mediterranean bronze types, that after Susan Sherratt (2000:88), was common in the Eastern Mediterranean from the end of the 13th BC onwards, could have overflowed Eastern palace-controlled markets and provoked the system to collapse (ibidem 1998 & 2000a). In such an international,

Table 4.

Hoard	Inv. nº	Item	Weight (gr.)	Aegean unit 6.5/6.8	Canonical weight (gr.)	Microasiatic unit 11.75	Canonical weight (gr.)
Flumenelongu	10296	bronze bracelet	142	----		12	141
Flumenelongu	10270	bronze bracelet	188	----		16	188
Flumenelongu	10271	double looped flat palstave	878	130 or 13 (6.8x10x13)	884	75	881,2
Flumenelongu	10251	plano-convex bronze ingot	1044	154 or 15 1/4 (6.8 x 10 x 5,4)	1047,2	90	1057,5
Flumenelongu	10250	plano-convex bronze ingot	1270	186 or 18 2/3 (6.8 x 10 x 18,6)	1264,8	108	1269
Flumenelongu	10244	plano-convex bronze ingot	1288	190 or 19 (6.8 x 10 x 19)	1292	110	1292,5
Flumenelongu	10238	plano-convex bronze ingot	1312	193 or 19 1/3 (6.8 x 10 x 19,3)	1312,4	112	1316
Flumenelongu	10237	plano-convex bronze ingot	1390	205 or 20 1/2 (6.8 x 10 x 20,5)	1394	118	1386,5
Flumenelongu	10246	plano-convex bronze ingot	1466	215 or 21 1/2 (6.8 x 10 X 21,5)	1462	125	1468,7
Flumenelongu	10239	plano-convex bronze ingot	1474	217 or 21 3/4 (6.8 x 10 x 21,70)	1475,6	126	1480,5
Flumenelongu	10255	plano-convex bronze ingot	1598	235 or 23 1/2 (6.8 x 10 x 23,5)	1598	136	1598
Flumenelongu	10254	plano-convex bronze ingot	1644	242 or 24 1/4 (6.8 x 10 x 24,2	1645,6	140	1645
Flumenelongu	10245	plano-convex bronze ingot	1652	243 or 24 1/3 (6.8 x 10 x 24,3)	1652,4	140	1645
Flumenelongu	10258	plano-convex bronze ingot	1672	246 or 24 2/3 (6.8 x 10 x 24,6)	1672,8	142	1668,5
Flumenelongu	10249	plano-convex bronze ingot	1712	252 or 25 1/4 (6.8 x 10 x 25,2)	1727	146	1715

free lance trade, Nuragic people[9] may have acted as middlemen between the more developed Eastern Mediterranean political systems, and the less developed Western Mediterranean and Atlantic ones. The fact is, that from Postmycenean times (Nuragic Final Bronze Age and Early Iron Age and Atlantic Late Bronze Age II and III), the amount of bronze items of *Iberian* type in Sardinian scrap hoards is not negligible. Neither is the presence of bronzes of Sardinian/Central Mediterranean typology in Iberian hoards and sites (Ruiz-Gálvez 1998: chapter 6). Perhaps it is not by accident, that *Portuguese* Late Bronze Age obeloi are to be found both in the Sardinian Monte Sa Idda scrap hoard, and in the Cypriot necropolis of Amathus (Karagheroghis & Lo Schiavo 1989). Perhaps it is not by accident either, that the Microasiatic standard was then common in Sardinia, the Iberian Peninsula and some areas of Atlantic Europe (Ruiz-Gálvez 1998: chapter 7 and 2000 a & b).

I wish to conclude my contribution with some comments on a cluster of weights and numerical combinations coming from Slovenian hoards recently published by Turk (2001). These are on the one hand, a hollow bronze sphere made of two halves, and one perforated bronze bead from the Grabe Hoard in Styria, and on the other hand, series of objects from the Kalnaski Vrh I hoard. The two of them are datable at the Late Bronze Age or Early Iron Age (ibidem: 251 & 266). The two hemispheres from Grave weigh respectively 54.875 and 32.925 gr. and the perforated bead 34.9 gr. The sum of the two spheres is 87.8 gr. and the perforated bead weighs exactly half of this (87.8:2=34.9) and the estimated ratio among the three weights could be 5:4:3. Concerning the Kalnaski Vrh I hoard, it contains a set of wheel-shaped pendants, tubulets, bracelets with spiral terminals, winged axes, necklaces etc. Again, the ensemble of items in the hoard seem to have a numeral relation, since, according to Turk (ibidem:266) *the sum of pendants, tubulets and ringlets from the hoard (41+41+63=145) almost corresponds with the sum total of ingots and plano-convex ingots in the hoard (144).*

In both, the Grabe and the Kalnaski Vrh I hoards, a kind of additive system of calculation seems to be operating, similar, although not exactly the same, to that suggested by Petruso (1984) for the Cape Gelidonya sphendonoid weights. According to the mentioned author (ibidem:302), the shapes are significant weight indicators, as also happens with Grabe and Kalnaski Vrh 1 hoards. On the basis of the Gelidonya wreck we can assume, that additive systems of calculation were operating in the Eastern Mediterranean at least from the 13th BC. Precisely Turk (ibidem:268), based on the presence of a 10.5 gr. weight in Cape Gelidonya, suggests the 10.5 gr. unit for Grabe and Kanalski Vrh 1 hoards. Yet,

9 The so-called Shardana? See other views on the origin of the Shardana in Lo Schiavo 2001.

Table 5.

Hoard	Item	Weight (gr.)	Aegean unit 6.5/6.8	Canonical weight (gr.)	Microasiatic unit 11.75	Canonical weight (gr.)
Arrubiu	*Plano-convex lead ingot*	1.748,9	26 (6.8x10x26)	1768	150	1762,5
Arrubiu	*Plano-convex lead ingot*	2.408,8	36	2448	200	2350
Arrubiu	*Plano-convex lead ingot*	4.099,4	60	4080	350	4112,5

Table 6.

Hoard	Inv. nº	Item	Weight (gr.)	Aegean unit 6.5/6.8	Canonical Weight (gr.)	Microasiatic Unit 11.75	Canonical Weight (gr.)
Olmedo	502	*Plano-convex copper ingot*	359,28	5 1/2 (65x10x5 1/2)	357,5	30 1/2	358,3
Olmedo	504	*Plano-convex copper ingot*	358,15	5 1/2	357,5	30 1/2	358,3

Lassen states (2000:235), that there are very few known weights from Late Mycenaean period, and all unmarked. Despite this, Petruso (quoted by Lassen 235), suggested a Mycenaean unit of 5.5 gr. based on some few findings as i.e two lead and one stone weights found at Late Bronze Age Mycenae weighing 1089.5 gr. and 1120 gr. respectively, eight stone weights weighing between 545-560 gr. found at Phokis, and two others from Tyrins weighing 541.4 gr. and from Magoula, weighing 560 gr. (Lassen 2000 235). The same 5.5 gr. unit has been proposed for some stone weights coming from Northern Italy and Switzerland, dating from Later Bronze Age (*Bronzo Recente*) to Early Iron Age, and that would link the area with the Fratesina gateway community (Cardarelli et allii 1997:632; Peroni 1998: 218-22).

So, the 5.5 gr. unit working on Slovenian Late Bronze Age/ Early Iron Age contexts, may perfectly be a Late Mycenean one. But it could also be the Microasiatic one, whose half unit is 5.7/5.8 gr. (Zaccagnini 1991:344). It is quite interesting, that the set of weighs from the 9th century BC. warrior trader grave of Lefkandi published by Popham and Lemos (1995) included a group of almond-like weights, weighing respectively 4.1 gr. (1/3 of the Microasiatic standard), 5.9 gr. (1/2 of the Microasiatic standard), 21.9 gr. and 25.0 gr. (approximately 2 microasiatic shekels, being the mean weight 23.45 gr. = 2 Microasiatic shekels); 44.9 gr. (4 Microasiatic shekels); 89.9 gr. (8 Microasiatic shekels) and 159.7 gr.(14 Microasiatic shekels). A second hexagonal set of weights of the same grave weigh respectively 10.5 gr. and 30.1 gr. and they could fit with 2 and 5 1/2 times the Late Mycenaean unit of 5.5 gr. or either 2 times and 5 times a 1/2 Microasiatic shekel of 5.7/5.8 gr. The same would be applicable under my view, to the Central European bi-conical ingots, for which Pare calculated a unit of ca. 475 gr. that he assumed, was a multiple of the Cypriot 9.3 gr. unit (Pare 1999; Petruso 1984). And, indeed 475 is ± five times 93.But it is also very close to the western mina of 470 gr. (see footnote 5), which is divisible among 70 Aegean units, 60 Eblaitic or Phoenician units, 50 Egyptian units or 40 Microasiatic units. Anyway I do not find any archaeological justification for the adoption of a Cypriot standard in Late Bronze Age Central European contexts. Quite on the contrary, and as I have tried to convey in the previous pages, there are archaeological arguments that support the hypothesis of the adoption in wide areas of Europe of firstly, the Aegean unit of 6.5/6.8 gr., perhaps also the so-called Late Mycenaean unit of 5.5 gr., and thereafter, the Microasiatic unit of 11.45/11.75 gr.

Acknowledgements

My gratitude goes to Mariano Torres for reviewing the text and for doing so many brilliant suggestions and critics to it, and to Maria Angeles Pérez and Oscar López for doing the statistics. An SPSS/PX+ programme was used for the analysis.

I also wish to express my gratitude to Dr Fulvia Lo Schiavo for friendly providing data and support.

The search for weight systems in Sardinia is one part of a wider research project called *Landscape and Nuragic territory at the Pranemuru plateau (Orroli, Nuoro)*, funded by the Spanish Ministry of Culture and Education (DGES project PB98-0840), the Spanish Institute for the Heritage (years 1999, 1990 and 1991), and the Madrid Complutense University (project PR269/98-0196), and conducted by M. Ruiz-Gálvez.

Author's address

Marisa RUIZ-GÁLVEZ
Universidad Complutense
E-28040 Madrid SPAIN

Bibliography

ARTZY, M. 1994: Incense, camels and collared rim jars: desert trade routes and maritime outlets in the Second Millennium. *Oxford Journal of Archaeology* nº13 (2). pp 122-44

ARTZY, M. 1998: Routes, trade, boats and nomads of the Sea. In Gitin, S./A Mazar/E. Stern (eds): *Mediterranean peoples in transition. Thirteenth to Tenth Centuries BCE. In Honor of Professor Trude Dothan.* Jerusalem, Israel Exploration Society. pp 439-48.

AUBET, Mª E. 1994: *Tiro y las colonias fenicias de Occidente.* Barcelona, ed. Critica. 2ª ed.

BAUER, A. 1998: Cities of the Sea: maritime trade and the origins of Philistine settlement in the Early Iron Age Southern Levant. *Oxford Journal of Archaeology* nº 17 (2) pp 149-68

BEGEMANN, F./SCHMITT-STRECKER, S./PERNICKA, E./LO SCHIAVO, F. 2000: Chemical composition and lead isotopy of copper and bronze from Nuragic Sardinia. *European Journal of Archaeology* vol. 4 (1) pp 43-85

BERNARDINI, P. 1989: Tre nuovi documenti di importazione dalla colina di Muru Manu. *Rivista di Studi Fenici* XVII (2). pp 285-90

BERNARDINI, P. 1991: *Micenei e Fenici. Considerazioni sull'età precoloniale in Sardegna.* Roma. Instituto per l'Oriente C.A. Nallino. Orientis Antiqui Collectio- XIX

BERNARDINI, P. 1991: Considerazioni sui raporti tra la Sardegna, Cipro e l'area Egeo-Orientale nell'età del Bronzo. *Quaderni della Soprintendenza Archeologica per le provincie di cagliari e Oristano* nº10. pp 29-67.

BLOCH, M. & PARRY, J. (eds) 1982: *Death & the regeneration of life.* Cambridge, Cambridge Univ. Press.

CACERES, Y. 1997: cerámicas y tejidos: sobre el significado de la decoración geométrica del bronce Final en la Península Ibérica. *Cómplutum* nº8. pp 125-40

CAMPUS, F. & LEONELLI, V. 1999: *La tipollogia della ceramica Nuragica I. il materiale edito.* Sassari Beta Gamma ed.

CARDARELLI, A./PACCIARELLI, M./PALLANTE, P. 1997: Pesi di bilancia dell'étà del Bronzo?. In M. Bernabò Brea/A. Cardarelli/ M. Cremaschi (eds): *Le Terramare. La più antica civiltà padana.* Modena, ed. Electa pp 629-42.

CARDINI, F. 1983: *I giorni del Sacro.* Novara, ed. Nova.

CURTIN, PH. 1984: *Cross-cultural trade in World History.* Cambridge, Cambridge Univ. Press.

COURTOIS, j-c. 1983: Le trésor de Poids de Kalavasoss-Ayios Dhimitrios 1982. *Repports of the Department of Antiquities, Cyprus.* pp 117-30

EARLE, TH. 1997: *How chiefs come to power. The political economy in Prehistory.* Stanford, Satanford Univ. Press

FERRARESE CERUTI, Mª LUISA 1997: *Archeologia della Sardegna Preistorica e Protoistorica.* Cagliari, ed. Poliedro

GALAN, E. & RUIZ-GÀLVEZ. M. 1996: Divisa, dinero y moneda. Aproximación al estudio de los patrones metrológicos prehistóricos peninsulares. In Mª A. Querol & T Chapa (eds): *Homenaje al profesor Manuel Fernández-Miranda.* madrid, Cómplutum extra nº6. Tº II. pp 151-65.

GENNEP, A. VAN 1986: *Los ritos de paso.* Madrid, ed. Taurus

GOMEZ DE SOTO, J. 1990: "Pictogrames", figurations anthropomorfiques et zoomorfiques sur les céramiques de la fin de l'Âge du Bronze, un révision. *Actes du 115è Congrès National des Sociétés Savantes. Pré.et.Protohistoire.* pp 149-62

HELMS, M. 1988: *Ulysses' sail. An ethnographical Odyssey of 'power, knowledge and geographical distance.* Princeton, Princeton Univ. Press

KARAGEORGHIS, V. & LO SCHIAVO, F. 1989: A West Mediterranean obelos from Amathus. *Rivista di Studi Fenici* XVII, 1. pp 15-29

KNAPP, B. 1990: Entrepreneurship, ethnicity and exchange: Mediterranean inter-island relations in Lae Bronze Age. *Papers of the British School at Rome* nº85. pp 115-53

IFRAH, G. 1997: *Historia Universal de las cifras.* Madrid. ed. Espasa Calpe.

LASSEN, H. 2000: Introduction to weight systems in Bronze Age East Mediterranean: the case of Kalavasos-Ayios Dhimitrios. In CH Pare (ed): *Metals make the World go round. The supply and circulation of metals in Bronze Age Europe.* Oxford, Oxbow. pp 233-245

LILLIU, G. 1982: *La civiltà Nuragica.* Sassari

LO SCHIAVO, F. 1982: Copper metallurgy in Sardinia during the Late Bronze Age. New prospects on its Aegean connections. In V. Karageorghis (ed): *Early metallurgy in Cyprus 4000-3000 BC. Acta of the International Archaeological Symposium Larnaca 1981.* Nicosia pp. 271-82.

LO SCHIAVO, F. 1976: Il ripostiglio del nuraghe Flumenelongu (Alghero-Sassari). Considerazioni preliminari sul commercio marittimo nel Mediterraneo occidentale in età protostorica. *Quaderni* nº2. pp 7-45.

LO SCHIAVO, F. 1988: Il ripostiglio della capanna I e gli altri bronzi Nuragici. In *Il nuraghe S. Antine nel Logudoro-Meilogu.* Carlo Delfine ed. pp 207-41.

LO SCHIAVO, F. 1989-90: Le più antiche asce a margini rialzati della Sardegna. *Rivista di scienze prehistoriche.* nº XLII, 1-2. pp 241-70

LO SCHIAVO, F. 1997: Sardegna. *Enciclopedia dell'Arte Antica. Classica e orientale.* Roma. Enciclopedia Italiana. pp 141-58

LO SCHIAVO, F. 2000: Sea and Sardinia. In D. Ridgway et allii (ed): *Ancient Italy in its Mediterranean Setting. Studies in honour of Ellen Macnamara.* Accordia Research Institute vol 4. pp 141-58

LO SCHIAVO, F. 2001: Late Cypriot Bronzework and bronzeworkers in Sardinia, Italy and elsewhere in the West. in L. Bonfante & V. Karageorghis (eds): *Italy and Cyprus in Antiquity 1500-450 BC.* Nicosia pp 131-52.

LO SCHIAVO, F./MACNAMARA, E./VAGNETTI, L. 1986: Late Cypriot Imports to Italy and their influence on local bronzework. *Papers of the British School at Rome* LIII pp 1-71

LO SCHIAVO, F. & USAI, L. 1995: Testimonianze cultuali di età Nuragica: La grotta Pirosu in località Su Benatu di Santadi. In V. Santoni (ed): *Carbonia e il Sulcis. Archeologia e territorio.* Cagliari, ed. S'Alvure. pp 147-476.

LO SCHIAVO, F. & VAGNETTI, L. 1993: Alabastron miceneo dal nuraghe Arrubiu di Orroli (Nuoro). *Rend della. Accademia dei Licei* IX, 1 pp 121-48

MEDEROS, A. &. HARRISON, R. J :1996: Placer de dioses. Incensarios en soporte de ruedas del Bronce Final de la Peninsula Iberica. In Mª A. Querol & T Chapa (eds): *Homenaje al profesor Manuel Fernández-Miranda.* madrid, Cómplutum extra nº6. Tº I. pp 237-53

NEGBI, O. 1991: Were there Sea Peoples in the Jordan valley at the transition from the Bronze Age to the Iron Age?. *Tel Aviv* 18 pp 205-43.

PARE, CH. 1999. Weights and weighing in bronze Age Central Europe. In *Eliten in der Bronzezeit. Ergebnisse zweier Kolloquien in Mainz und Athen.* Mainz. Römish-Germanischen ZZentralmuseum. Monographien nº43 pp 421-514.

PARISE, N. 1993: Forme della circolazione metallica fra Tarda Età del brono ed Età Geometrica nel Mediterraneo Orientale. *Moneta e non moneta. Atti del Convegno Internazionale di Studi Numismatici. (Milan).* Rivista Italiana di Numismatica e Scienze Affini nº95. pp 53-8

PERONI, R. 1998: Bronzezeitliche Gewichtssysteme in Metallhaldel zzwischen Mittelmeer und Ostsee. In B. Hänsel (ed): *Mensch und Umwelt in der Bronezeit Europas.* Kiel Oetkes-Voges Verlag pp 217-24

PETRUSO, K.M. 1979: Reflections on Cycladic and Minoan metrology and trade. In J.L. Davis/J.F. Cherry (eds): *Papers in Cycladic Pregistory.* UCLA: Institute of Archaeology. Monograph XIV. pp 135-42.

PETRUSO, K.M.1984: Prolegomena to Late Cypriot weight metrology. *American Journal of Archaeology* n°88 pp 293-304

POPHAM, M. & LEMOS, I.S. 1995: a Eubean warrior trader. *Oxford Journal of Archaeology* n°14 (2). pp 151-57

PULAK, C. 2000: The balance weights from the Late Bronze Age shipwreck at Uluburum. In CH Pare (ed): *Metals make the World go round. The supply and circulation of metals in Bronze Age Europe.* Oxford, Oxbow. pp 247-66

RIVALLAIN, J. 1994: *Échanges et practiques monétaires en Afrique du XVè au XIXé siècles á travers les récits des voyageurs.* Lyon

RUIZ-GÁLVEZ, M. 1998a: *La Edad del Bronce en la Europa Atlántica. Un viaje a los orígenes de Europa Occidental.* Barcelona, ed. Crítica.

RUIZ-GÁLVEZ, M: 1998b: Peripheral, but not that much. In S. Jorge (ed): *Existe uma Idade do Bronze Atlântico?.* Lisboa. Trabalhos de Arqueologia n°10. pp 101-12

RUIZ-GÁLVEZ, M. 2000a: Weight systems and exchange networks in Bronze Age Europe. In CH Pare (ed): *Metals make the World go round. The supply and circulation of metals in Bronze Age Europe.* Oxford, Oxbow. pp 267-79

RUIZ-GALVEZ, M. 2000b: La precolonización revisada: de los modelos del s. XIX al concepto de interacción. En P- Fernández-Uriel/C. G.Wagner/F. López Pardo (eds): *Intercambio y comercio preclásico en el Mediterráneo.* Madrid, Centro de estudios fenicios y Púnicos. pp 9-25.

SHALINS, M. 1988: *Islas de Historia.* Barcelona, ed. Gedisa.

SHERRATT, A. 1991: Sacred and profane substances: the ritual use of narcotics in later Neolithic Europe. In P. Garwood et allii (eds): *Sacred and Profane: proceedings of a conference on archaeology, ritual and religion.* Oxford, OUCA Monograph n°32. pp 50-64.

SHERRATT, S. 1994: Commerce, Iron and Ideology: Metallurgical innovation in 12th-11th century Cyprus. In V. Karageorghis (ed): *Cyprus in the 11th Century BC.* Nicosia. pp 59-107.

SHERRATT, S. 1998: "Sea Peoples" and the economic structure of the Late Second Millennium in the Eastern Mediterranean. In S. Guitin/A Mazar/E. Stern (eds): *Mediterranean peoples in transition. Thirteenth to Tenth Centuries BCE. In Honor of Professor Trude Dothan.* Jerusalem, israel Exploration Society. pp 292-313.

SHERRATT, S. 2000: Circulation of metals and the end of the Brone Age in the Eastern Mediterranean. In CH Pare (ed): *Metals make the World go round. The supply and circulation of metals in Bronze Age Europe.* Oxford, Oxbow. pp 82-98

SHERRATT, A. & S. 1991: From luxuries to commodities: the nature of Mediterranean Bronze Age trading XC. pp 351-86systems. In N.H. Gale (ed): *Bronze Age trade in the Mediterranean.* Göteborg, Studies in Mediterranean Archaeology.

SHERRATT, A. & S. 1993: The growth of the Mediterranean economy in the early first Millenium BC. In J. Oates (ed): *Ancient trade, new perspectives.* World Archaeology n°24(3) pp 361-78

TORELLI, M. 1996: *Historia de los Etruscos.* Barcelona, ed. Critica.

TORRES, M. 1998: La cronología absoluta europea y el inicio de la colonización fenicia en Occidente. Implicaciones cronológicas en Chipre y el Próximo Oriente. *Complutum* n°9 pp 49-60

TURK, P. 2001: The weight of objects in Late Bronze Age hoards in Slovenia and possibilities for determining weight standards. *Arheološki vestnik* 52. pp 249-79

TYKOT, R.H. 1994: Radiocarbon dating and absolute chronology in Sardinia and Corsica. In R. Skeates & R Whitehouse (eds): *Radiocarbon dating and Italian Prehistory.* Accordia Research Centre, Univ. of London. & British School at Rome. vols 3 6 8. pp 115-145

UGAS, G. 1985: La produzione materiale Nuragica. Note sull'apporto etrusco e greco. Cagliari. *Società e cultura in Sardegna nei periodi orientalizzante e archaico. Rapporti tra Sardegna, fenici, Etruschi e Greci.* pp.41-146

UGAS, G. & USAI, L. 1986: Nuovi scavi nel santuario Nuragico di S. Anastasia di Sardara. Cagliari, *La Sardegna nel Mediterraneo tra il Secondo e il Primo Millennio a.C.* pp 167-218.

VAGNETTI, L. 1993a: Mycenean pottery in Italy: fifty years of study. In C Zerner & J. Winders (eds): *Proceedings of the International Conference "Wace and Blegen".* Amsterdam. J.C. Gieben publ. pp 143-57

VAGNETTI, L. 1993b: Aspetti della presenza micenea nel Sud-est Italiano. *Atti del trentesimo convegno di studi della Magna Grecia. taranto- Lecce 4-9 ottobre 1990* Napoli. pp 363-82

VAGNETTI, L. 1998: Variety and function of the Aegean derivative pottery in the central Mediterran in the Late Bronze Age. In S. Guitin/A Mazar/E. Stern (eds): *Mediterranean peoples in transition. Thirteenth to Tenth Centuries BCE. In Honor of Professor Trude Dothan.* Jerusalem, israel Exploration Society. pp 66-79.

VAGNETTI, L. & LO SCHIAVO, F. 1989: Late Bronze Age long distance trade: the role of Cypriots. In E Peltenburg (ed): *Early Society in Cyprus.* Edinburgh Univ. Press. pp 217-43.

VILAÇA, R. 1995: *Aspectos do povoamento da Beira Interior (Centro e Sul) nos finais da Idade do Bronze.* Lisboa, Trabalhos de Arqueologia n°9. 2 vol.

WEBSTER, G. 1996: *A Prehistory of Sardinia 2300 to 500 BC.* Sheffield, Monographs in Mediterranean Archaeology n°7.

ZACCAGNINI, C. 1986: Aspects of copper trade in the eastern Mediterranean during the Late Bronze Age. In M Marazi/S. Tusa/ L. Vagnetti (eds): *Traffici micenei nel mediterraneo.* Taranto. Instituto per la Storia e l'Archeologia della Magna Grecia. pp 413-24.

ZACCAGNINI C. 1991: "Nuragic" Sardinia: Metrological notes. *Atti del II Congresso Internazionale di Studi fenici e Punici.* Roma, Consiglio Nazionale delle Ricerche. Vol. I. pp 343-7.

ANCIENT POTENTIAL TIN SOURCES IN THE AEGEAN

Nikos SKARPELIS

Abstract: This paper is designed to introduce data to archaeologists on Aegean ores comprising tin-bearing minerals. The minerals were identified mainly during exploration works for precious metals ores. In some cases tin grades proved to have been reasonably high so as to have been used for tin bronze production in antiquity. The chemistry of metal objects analysed by many scholars is compatible to the geochemical signature of the mineralizations. It is concluded that some outcrops of mineralizations should be considered as ancient potential tin sources.

INTRODUCTION

Tin had been used as an alloying material with copper for bronze production since the third millennium BC, having being one of the most important components of the "high-tech" industry of bronze age. This improved the mechanical properties of the bronze objects produced. Tin contents of these objects ranged from 1.2 to 13.5% (McGeehan Liritzis, 1996). It is known that the Aegean "tin problem" remains a crucial one for the Bronze Age. An overview of the problem was addressed by Muhly (1999). Despite the economic and strategic importance of tin, sources of the metal easily accessible in the Aegean and the adjacent areas have until recently remained unknown. Due to the fact that the metallogenetic setting of the mineralizations exposed in the aforementioned areas was not clarified and the corresponding ores were not entirely known in terms of mineralogy and geochemistry, it was believed that tin was imported in the region.

Extended exploration in the Aegean, been carried out over the last years from Milos to Thrace and from western Cyclades to Minor Asia, resulted in discoveries and re-evaluation of precious and base metals mineralizations. All these mineralised outcrops exhibit some common mineralogical and geochemical features and several of them comprise tin-bearing minerals. Since the Aegean region was important for its lead, zinc, silver, copper, gold and iron mineralizations, earlier chemical analyses of ores were limited to these elements. Skarpelis (1995) presented data on the geochemistry of the epithermal ores of Thrace and found tin grades up to 0.34% in copper rich portions of the Agios Filippos high-sulfidation epithermal ore deposit.

It is known that the hypogene tin and tin-bearing minerals are cassiterite and complex associations of copper, iron and tin (as major elements participating in the structure of sulfosalts), along with arsenic, zinc, lead, bismuth, antimony and rarely vanadium. On oxidation, these minerals are transformed into fine grained aggregates of oxides, hydroxides and hydrous carbonates of single metals or metal associations. A "tin oxide" or "hydrous tin oxide" compound "– corresponding to the minerals romarchite and hydroromarchite respectively - is produced at the oxidation zone of sulfide mineralizations, comprising tin-bearing mineral species of the sulfosalt group (e.g stannite, koesterite). In the archaeological literature there is a common view that tin metallurgy was solely based on smelting of oxide ores. Most scholars suggest that "tin oxide" was the single chemical substance smelted to provide tin for the manufacture of tin bronze. In some cases, a chemical compound dominated by tin-oxide was referred to as the mineral cassiterite. This is highly confusing for economic geologists. If an accurate mineralogical investigation lacks, the occurrence of "cassiterite" in the protore is in doubt.

SOURCES OF TIN: RECENT ORE DEPOSIT MODELS

According to the recent ore deposit models, cassiterite and tin-bearing sulfosalts occur as the main or subordinate ore minerals in the following deposit types:

- **Tin in granitoids**: Cassiterite is the predominant ore mineral in *"Manto and stockwork"* ore deposits genetically associated with granitic plutons and associated felsic dikes (e.g. Renison Bell, Mt. Bischoff and Cleveland, Australia; Dachang and Gejiu, China). Ore deposits of this type are large and high grade, containing millions to tens of millions of tonnes averaging about 1% Sn. *"Tin skarn"* is a cassiterite-dominant mineralization genetically associated with skarn gangue minerals, developed in reduced and deep-level environments, possibly associated with greisen alteration. *"Porphyry tin"* mineralizations comprise disseminated cassiterite and stannite minerals and are genetically associated with porphyritic intrusions. They are large, containing tens of millions of tonnes at grades of 0.2 to 0.5% of tin (e.g. Cerro Rico stock, Bolivia; Yinyan, China; Taronga, Australia; East Kemptville, Nova Scotia). Copper grades are very low in the above mineralization types.

- **Tin-Silver veins**: Polymetallic tin veins or tin-silver veins are hosted in various rock lithologies and are genetically associated with quartz-bearing and peraluminous intrusive rocks. They comprise variable amounts of cassiterite, stannite, chalcopyrite, galena and various lead sulfosalts and silver minerals. Potentially, bulk-mineable bedrock deposits contain in the order of 0.2% tin with 70-200 gr/tn silver (e.g. Potosi, Bolivia). These veins are a source of cassiterite for economic placer deposits.

- **Epithermal-type ores**: (according to the model proposed by Buchanan (1981) as it was later developed by Heald et al. (1987). This class of ores has recently become a focus for exploration throughout the world, due to the very attractive gold, silver and copper grades. Based on the hydrothermal alteration characteristics of the host rocks and the sulfidation state of ore minerals, two major types of epithermal ores are identified: High-sulfidation (HS) and low-sulfidation (LS). It is worth mentioning that the high-sulfidation epithermal mineralizations - in addition to base metals sulfides - comprise a wealth of compositionally complex sulfosalts. Among them the occurrence of copper-arsenic- and copper-tin-sulfosalts must be emphasised. The epithermal high sulfidation mineralizations are characterised by a geochemical signature which in addition to gold, copper and arsenic is marked by silver, lead, zinc, tin, antimony and mercury. Tin-bearing sulfosalts become easily oxidized under supergene conditions. Hence epithermal type mineralizations should not become a source for tin-minerals for placer deposits. High-sulfidation epithermal type mineralizations have lower tin and higher copper and arsenic grades in comparison to granite-hosted tin mineralizations and tin-placers.

- **Carbonate-hosted massive sulfide ores**: These are base- and precious-metal sulfide ores, hosted by carbonate rocks. Most are pyritic with zinc-lead-silver-copper and gold metals. Arsenic content is rather high depending on the modal percentage mainly of arsenopyrite. Tin has been produced in a few such deposits (e.g. Santa Eulalia, Arizona).

- **Placer tin deposits**. They are formed by weathering and degradation of the host rocks, followed by removal and redeposition of cassiterite downslope or along the streams.

AEGEAN MINERALIZATIONS COMPRISING TIN-BEARING MINERALS

Exploration works been carried out since 1985 in the Aegean resulted in discoveries of ore deposits and mineralised outcrops. Several outcrops of base metal sulfides, designated in the past as "mixed sulfides", were "re-interpreted". Several of them were classified into the epithermal type ores (Skarpelis, 2002). Epithermal mineralizations are widespread in the Aegean and the adjacent areas. They are part of a Tertiary metallogenetic belt extending from the Southern and NE Aegean islands (Cyclades, Lemnos, Lesbos, Chios) and Minor Asia to Rhodope, Srednogorie, Apuseni Mts and the Carpathians. HS and LS epithermal systems, hosted mainly in volcanic sequences, were discovered. The geology, mineralogy and the available geochemical data of Aegean mineralizations comprising tin-bearing minerals are briefly mentioned below. The location of mineralized outcrops is shown in Fig. 1, whereas the type and the ideal chemical composition of the minerals identified is given in Table 1.

a. **Western Thrace** (Agios Filippos, Pefka and Perama Hill High Sulfidation type epithermal ores). Although tin-

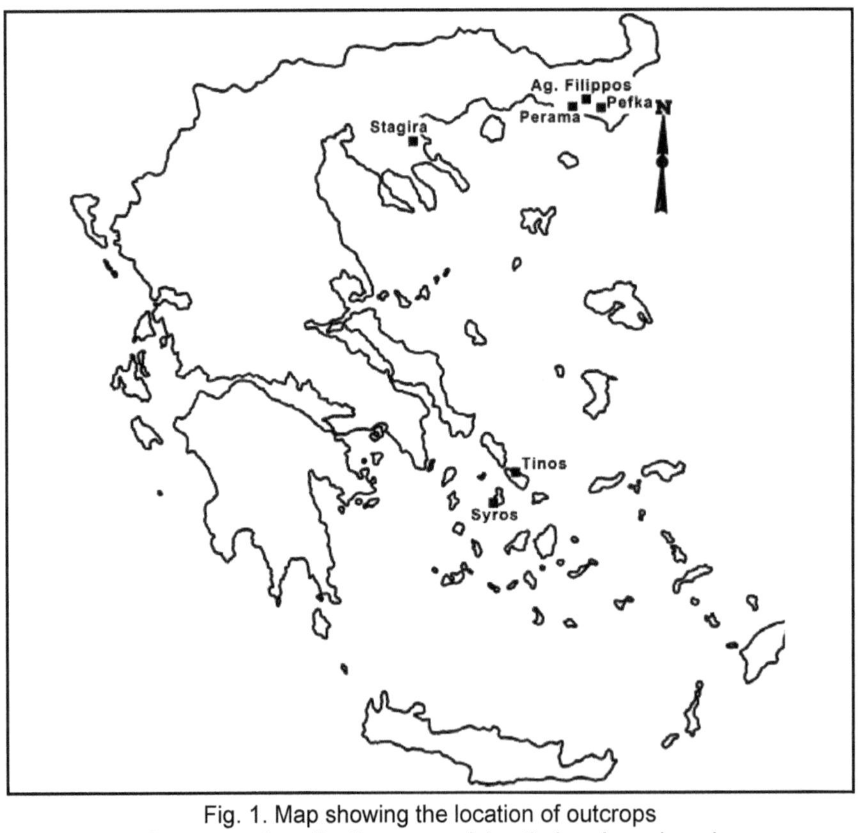

Fig. 1. Map showing the location of outcrops of Aegean mineralizations comprising tin-bearing minerals

Table 1. Tin and Tin-bearing minerals identified in the Aegean Mineralizations

	Stagira (NE Halkidiki)	Agios Philippos	Pefka	Perama Hill	Tinos	Syros
Koesterite $Cu_2(Zn,Fe)SnS_4$	+	+				
Stannite Cu_2FeSnS_4	+		+	+		+
Levyclaudite $Pb_8Sn_7Cu_3(Bi,Sb)_3S_{28}$		+				
Mawsonite $Cu_6Fe_2SnS_8$			+		+	
Stannoidite $Cu_8(Fe,Zn)_3SnS_{12}$			+			
Sn-Germanite (Cu,Ge)(S,Sn,As)			+			
Cassiterite SnO_2					+	
Colusite $Cu_{26}V_2(As,Sn,Sb)_6S_{32}$	+		+			
Canfieldite Ag_8SnS_6					+	
References	1, 2	3	4	5	6	7
1. Dimou et al. (1986); 2. Nebel et al. (1991); 3. Skarpelis (1999) and references therein; 4. Dimou et al. (1994); 5. Skarpelis (unpubl.); 6. Melidonis (1980); 7. Melidonis & Constandinides (1983)						

bearing sulfosalts are subordinate minerals in the epithermal mineralizations, especially in the copper-rich portions of the ore bodies, still they contribute to tin grades which in some cases are comparable to tin grades of granite-hosted cassiterite bearing mineralizations. Copper, arsenic and tin contents of the Agios Filippos ore yielded values up to 22.7, 9.2 and 0.34% respectively (Skarpelis, 1995, 1999). It is worth mentioning that the High Sulfidation type epithermal ore of the Chelopech deposit (Srednogorie - Bulgaria) comprises stannite, mawsonite, cassiterite, stannoidite, colusite, nekrasovite ($Cu_{13}VSn_3S_{16}$), hemusite (Cu_6MoSnS_8) (Petrunov, 1995; Popov & Kovachev, 1996). Copper and arsenic grades are around 19.0 and 4.5% respectively.

b. **Stagira** (NE Halkidiki): The tin-bearing mineral colusite has been identified in the carbonate-hosted lead-zinc mineralizations in Madem Lakkos and Mavres Petres (Nebel et al., 1991). A paragenesis of sulfide minerals comprising many sulfosalts has been found in the sulfide mineralization at Pinaki Lakkos near Stagira. The mineralization at Stagira consists of veins and irregular impregnations with a wide variety of sulfide minerals. It is enriched in copper, arsenic, antimony, manganese and bismuth. Dimou et al. (1986) describe enargite and the tin-minerals koesterite and stannite. Vavelidis et al. (2000) reported findings of litharge slags and copper and lead objects belonging to the Archaic, Classic and Hellenistic Eras in the ancient site of Stagira. The copper objects contain up to 11.95% tin. It is possible that copper and tin were extracted by processing ore from the nearby exposures of tin-bearing mineralization.

c. **Tinos island**: Outcrops of an epithermal polymetallic vein-type mineralization associated to Miocene back arc extension occur at the NE and W part of the island. The mineralization comprises subordinate tin-bearing sulfosalts and minor cassiterite (Melidonis, 1980; Tombros & Seymour, 1998). Due to their very low modal percentage of cassiterite and tin-bearing sulfosalts, these mineralised outcrops are considered to be subeconomic. Samples from the oxidized parts of the mineralization contain very low tin (<10 gr/tn, this study), a result indicating they were not the suitable material to be smelted and processed for bronze production.

d. **Syros island**: The sulfide mineralization occurs as replacement bodies hosted in marbles. Oxidation is pronounced so that only relics of sulfide minerals are rarely found. Stannite was identified by Melidonis & Constantinides (1983) as a minor mineralogical constituent.

DISCUSSION

The Aegean region is characterised by complex geologic and consequently metallogenetic evolution. A large number of outcrops of various types of rocks and ores are exposed. Recent studies shed light on the mineralogy and geochemistry of ores as far as arsenic, copper and tin metals are concerned. Although intensive geological surveys are necessary to evaluate the concentrations of tin accessible for exploitation in antiquity, we think that the available data allow identification of the potential tin sources in the Aegean and the adjacent areas. Furthermore they indicate that several ore deposits of high significance as metal sources in antiquity cannot be classified into the tin- or potential-tin sources. Besides recent studies of the mineralogy and chemistry of the Laurium (Skarpelis, unpublished) and NE Halkidiki (Nebel et al., 1991) mineralizations indicated very low tin grades in the Cu-rich sulfide ore and also in sphalerite concentrates. Furthermore, the petrological and geochemical characteristics of extensive outcrops of Tertiary granitoids of Rhodope (northeastern Greece and south Bulgaria) and the Cyclades, indicate they are not compatible with granitic rocks capable of hosting tin mineralization. But the leucogranites exposed in the Cyclades (Altherr, 1981) are of special interest, since they have gross mineralogical and geochemical features in common with tin-bearing granites. The occurrence of cassiterite as a subordinate mineral in the Ikaria S-type granite (Altherr, op. cit.) is worth to be mentioning.

Metalworking based on local ore sources other than tin took place in the Aegean in the Early Bronze Age (Renfrew, 1967; Branigan, 1974). There is sufficient evidence now that bronze production and significant bronze use took place in the Aegean and the adjacent areas:

- at the sites of Troy and Beshiktepe (Pernicka et al. 1984; Muhly, 1993)
- at Poliochni on Lemnos (Pernicka et al., 1990)
- at Kastri on Syros (Stos-Gale et al., 1984)
- at Thermi on Lesbos (Begemann et al., 1992; Stos-Gale, 1992)

The earliest example of metallic tin in the region comes from Sitagri (McGeehan Liritzis, 1996).

The copper - arsenic alloy – referred to commonly as arsenical copper – was amongst the earliest and most widely used alloys. Concerning ancient exploitation of arsenic, it is well known that arsenical copper was the alloy characteristic of the Aegean EBA (Renfrew, 1967; Branigan 1974). Arsenical copper might be an accidental product of the smelting of arsenical oxidized copper ores as suggested by Gale & Stos-Gale (1982) or copper-arsenic rich sulfide ores, which are abundant in Aegean. Tin bronze became the dominant alloy in the LBA. Cycladic and other bronzes sometimes contain a high proportion of arsenic, rather than tin. It should be mentioned that - as Renfrew (1967) stated - tin-bronze was more common in the East Aegean.

It is known that the main parameters taken into consideration for the assessment of an ore mineralization are: tonnage - grade and mineralogy. A rough estimation of the tonnage of high-in tin ore is impossible on the basis of the available data. Tin-grades of epithermal mineralizations are in some cases high enough.

An accurate estimate of the demand for tin at the bronze ages is impossible. The question that arises however is whether the quantities of tin in the various mines were sufficient so as to supply the increasing demands for this metal at those times. It is possible that only small-scale mining existed, as a result of low tonnage of various ancient tin deposits plus the difficulties in transportation of raw material.

Production of bronze requires a particular smelting technology, which might have been applied to tin-bearing ores. Archaeologists continue searching for the major sources of tin in the Mediterranean Sea. Most agree that cassiterite was the single mineral smelted to provide tin for the manufacture of tin bronze (Lechtman, 1991) and that bronze was produced, under reducing physical conditions, by the addition of a cassiterite concentrate to molten copper, necessary for the production of metallic tin. Then it was alloyed with copper in crucibles. The hypothesis of deliberately produced bronzes, requires mixing of two metallurgical products: tin produced by smelting cassiterite concentrates and copper produced by smelting copper oxides or hydrous copper carbonates. As it is suggested by Muhly (1993) such a process "would have been erratic and difficult to regulate so that it would have been almost impossible to maintain a good control over the copper:tin ratio". Thus it is reasonable to assume that the shift from "arsenical copper" to "tin bronze" was possibly a result of accidental exploitation of copper- and tin-rich ores at sectors of the ore bodies comprising copper- , tin- and arsenic-bearing minerals.

The chemistry of various objects can be explained by the common association of Cu, Sn and As in mineralizations. It is true that most mineralized outcrops in the Aegean comprise Cu-As sulfosalts and several of them - in addition - Cu-Sn minerals. They also contain bismuth, nickel, cobalt, antimony, zinc, silver and gold. That polymetallic character of the mineralizations can evidently be traced in the chemistry of the various objects. It is remarkable to see the analysis of a metal object from Chios island given by Mangou & Ioannou (1997). They determined high antimony in that object. If we take into consideration that Chios island hosts epithermal ores rich in antimony (Kerames) and base metals (Kambia area), it is rather reasonable to propose local exploitation and processing of ore than import of raw material, unless we exclude local smelting. Furthermore it is worth mentioning that there is a wide range of tin contents of the objects analysed (see analyses given by McGeehan Liritzis, 1996; Mangou & Ioannou, 1999). This variability of the metal contents can be explained either as a result of large variation of Sn/Cu ratio in the ore or as a result of application of non-strict rules regulating the quality of copper and tin mineral concentrates used for smelting and production of bronze. It should be also a result of non appropriate application of metallurgical techniques for the production of high quality alloy. Although linking groups of compositionally similar artefacts with particular ore sources without detailed understanding of production processes may lead to erroneous conclusions, we think that the geochemical signature of the HS epithermal deposits, if compared to concentrations of various elements analysed in tin bronzes, is a strong indication that they may typify the perfect "canditate" for having been a source of tin, copper and arsenic in ancient times. It is worth mentioning that, while browsing among the literature, we realised that detailed studies of the relics of primary minerals and alloys produced after smelting of ore, lack. It seems that the models on ancient methods of ore processing are based on limited data.

The definitive determination of Bronze Age tin sources is a challenge which still faces the ore deposit geologist, the archaeologist and the metallurgist. In the future the archaeometallurgists dealing with problems on the provenance of metals in the Aegean and the adjacent areas, have to take into consideration that sources of tin-bearing minerals do exist in the area.

CONCLUSIONS

a. The geological data indicate that tin participates in the chemistry of some of the Aegean ores at reasonable grades to be exploited in antiquity.

b. Tin- and tin-bearing minerals are found in the mineralizations, allowing their characterization as potential sources of tin in antiquity. Their geochemical signature is compatible with the chemistry of metal objects analysed.

c. In searching for tin sources during the bronze age it is necessary to think in terms of very much smaller deposits in comparison to tin deposits operated in historical times. The demand was low and the reserves of the deposits could be much smaller than those in historic periods.

d. Intensive geological and archaeometallurgical surveys are needed to evaluate the concentrations of tin accessible for exploitation in antiquity in the Aegean and the adjacent areas, and also to investigate the method applied for bronze production by processing Cu-Sn-As bearing sulfosalts.

e. Further mineralogical and geochemical research – especially when we deal with high sulfidation epithermal ores - is necessary in order the "tin problem" to be settled.

Acknowledgements

Funding of research projects dealing with the study of epithermal mineralizations in the Aegean by the Greek General Secretariat for Research and Technology and the Special Account for Research Grants of the University of Athens is acknowledged.

Author's address

Nikos SKARPELIS
Department of Geology
University of Athens
Panepistimioupoli
157 84 Zografou, Athens, GREECE
E-mail: skarpelis@geol.uoa.gr

Bibliography

ALTHERR, R., 1981, Zur Petrologie der Miozanen Granitoide der Zentral-Agais (Griechenland). *Dr. habil. Thesis*, Univ. Braunschweig, Germany, 218p.

BEGEMANN, F., SCHMITT-STRECKER, S. & PERNICKA, E., 1992, The metal finds from Thermi III-V: a chemical and lead-isotope study. *Studia Troica*, 2, p. 219-239.

BUCHANAN, L.J., 1981, Precious metal deposits associated with volcanic environments in the Southwest. In *Relations of Tectonics to Ore Deposits in the Southern Cordillera*, edited by W.R. Dickson & W.D. Payne. Ariz. Geol. Soc. Dig. 14, p. 237-262.

BRANIGAN, K., 1974, Aegean metalwork of the Early and Middle Bronze Age. Oxford Monographs on Classical Archaeology, edited by B. Ashmole, C. M. Robertson & J. Boardman. Oxford at the Clarendon Press.

DIMOU, E., DIAKAKIS, M. & ECONOMOU, G., 1986, La mineralisation sulfuree de la region de Stagira (site Piavitsa) de Chalkidiki et la presence des mineraux rares d' alabandite et de Mn-sphalerite. *Geol. & Geoph. Res.*, Special Issue, Athens, I.G.M.E., p. 73-87 (In Greek).

DIMOU, E., MICHAEL, K. & SERMENT, R., 1994, Mineralogy of the Epithermal – polymetallic mineralization of Pefka – Rhodope. *Bull. Geol. Soc. Greece*, 30, 1, p. 533-550 (In Greek).

HEALD, P., FOLEY, N.K. & HAYBA, D., 1987, Comparative anatomy of volcanic-hosted epithermal deposits: Acid-sulphate and adularia-sericite types. *Econ. Geol.*, 82, 1, p. 1-26.

GALE, N.H. & STOS-GALE, Z.A., 1982, Bronze Age Copper Sources in the Mediterranean: A New Approach. *Science* 216, no 4541, p. 11-19.

LECHTMAN, H., 1991, The Production of Copper-arsenic Alloys in the Central Andes: Highland Ores and Coastal Smelters? *Journal of Field Archaeology*, 18, 1, p. 43-76.

MANGOU, H. & IOANNOU, V.P., 1997, On the chemical composition of prehistoric Greek coperr-based artefacts from the Aegean region. *Annual of the British School at Athens*, 92, p. 59-72.

MANGOU, H. & IOANNOU, V.P., 1999, On the chemical composition of prehistoric Greek coperr-based artefacts from the Mainland Greece. *Annual of the British School at Athens*, 94, p. 81-100.

McGEEHAN LIRITZIS, V., 1996, The role and development of metallurgy in the Late Neolithic and Early Bronze age of Greece. *Studies in Mediterranean Archaeology and Literature, Pocket-Book* 122, edited by Paul Astrom, Paul Astroms Forlag.

MELIDONIS, N., 1980, The geological structure and ore deposit geology of Tinos island, Cyclades, Greece. I.G.M.E., *Geol. & Geoph. Res.* 13, p. 1-80, Athens.

MELIDONIS, N. & CONSTANTINIDES, D., 1983, The Stratabound Sulphide Mineralization of Syros (Cyclades, Greece). *Z. dt. Geol. Ges.* 134, p. 555-575.

MUHLY, J.D., 1993, Early Bronze Age tin and the Taurus. Amer. *J. Archaeology*, 97, p. 239-253.

MUHLY, J.D., 1999, Copper and Bronze in Cyprus and the Eastern Mediterranean. In "The Archaeometallurgy of the Asian old world". edited by V.C. Pigott. *University Museum Symposium Series*, vol.VII, The University Museum, University of Pennsylvania, p. 15-25.

NEBEL, M.L., HUTCHINSON, R.W. & ZARTMAN, R.E., 1991, Metamorphism and Polygenesis of the Madem Lakkos Polymetallic Sulfide Deposit, Chalkidiki, Greece. *Econ. Geol.*, 86, 1, p. 81-105.

PERNICKA, E., BEGEMANN, S., SCHMITT-STRECKER, S. & GRIMANIS, A.P., 1990, On the composition and provenance of metal objects from Poliochni on Lemnos. *Oxford Journal of Arcaeology*, 9, p. 263-298.

PERNICKA, E., SEELIGER, T.C., WAGNER, G.A., BEGEMANN, F., SCHMITT-STRECKER, S., EIBNER, C., OZTUNALI, O. & BARANYI, I., 1984, Archaeometallurgische Untersuchungen in Nordwestanatolien, *Jahrbuch des Romisch-Germanischen Zentralmuseums*, Mainz 31, p. 533-599.

PETRUNOV, R., 1995, Ore mineral parageneses and zoning in the deposit of Chelopech. *Geochem. Mineral. and Petrol.*, 30, p. 89-98 (In Bulgarian).

POPOV, P. & KOVATCHEV, V., 1996, Geology, composition and genesis of the ore mineralizations in the central and southern part of Elatsite – Chelopech ore field. In *Proc. Annual Meeting, UNESCO-IGCP Project No 356*, Sofia, vol. 1, p. 159-170.

RENFREW, C., 1967, Cycladic Metallurgy and the Aegean Early Bronze Age. *Aegean J. Archaeology*, 71, p. 1-20.

SKARPELIS, N., 1995, Minor elements in the base metals part of an epithermal system: the Kirki (St. Philippe) mine, Thrace, Northern Greece. *EUG 8, Strasbourg, Abstract Supplement No 1 to TERRA Nova*, vol. 7, p. 293.

SKARPELIS, N., 1999, The Agios Filippos ore deposit, Kirki (Western Thrace). A base metal part of a high sulfidation epithermal system. *Bull. Geol. Soc. Greece*, 33, p. 51-60.

SKARPELIS, N., 2002, Geodynamics and evolution of the Miocene mineralisation in the Cycladic – Pelagonian belt. *Bull. Geol. Soc. Greece*, 34, 6, p. 2191-2206.

STOS-GALE, Z.A., 1992, The origin of metal objects from the Early Bronze Age site of Thermi on the island of Lesbos. *Oxford Journal of Archaeology*, 11, 2, p. 155-177.

STOS-GALE, Z.A., GALE, N.H. & GILMORE, G.R., 1984, Early Bronze Age Trojan metal sources and Anatolians in the Cyclades. *Oxford Journal of Archaeology*, 3, 3, p. 23-37.

TOMBROS, S. & SEYMOUR, St. K., 1998, Applied geothermometry of the hydrothermal Au-Ag vein mineralization, Apigania, Tinos. *Bull. Geol..Soc. Greece*, 32, 3, p. 165-172.

VAVELIDIS, M., KIOURTZOGLOU,E. & SISMANIDIS,K., 2000, Archaeometallurgical investigations at ancient Stagira, NE Chalkidiki, Macedonia, Greece. DMG Meeting 2000, *Eur. J. Miner.* 12, Bh 1, p. 220.

THE PROBLEM OF TIN IN THE ANCIENT WORLD
(Part 2*)

John E. DAYTON

Abstract: The sources of the tin used to make the bronzes of the Bronze Age has been a matter of much speculation since the writer's paper of 1971. Obvious and well-known tin deposits in Europe (Cornwall, Spain and Bohemia have been ignored by Middle Eastern archaeologists in a frantic search for an eastern source. (The idea that bronze and technology could have spread from Barbarian Europe being anathema to the belief in the superiority of the not so very 'Fertile Crescent'). Minute traces of tin at the ppm level have been hailed as the source of all Bronze Age bronze. e.g. at Kestel in Anatolia, and traces in eastern Iran and Afghanistan.

Recent lead isotope analysis of ancient tin ingots (Begemann et all 999) has shown that some tin in Phoenician times was coming from Central Africa, and the area known to the Ancient Egyptians as 'The Land of Punt." This has confirmed the ignored work of Brill et a] (1974) who analysed lead from 12th Dynasty tombs and of the writer (1971, 1978, and 1986) who analysed lead ores from the area.

Résumé: Les sources d'etain entrant dans le réalisation des bronzes de l'Age de Bronze ont fait l'objet de beaucoup de spéculation depuis l'article de l'auteur de 1971. Les gisements d'Europe (de Cornouailles, d'Espagne et de Bohême) évidents et bien connus, ont été ignorés par les archéologues du Moyen-Orient dans leur recherche frénétique à trouver une source orientale. (L'idée que le bronze et la technologie puissent être propagés par l'Europe Barbare était anathème à la croyance en la supériorité du pas très 'Croissant Fertile'). D'infimes traces d'étain (quelques particules parmi des millions) ont été saluées comme étant la source de tout le bronze de l'Age de Bronze (à Kestrel en Anatolie, ainsi que des traces relevées en Iran Oriental et en Afghanistan.

La récente analyse de lingots d'étain pour des isotopes de plomb (Begemann et al 1999) à démontré qu' au temps des Phéniciens certains étains provenaient d'Afrique Centrale et d'une contrée connue aux Anciens Egyptiens sous le nom de ' La Terre de Punt'. Ceci à confirmé les travaux ignorés de Brill et al (1974) qui analyse le plomb des tombes de las 12ème dynastie et de l'auteur (1971, 1978, et 1986) qui analyse le minerai de plomb de cette région.

THE OCCURENCE OF TIN

Tin is a high temperature metal, which originated from the Earth's magma at about 600°C and about *350* million years ago. It is of igneous origin and occurs with quartz and acid granites, but most granites do not contain tin. The world's tin deposits are thus well defined. Tin does not occur anywhere in the Near East, nor in Anatolia, Iran, and the whole of India. In the whole of the Soviet Union workable deposits only occur in far eastern Siberia, discovered in 1933. During the last war, the Soviets were desperate for tin, and it is surprising that they did not find the Zeravshan deposits, an area famous for its gold. Strangely, there is no tin deposit in either the USA or Canada, and there is only a little tin in South America, which until recent times was a completely separate continent. In China there is tin in the far south along the Red River, and stannite is smelted in Yunnan where the final product, (pey tung) assays at 44% tin, 52% copper and 1.8% lead. Rich tin deposits exist of course in Burma, Malaya, and Indonesia, but curiously in Roman times tin was exported from Rome to India (Charlesworth 1924).

The countries of Central Africa, now called Uganda, Zaire, Rwanda, Tanzania, Buganda, Urundi, the Congo, and northern Zimbabwe are rich in copper, gold, cobalt and tin deposits. The rifting of the valley allowed mineral rich magma to come to the surface. There appears to have been three orogenic cycles at 2,760 m.yrs., 2,100 m.yrs, and 700 m.yrs. with considerable later volcanic activity. Crystalline schists, gneiss, quartzites, and the important pegmatites are features of the area. Tin was emplaced at a late stage.

One of the main tin belts 150km by 90km between 29° and 32° East, an area bigger than the whole of Cornwall, is the Ankole-Karagwe system, west of Lake Victoria which extends to Lake Kivu and the Congo. The tin belt also extends southwards through Rwanda, Buganda and into north-west Tanzania. The tin occurs with native copper, iron pyrites, gold, and argentiferous galena. Wolfram and bismuth are also found. (Bismuth is a good analytical mineral, as it cannot be removed in smelting and remains in the final metal). The tin occurs in coarse-grained intrusions of muscovite-biotite granite, and some of the veins are 60 cms. wide (see Map 1).

There are significant deposits of tin, and more importantly, actual tin mines, in northern Nigeria on the Jos Plateau, and also along the African Rift Valley in Rwanda, Tanganyika, The Congo and Northern Rhodesia. Rich deposits of tin exist near the ruins of Zimbabwe and were probably exploited after 1,000 A.D by Arab slave traders who knew the area well.

Now these are no mythical deposits with a few parts per million of tin, but real mines with real records of production. In 1952 Maniema and Kivu produced 10,436 tons of cassiterite, Katanga produced 1,645 tons, Rwanda and Urundi 2,645 tons, while Tanzania produced 295 tons in 1966.

* Part 1 of this paper was published in 1971. Part 2 has been a long time coming.

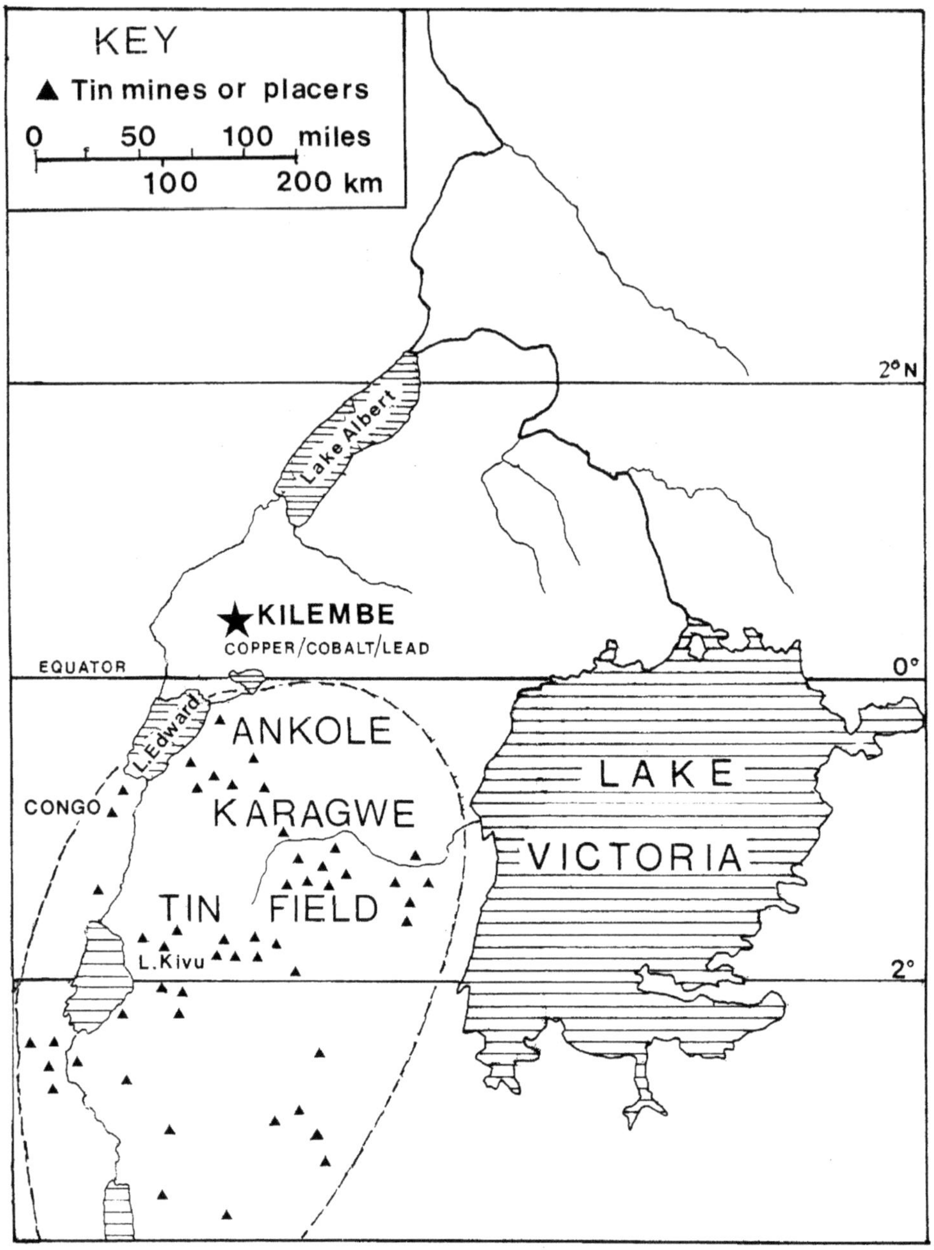

Map 1. The Ankole-Karagwe tin field. Rich deposits of cassiterite were discovered in the African Lakes region in 1924 and mapped by A.D. Combe for the Geological Survey of Uganda. The tin field continued into Tanzania, Ruanda, and the Belgian Congo. The Ankole-Karagwe system is associated with granites, gneisses, and schist's as one would expect in the great western Rift Valley.

All these tin deposits were mentioned by the writer in 1971 in a paper, which aroused great anger for some strange reason in Middle Eastern archaeological circles, which strongly defended their 'superior' civilisation against that of barbarian Europe. As tin bronzes are notoriously late in appearing in Egypt, it did not appear possible that the Ancient Egyptians were trading so far south for tin. However, from the time of the 5th Dynasty c. 2500 B.C., the Egyptians were trading with the Land of Punt, in Central Africa, for ivory, ebony, leopard skins, gold, and pygmies. These expeditions continued until the time of Ramses III c. 1198-1166 when the Phoenicians appear on the scene, and as we shall see, are actually obtaining tin from Central Africa. Nonetheless, all the evidence still points to the European Beaker Folk, the Unetice Culture and the Mycenaeans as the originators of the Bronze Age.

As stated by the writer in 1971, the large amounts of tin that heralded in the Bronze Age in the Near East sometime after 1800 B.C. (or even later at c.1650 B.C. on the evidence of the Shaft Graves at Mycenae that are very rich in bronze swords) must have come with the new peoples from Central Europe via the Brenner Pass and the Adriatic. It is no coincidence that the first known finds of metallic tin occur as plating on Mycenaen goblets. (Higgins 1967:76) A glance at the mass of bronze torques in the Munich Prehistoric Museum would convert the most doubtful Near Eastern archaeologist, while the lead isotope analysis (Dayton 1978 and 1986) confirm the European provenance of much Aegean metallurgy.

The problem of the source of tin was complicated firstly by James Muhly (an Akkadian scholar) who asserted that the

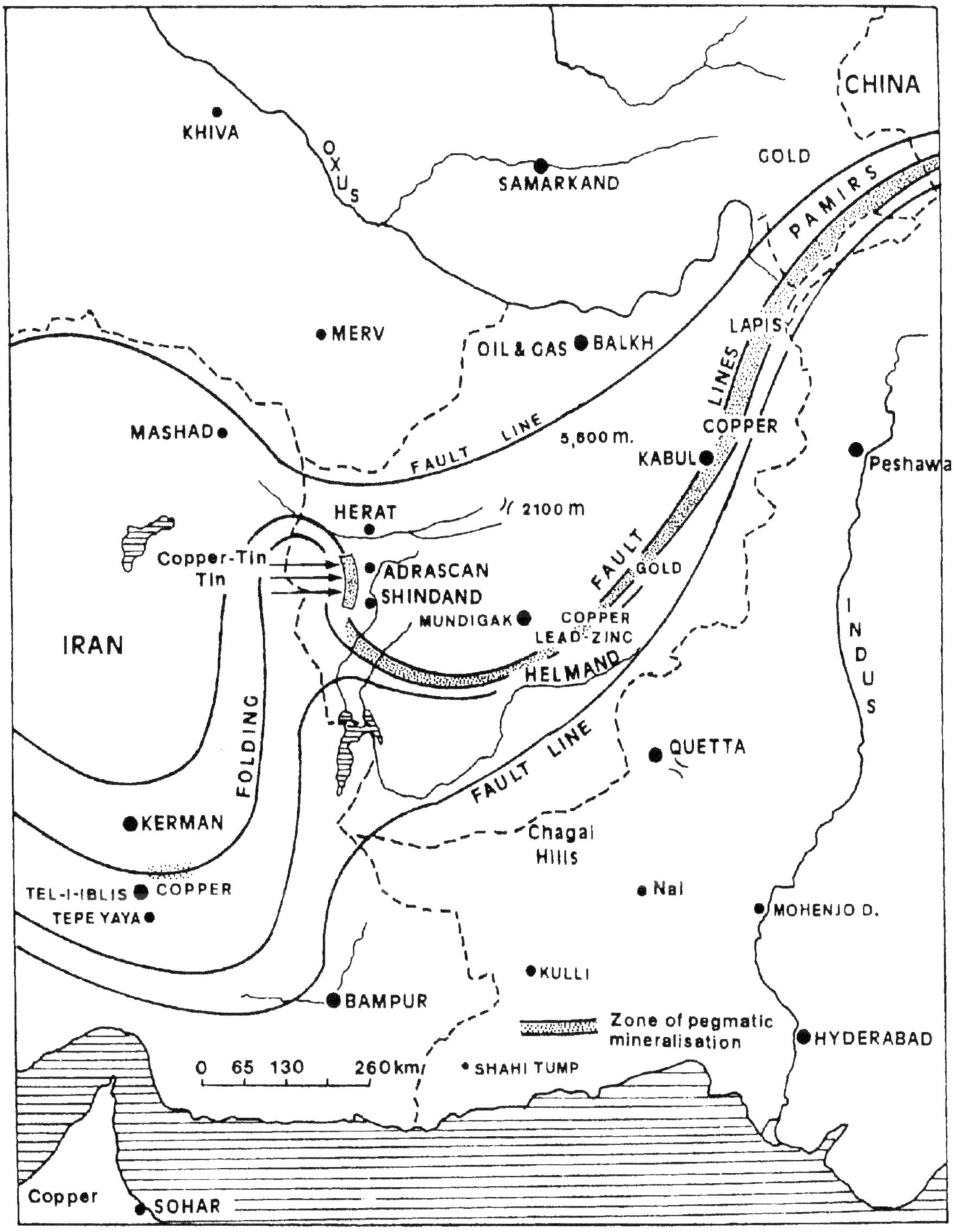

Map 2. The mineralized zone of Afghanistan (shaded) and south-eastern Iran. Afghanistan consists of arid deserts in the east, and a succession of very high mountain chains running from the south-east to the north-east where the Pamirs reach a height of 8,000 meters.

word *annaku* means tin. Scholars have argued over its meaning for years, and the writer thinks it means a bronze ingot torque that would have made a spear. It is doubtful if the Babylonians ever knew tin as a metal. Muhly can now state dogmatically that tin was imported into Cyprus from Afghanistan via Anatolia, while Tamara Stech, also in Piggott (1999), suggests that the anachronistic true tin bronzes of Ur and Troy come from Afghanistan as well. Clerziou and Berthoud (1982) found traces of alluvial tin south of Herat on the Persian/Afghan border. The writer, originally a geologist, studied these seven deposits, of which only that of Shindand was significant. The deposits showed no sign of ever having been worked but Shindand could have provided the semi-bronze of nearby Mundigak (see Map 2 and Dayton 1996). Cierny and Weisgerber (personal communication) have located considerable deposits of stannite and cassiterite in the Zeravshan valley, south of Samarkhand. Stannite, (Bell metal ore, Cu_2FeSnS_4 is difficult to smelt because of the sulphur content, but will produce a natural bronze. However, it is impossible that

LEAD ISOTOPE PLOTS FOR MAJOR ORE DEPOSITS

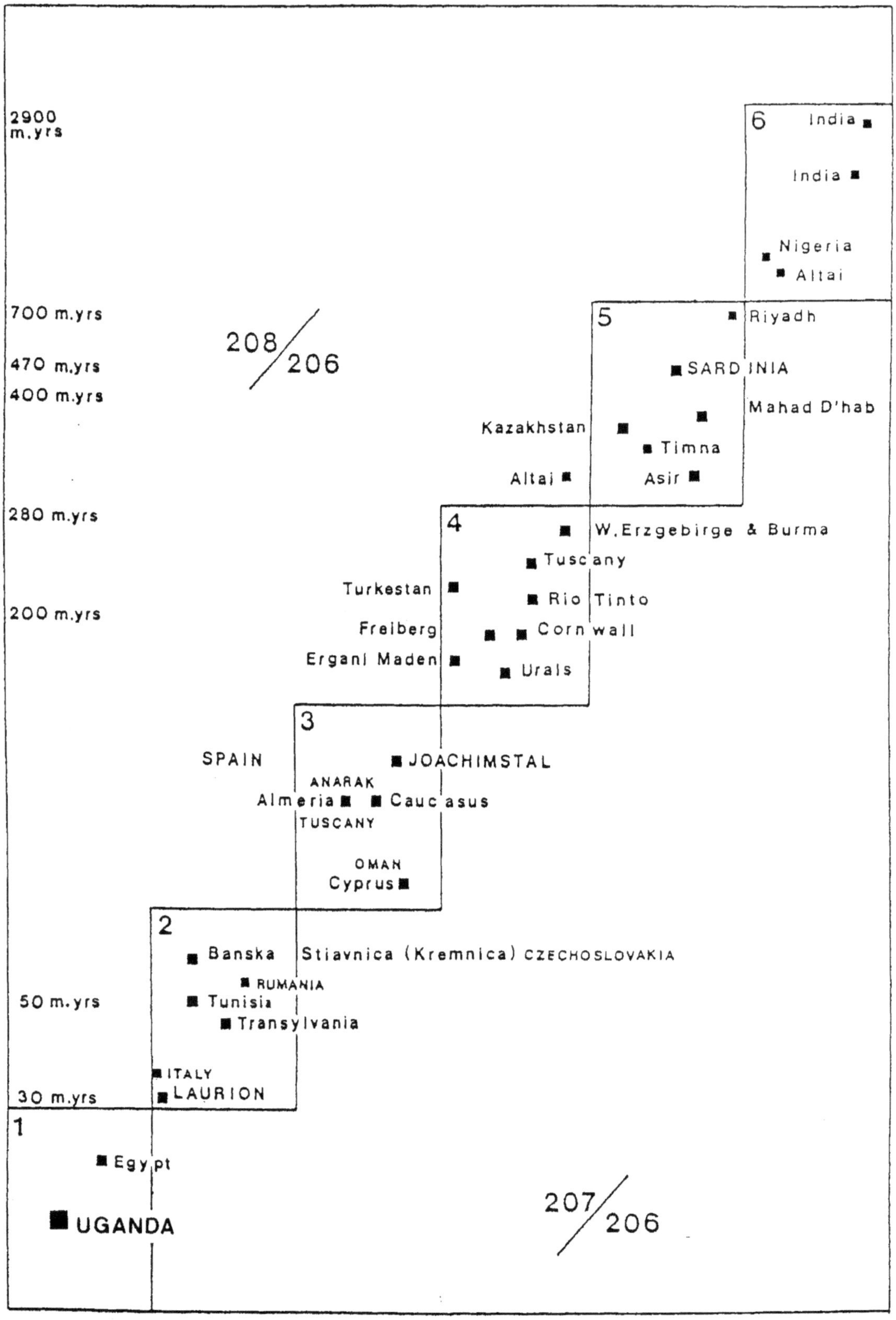

Figure 1. Key to lead isotope plots of ores and artifacts with a geological age of 30 to 3000 million years. Graphs 1 through 6 (Figures 2.2-2.7) can best be understood by plotting them along a line from youngest to oldest deposit. Although many deposits would not be confused analytically with others, some deposits are of the same age and thus cannot be differentiated for archaeological purposes. Here, however, other methods such as trace analysis may help, while other diagnostic differences on which archaeologists traditionally have had to rely are still valid.

these small deposits could have produced all the tin for the Aegean and European Bronze Ages. An enormous amount of wood would also have been required for charcoal and smelting.

ARCHAEOLOGICAL FINDS OF TIN

Fragments of tin were found in the Tomb of Tutankhamun (Lucas 1933:176-177). A Mycenaean LH1 goblet from Ialysos on Rhodes was plated with metallic tin, as was one from Mavro Spelio. Karo (1931) mentions the existence of tin foil in the Shaft Graves of Mycenae. An 18th Dynasty tin ring was in the Petrie collection. However, finds of tin have had an unlucky life. A bangle of tin from Thermi, on Lesbos vanished (Lamb 1936). Tin found in the Cape Gelidonya wreck was lost in the post. Finally in 1984, tin ingots were found in the Ulu Burun (Kas) shipwreck and are now mouldering in the Bodrum museum although Bob Maddin was able to obtain samples from them. The writer is firmly of the opinion that the Kas ship was a Mycenaean trading vessel with a cargo of 80 oxhide copper ingots, ingots of tin, and also round cakes of cobalt blue glass which the writer considers were the slag produced by silver smelting in Bohemia (Dayton 1993). The wreck contained a bronze double-axe blade, a bronze sword, a Mycenaean seal, amber beads, and a faience rhyton. The ship was obviously a Mycenaen trading vessel, although Maddin, who is not an archaeologist, insists it was Canaanite while Muhly refuses to believe the ingots were of tin. (In 1984 the writer analysed isotopically the copper ingots from Haghia Triada and found they came from Sardinia, where many Mycenaean sherds have been found, and it is clear that the Mycenaeans traded far and wide.) See Figs. 1 and 2.

THE SCIENTIFIC BREAKTHROUGH

In 1974 Bob Brill, Lynas Barnes and Barbara Adams analysed Egyptian 12th Dynasty eye paint (kohl) from Abydos, Lahun, Kahun, and Qau. Brill and the writer independently analysed yellow glass, coloured with lead/antimony from Tel Amarna. All the analysis showed that the lead came from the Central African lakes area (see Brill et al 1974 and Dayton 1978:443). In 1986 the writer compiled a world plot of some 5000 lead isotope ratios from the geological literature including 695 analysis from his own sources which included lead samples from the Kilembe mine in Uganda (see Figs. 1 and 2).

In 1981, eleven tin ingots were found in the sea off Haifa. (Wachsmann and Raveh 1981:160) and there they remained until the indefatigable Bob Maddin obtained samples. Seventeen years later the lead in the Haifa ingots was analysed by Begemaun *et* al (1999) at Bochum. The publication was kindly brought to my attention by Professor Thilo Rehren and the analysis were remarkable, and as the plot in fig.2 shows, unique in falling in the Uganda zone. Lead in the Kas tin ingots was also analysed. These analysis showed that seven of the Haifa ingots came from the western Rift Valley mineral field, and two had been contaminated with lead from Laurion. The Kas tin ingots had clearly been contaminated, or adulterated, with lead from Laurion. The 12th Dynasty kohls had clearly come from the Kilembe mine. The copper ore at Kilembe is rich in cobalt and may have been used for the crude black and blue faience of the Middle Kingdom, for there is no cobalt in the copper of Cyprus and of Laurion. The writer found copper and cobalt in an 11th Dynasty bead. Next to the copper mine at Kilembe is the lead mine from where the kohl came.

CONCLUSIONS

It is clear that the Ancient Egyptian were getting kohl and probably cobalt- rich copper ore from Kilembe in Uganda as early as the 11th Dynasty sometime after 2000 B.C.

There is no evidence that they were obtaining tin from this area, but they were certainly getting gold as well as ivory, ebony, and pygmies from the area. Black cassiterite would have had no obvious value to them.

Much later, the Phoenicians, not the Canaanites, were definitely obtaining tin from the area at about 1000 B.C, and we know that King Soloman had a fleet manned by Phoenician sailors trading down the Red Sea from Eilat and probably calling at the island of Suakin to barter.

The Haifa ingots were probably jettisoned from a Phoenician ship, during a storm, which was on its way to the industrial island of Cyprus with its copper mines.

The provenance of the tin ingots in the Kas shipwreck remains a mystery. However, the analytical results show what can be achieved by modern analysis if only museum curators would provide the very tiny samples now required to enable archaeology to be based on scientific facts rather than mythology and doubtful king lists.

References

BEGEMANN, F., HAUPTMANN A., PERNICKA E., REHREN T., & YALCIN Ü, (eds.) 1999, The Beginnings of Metallurgy. *Der Anschnitt, Beiheft 9,* Bochum.

BRILL, R.H., BARNES, I.L., ADAMS, B., 1974, Lead isotopes in Ancient Egyptian Objects. In *Proceedings of the Third Solid State Conference, Cairo* edited by A. Bishay. Plenum Press, N.Y.

CHARLESWORTH, M.P., 1924, *Trade Routes and Commerce of the Roman Empire.* Cambridge.

CIERNEY & WEISBERGER, 2001 Personal Communication.

CLERZIOU, S. and BERTHOUD, T., 1982, Early Tin in the Near East. In *Expedition 9,* p.21

DAYTON, J.E., 1971, The Problem of tin in the Ancient World. In *World Archaeology 3.* p.49

DAYTON, J.E., 1978, *Minerals, Metals, Glazing & Man* Geo.Harrap, London.

DAYTON, J.E. & DAYTON, A.J., 1986, The uses and limitations of Lead Isotope Analysis. In *24th Archaeometry Symposium, Washington D.C. 1984,* edited by J.S.Olin & M.J.Blackman.

DAYTON, J.E., 1996, Recent mineral discoveries in Central Asia. In *Annali Instituto Universitario Orientail, Napoli, Vol.56.*

HIGGINS, R., 1967, *Minoan and Mycenaean Art.* London.

KARO, G., 1930, *Die Schachtgräber von Mykenai.* Bruckmann, Munich.

LAMB, W., 1936, *Excavations at Thermi in Lesbos.* London.

LUCAS, A., 1962, *Ancient Egyptian Materials and Industries.* London.

LO SCHIAVO, F., MADDIN, R., MUHLY J.D., and STECH, T., 1985, Preliminary Research on Ancient Metallurgy in Sardinia 1984. In *The American Journal of Archaeology* 2nd series 89: p. 316-318.

WACHSMANN, S. & RAVEH, K., 1981, An underwater salvage excavation near Kibbutz-Hotrim, Israel. In *International Journal of Nautical Archaeology* 10: p.160.

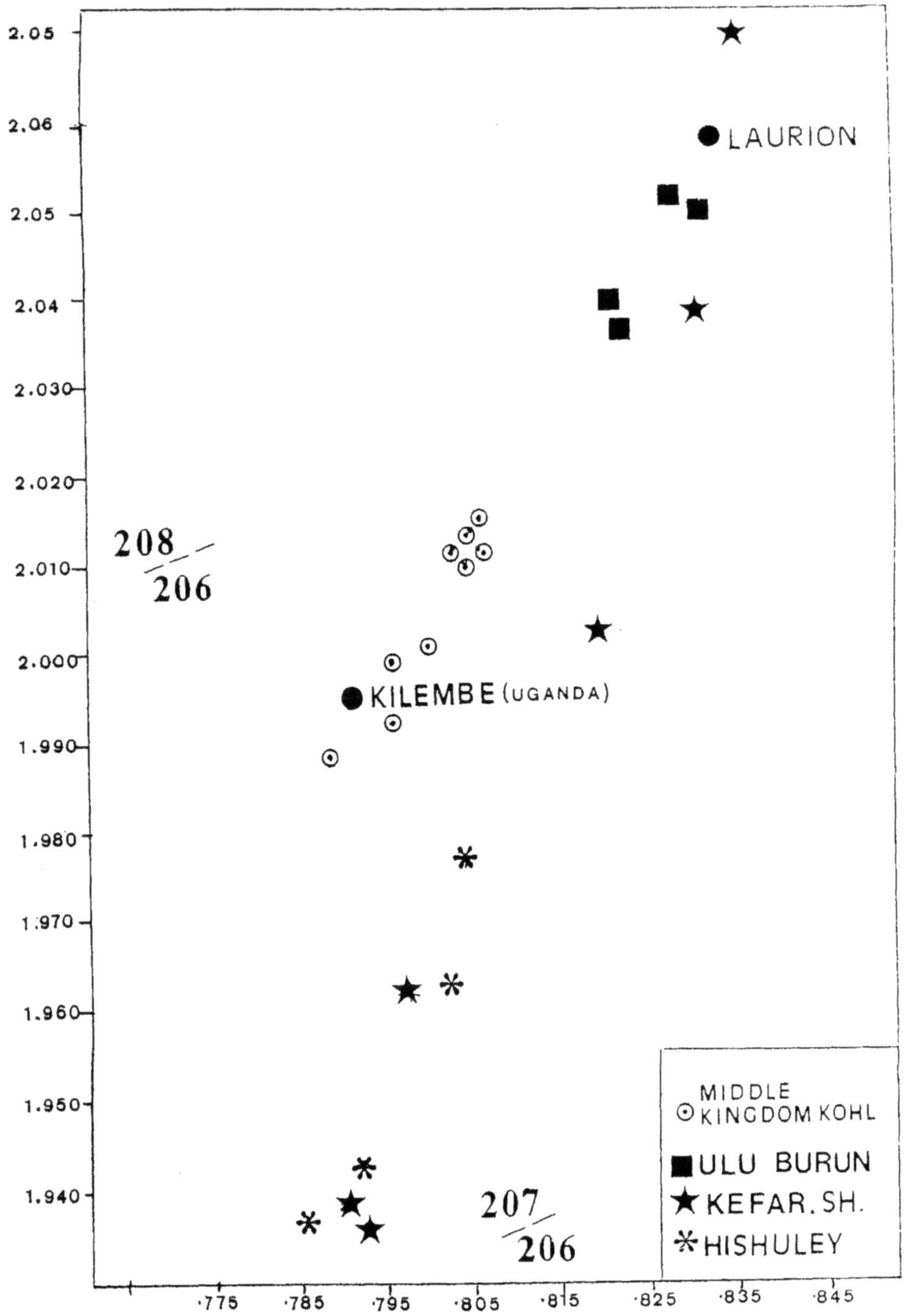

Figure 2. Detail of Fig. 1 (graph 1, Uganda). Isotope plot of the Haifa and Ulu Burun tin ingots, showing that the Haifa ingots derived from the Central African tin belt. (Laurion was a rich Mycenaean copper mine before it was exploited for galena and silver after 500 B.C.).

www.ingramcontent.com/pod-product-compliance
Lightning Source LLC
LaVergne TN
LVHW071212210926
848482LV00015B/793